MAN IN INNER AND OUTER SPACE

Selected Lectures on the U.S. Manned Moon Landing Programme, the Sun and our own Planet

Edited by

S. T. BUTLER, M.SC., PH.D., D.SC.

Professor of Theoretical Physics, University of Sydney

and

H. MESSEL, B.A., B.SC., PH.D.

Professor of Physics and Head of the School of Physics, University of Sydney

A course of lectures contributed in the 11th International Science School for High School Students, sponsored by the Science Foundation for Physics within the University of Sydney, at the University of Sydney, August 26–September 6, 1968

1968
THE QUEEN'S AWARD
TO INDUSTRY 1968

PERGAMON PRESS

OXFORD · LONDON · EDINBURGH · NEW YORK
TORONTO · SYDNEY · PARIS · BRAUNSCHWEIG

Pergamon Press Ltd., Headington Hill Hall, Oxford
4 & 5 Fitzroy Square, London W.1

Pergamon Press (Scotland) Ltd., 2 & 3 Teviot Place, Edinburgh 1

Pergamon Press Inc., Maxwell House, Fairview Park, Elmsford,
New York 10523

Pergamon of Canada Ltd., 207 Queen's Quay West, Toronto 1

Pergamon Press (Aust.) Pty. Ltd., 19a Boundary Street,
Rushcutters Bay, N.S.W. 2011, Australia

Pergamon Press S.A.R.L., 24 rue des Écoles, Paris 5^e

Vieweg & Sohn GmbH, Burgplatz 1, Braunschweig

Printed in Great Britain by A. Wheaton & Co., Exeter

THE COMMONWEALTH AND INTERNATIONAL LIBRARY

Joint Chairmen of the Honorary Editorial Advisory Board

SIR ROBERT ROBINSON, O.M., F.R.S., LONDON

DEAN ATHELSTAN SPILHAUS, MINNESOTA

Publisher : ROBERT MAXWELL, M.C., M.P.

MAN IN INNER AND OUTER SPACE

PREFACE

The Science Foundation for Physics within the University of Sydney is honoured to present such a distinguished group of lecturers at its 11th International Science School for High School Students.

On behalf of the Foundation we wish to take this opportunity of thanking Professor Bracewell, Dr. MacDonald, Dr. May, Dr. Rees and the three American astronauts, for having given so generously of their time and effort.

We have chosen the general heading "Man in Inner and Outer Space" for the 1968 International Science School because its lecture topics range from America's manned moon landing programme to selected research fields, the sun and our own planet. In each of the fields discussed the lecturers are specialists of world renown and the material has been specially prepared, written and edited for fifth-year high school students. We therefore feel that the lectures will be of interest not only to the students, but to the widest sections of the public. We feel that the material presented will be generally appreciated by the increasingly more science-conscious layman in this scientific age and also, in fields other than his own, by the specialised scientist.

The Foundation's 1968 International Science School and, indeed, this book, are intended to stimulate and develop science consciousness, in Australia and throughout the world. The Foundation is therefore honoured that the United States President, for the second year running, has endorsed a special scheme under which 10 American students, designated by the President as "Lyndon B. Johnson Australian Science Scholars" are attending the Science School in Sydney.

The Foundation is similarly honoured that Prime Minister Eisaku Sato of Japan, and H.R.H. The Prince Philip, Duke of Edinburgh, as Patron of Britain's Royal Institution, have kindly consented to associate themselves with the scheme and, on behalf of the Foundation, to award five scholarships each to top students from their countries. These students have been named the "Sato Eisaku Australian Science Scholars" and the "Royal Institution Australian Science Scholars".

The 20 overseas students, as well as those selected from throughout Australia and New Zealand deserve the applause of all, and the Science Foundation for Physics within the University of Sydney is happy to honour and reward the ability and diligence of these young people.

H. MESSEL and S. T. BUTLER.

Sydney, August, 1968.

CONTRIBUTORS OF LECTURES

R. N. BRACEWELL

Professor of Electrical Engineering,
Stanford University, Stanford, California.

G. J. F. MacDONALD

Executive Vice-President,
Institute for Defense Analyses, Washington, D.C.

R. M. MAY

Reader in Physics,
School of Physics, University of Sydney.

E. F. M. REES

Deputy-Director, Technical,
NASA George C. Marshall Space Flight Center, Huntsville, Alabama, and
Director, Apollo Special Task Team at NR,
NASA Manned Spacecraft Center, Downey, California.

D. K. SLAYTON, A. B. SHEPARD, L. G. COOPER

NASA Astronauts.

CONTENTS

THE SPONSORS

The Science Foundation for Physics within the University of Sydney gratefully acknowledges the generous financial assistance given by the following group of individual philanthropists and companies, without whose help the 1968 International Science School for High School Students and the production of this book would not have been possible.

Ampol Petroleum Limited

The Sydney County Council

W. D. & H. O. Wills (Aust.) Limited

and

Birt & Co. (Pty. Ltd.)

A. Boden, Esq.

The Sun

(Five Chapters)

by

R. N. BRACEWELL

Professor R. N. Bracewell,
Professor of Electrical Engineering,
Stanford University, Stanford, California.

CHAPTER 1

Introduction to the Sun

Theories of the sun

It is very natural to wonder about the nature of the sun and we can be sure that in the course of man's time on earth many ideas have been entertained. One of the earliest opinions, of which we have a record, is that of Thales, a Greek astronomer who lived in the seventh century B.C. in Egypt, and who, according to accounts, gave a correct explanation for eclipses of the sun and eclipses of the moon. You will see from this that he must have understood that the moon shone by reflected light, having no light of its own, and since one could hardly stumble onto the geometrical explanation of eclipses without real understanding, it has to be admitted that in those days there were powerful intellects that could take in everyday experiences, handle them with rigorous logic, and make correct deductions as to the causes. This being so, it is certain that occasional men (and perhaps women?) over the past hundreds of thousands of years have gained some understanding of the sun and its system of planets. It would not be too much to suppose that some prehistoric men clearly grasped the notion that the sun is at the centre of our solar system, a point that was again reached, as we know, by some fine minds of the seventeenth century. I think this is a fascinating thought. Still, some problems are harder than others, and while from time to time, say once a century, there may have been a caveman with the brain of a Newton, for whom the motions of the moon and planets were an open book, I don't think he would have been able to say why the sun shines as it does. This is a problem whose explanation depends on a great deal of physics that no one person could discover.

For example, an understanding of how sunlight is caused requires a knowledge of spectrum analysis and the thermodynamics of radiation, subjects that were only beginning to be worked on towards the close of the nineteenth century. One of the principal contributors to the new science of chemical analysis by means of spectra was G. R. Kirchhoff (1824-1887), a German physicist whose name is also well remembered in connection with electric circuits, and it is interesting to look back now and note that Kirchhoff examined the sun spectroscopically and found reason to believe that it was in a solid or liquid state. Shortly afterwards, however, it became generally understood that the sun was in a gaseous condition, so we see that this is a relatively recent story. Consequently, we should not be surprised to hear that Thales thought the sun was solid, for Sir William Herschel (1738-1822) who was the first to turn a really large telescope on the sun thought so too. (In fact he said it was cool, possibly habitable, and had mountains 600 miles high, all underneath a bright outer envelope.)

In addition to these imperfect ideas, which nevertheless are typical of scientific progress as it advances through, and disposes of, one incorrect hypothesis after another (the incorrect always greatly outnumbering the correct), many notions have been recorded in literature whicn can be described as folklore. Everyone knows the story of the sun god driving his chariot across the sky and being rowed back at night through the underworld; but in Egypt he rowed his boat across the sky during the day. We may ask, in what way do these incorrect descriptions differ from the equally incorrect descriptions that we dignify by the name of scientific hypotheses? I think there is a lot in common, the difference being that in the one case the idea has been preserved by literature so that we no longer recognize it for what it originally may have been — a reasonable conjecture on the part of some bright individual with that rare curiosity to want to wonder about the causes of natural phenomena.

In any case I would like you to know that the discovery of new knowledge about the sun is proceeding vigorously, that there are many exciting suggestions now current as to why the sun does this or that. Of necessity, most of these suggestions will turn out to be partly or wholly wrong but gradually the body of generally agreed conclusions increases.

To engage in, or be aware of, the ongoing discussions is by far more stimulating intellectually than to learn about what has hitherto been established, yet on the other hand this poses a dilemma, because without knowing what has been established one cannot easily contribute to or follow the day by day discussions. In respect to physical science, the Greeks were in a much better position than we are because not much had been established, but in some subjects, especially geometry, their position was the same as ours.

So my plan will be to present as much background material as seems necessary in a variety of topics connected with the sun and to attempt to show what the questions are about at the present time and in some cases to show what the contending theories were at earlier stages.

First let's run over the principal regions into which the sun is divided to establish a framework that can be used later for reference when we come back to look at different items in more detail. *(Figure 1.)*

The photosphere

As everyone knows, the sun looks like a circular disc, but it is hard to look at it for long enough to decide whether there is any detail to be seen, and we all learn as small children to keep the

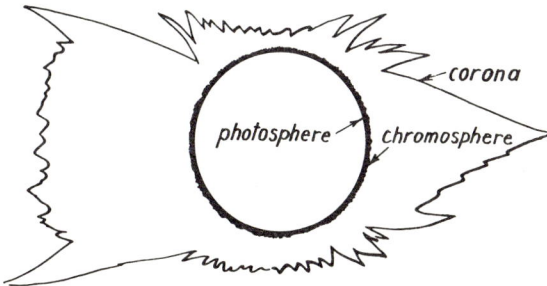

Figure 1. The main parts of the sun.

sun well out of our eyes. In fact it is extremely dangerous to look
at the sun. Your power to stand the pain is greater than the
power of your eye to protect itself, so that it is quite possible to
suffer eye damage leading to loss of sight without feeling serious
pain at the time. However, most people have seen the disc of
the sun through fog or through thick haze at sunset. Another
way is to let sunlight pass through a pinhole in a large card and
fall on another card to form an image, which you will find to
be quite round. *(Figure 2.)* This is an excellent thing to do

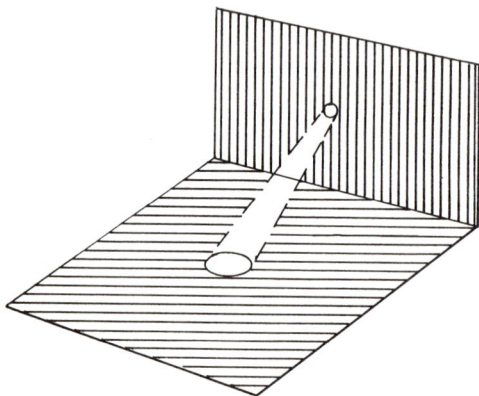

*Figure 2. A round pinhole throws a round pool of sunlight on the ground.
What happens if the pinhole is square?*

when a solar eclipse is coming on, because then the spot of
light has a bite out of it, which looks most unusual and can leave
no doubt that one is looking at a solar image. Even the round
pools of light seen dappling the shade beneath leafy trees become
crescent shaped at the time of an eclipse and I hope you will look
out for this interesting appearance. Unfortunately there will not
be a solar eclipse in Australia until 1974, and that will be invisible
from Sydney, so you can't check on me immediately. However,
if what I say is true, then a square pinhole will throw a round
patch of light, and you might find it interesting to see for yourself
whether it does or not.

This round disc, which is the source of sunlight, we call the **photosphere,** the sphere from which the light comes *(Figure 1).*

The chromosphere and corona

At a time of total solar eclipse, the sky darkens abruptly and during the period of totality, which may last only a minute or so, the sun may be looked at directly by eye, because the part of it that is not blocked by the moon is not bright enough to be harmful. Two distinct things are seen: a thin red band called the **chromosphere** and an extensive white **corona.** The word chromosphere was introduced precisely 100 years ago. You might be interested to look up the Philosophical Transactions of the Royal Society of London for the year 1868 (vol. 159, p. 430) to check the circumstances. It is based on the Greek word **chroma,** meaning colour, and refers directly to the red colour seen rimming the moon during an eclipse. The word **corona** is Latin for halo and has been in use since ancient times, with that meaning and thus was not specially coined to refer to the sun. On the contrary, various halos may be seen around both the sun and moon depending on meteorological conditions and the one that is visible during an eclipse was not especially distinguished from those that have their origin with water droplets or ice crystals in the earth's atmosphere. Indeed, long after it was established that the corona was an extraterrestrial phenomenon, opinion was divided over whether it belonged to the sun or to the moon. Of course it is obvious that the corona cannot belong to the moon, because the moon has no atmosphere. But wait a moment. How do we know the moon has no atmosphere. Can anyone recall the line of argument that leads to this conclusion, and does this argument say *no* atmosphere at all, or merely that the density of atmosphere cannot be greater than some small fraction of the density of the earth's atmosphere (at standard temperature and pressure air contains $2 \cdot 7 \times 10^{19}$ molecules per cubic centimetre). Even if the moon's atmosphere were at most a thousand million times less dense than the earth's atmosphere, it could still contain twenty thousand million particles per cubic centimetre. This is a lot, and perhaps with backlighting *(Figure 3)* from the sun such as a total eclipse of the sun provides, would look like the corona. Now here is a problem I would like

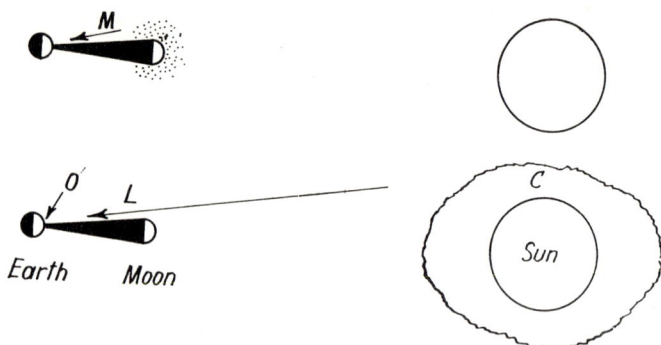

Figure 3. During a total eclipse of the sun, faint light L from the solar corona C can be seen by an observer in the moon's shadow at O. It was once thought that the corona belonged to the moon, being perhaps due to photospheric light scattered along M toward the observer by a lunar atmosphere. How would you have planned to settle this controversy with the resources available in 1850?

you to discuss over lunch. If you were back in 1850, when it was still being argued whether the eclipse corona belonged to the sun or to the moon, what would you plan to do if you were able to travel to a place where the next total eclipse of the sun would occur and where it would thus again become possible for you to view the corona?

In this survey of the main parts of the visible sun we have seen that the sun is bounded by an opaque white shell called the photosphere. Outside this is a thin transparent zone from which comes faint red light and outside this again is a transparent but very extensive region sending us faint white light.

Size and mass

Some numerical magnitudes which we shall need to call on from time to time are given in Table 1, where they are compared with corresponding values for the earth. Thus, from this table we see that the sun is 109 times bigger than the earth and consequently has more than a million times the earth's volume. Yet we see that its mass exceeds that of the earth by only a third of a million.

In other words, the density of an average sample of solar material is rather less than that of an average sample of earth. Remembering that the photosphere is believed to be in a gaseous state this sounds reasonable except for one thing. The sun's mean density is greater than the density of water, so we are dealing with a pretty dense gas. There is a big difference between the density of gases and liquids as we experience them in everyday life; thus the air in this room only weighs about one ton, but if the room were filled with water, the water would weigh one thousand tons. So when we talk about a gas that is denser than water, obviously something is going on that is out of our everyday experience.

TABLE 1
Numerical Magnitudes

	Sun	Earth	Ratio
Radius	$6 \cdot 96 \times 10^{10}$	$6 \cdot 37 \times 10^8$ cm	109
Mass	$1 \cdot 99 \times 10^{33}$	$5 \cdot 98 \times 10^{27}$ g	333,000
Mean density	$1 \cdot 41$	$5 \cdot 5$ g cm^{-3}	$0 \cdot 26$
Surface gravity	$2 \cdot 74 \times 10^4$	$9 \cdot 81 \times 10^2$ cm sec^{-2}	28
Escape velocity	6×10^7	$1 \cdot 1 \times 10^6$ cm sec^{-1}	55
Moment of inertia	6×10^{53}	8×10^{44} g cm^2	$7 \cdot 5 \times 10^8$
Angular momentum	$1 \cdot 7 \times 10^{48}$	$5 \cdot 9 \times 10^{40}$ g cm^2 sec^{-1}	$2 \cdot 9 \times 10^7$
Radiated power	$3 \cdot 9 \times 10^{33}$	2×10^{20} erg sec^{-1}	2×10^{13}
Sun-earth distance	$1 \cdot 496 \times 10^{13}$ cm		

It is not hard to see what this must be. As we know, every square inch of the earth's surface supports the weight of the air that rests on it, and this amounts to about 15 pounds. As we go into the interior of the earth the pressure increases in accordance with the additional weight of the overlying rock. But if the sun is roughly 100 times bigger than the earth, then from this fact of increased size alone we should expect much increased pressures and so in turn, because of the compressibility of gases, much increased densities. Dense gases, of course, mean even greater pressures. The overall result is that the density of the solar material increases towards the centre where it must reach a value around 100 grams per cubic centimetre. It is prevented from congealing into a solid by the high temperature.

I have not mentioned how the size of the sun is known and so we may pause here for a moment. As you may surmise this piece of information has been available for a long time, and yet this single dimension summarizes many fascinating chapters in the history of astronomy, going back to ancient times. We may pick up the thread here with the commission given to Captain Cook by the Royal Society to proceed to Tahiti to observe the transit of Venus in 1769. Hardly anyone remembers now why it was felt to be important to travel to the South Seas for this purpose, and to tell the truth the results did not prove to be world shattering; still, the expense of the expedition has generally been felt to be justified by the discovery of the east coast of Australia later on the same voyage. In the Venus transit observation Venus acts as a convenient marker against which the position of the sun in the heavens can be observed. If now the observation is made by two observers on opposite sides of the earth there will be an effect of parallax which causes the sun to appear displaced, relative to Venus, by several seconds of arc. Since the apparent displacement of the sun depends on how far away it is, one can work back to find the sun's distance. Generally speaking, it is difficult to measure angles as small as a few seconds of arc, but one of the attractive features of the method was that the two observers, one in Europe and one in Tahiti, needed only to time the moment when the planet crossed the sun's limb. The time discrepancy between their two observations, due to parallax, could then be converted to angular measure from the known angular velocity of Venus across the face of the sun. The basic idea, which is that of surveying from a known baseline, has since been applied in other ways involving asteroids, especially the asteroid Eros, and for many years now the distance to the sun has been known with an accuracy of about one part in one thousand. On multiplying this distance by the angular diameter of the sun, which can be measured to an accuracy of better than one in ten thousand, we obtain the linear diameter of the sun. The result still has only three figure accuracy, which hardly qualifies as astronomical. The fact is that the very precise knowledge about the solar system is related to timing (a sidereal year contains 365·256366 days) and to its configuration or shape, but its actual scale has not been as well defined.

This situation has now been revolutionized by the introduction of radar. By measuring the time taken by a radio signal to make the trip out to and back from a reflecting target one can calculate the distance by multiplying by the speed of light. Of course, the accuracy obtainable will be limited by the accuracy with which time intervals can be measured and by the accuracy with which the speed of light is known. I cannot go into the details of how the time of flight is measured, which is itself a very interesting subject involving the use of atomic clocks and digital computers to generate radio signals and modulate them with special codes; it is sufficient to say that time intervals can be measured better than we can measure anything else. As for the velocity of light, it is known now to about one part in a million (299,792·9 ± 0·4 kilometres per second).

Strange to say, the distance to the sun is not measured by observing radar reflections from the sun, and the reasons for this are rather interesting. First of all, the sun absorbs microwaves, which completely eliminates the radar technique in the shorter wavelength range. At longer wavelengths, around 10 metres, echoes can be obtained, but they are returned from levels high in the corona, sometimes from more than one place, and it is not very definite how far further down it is to the photosphere. The technique that has proved successful is to measure the range to Venus, which we see here again playing an important intermediary role, and to continue doing this as it makes its orbit around the sun. You will see that the average distance to the centre of Venus is equal to the distance to the centre of the sun, but that some tricky corrections may have to be made to allow for the radius of Venus and for the precise location of its elliptical orbit. The present position is that the distance to the sun is now known to within one part in 10^8, or about one mile. This is quite remarkable when you remember that the sun itself is 866,000 miles in diameter.

The mass of the sun is something that I shall not deal with. It depends on the famous experiment of weighing the earth and on the observed motions within the solar system, which you may care to study further if you are interested in celestial mechanics.

Power output

Returning now to the numerical table, we find that the sun radiates $3 \cdot 9 \times 10^{33}$ ergs per second. This is a very large power that is difficult to have a quantitative feel for at first sight, and it does not help much in this case to compare against the earth, which generates a trivial amount of heat, mainly from disintegration of radioactive nuclei in the rocks. It is, however, very easy to measure the sun's power output, or luminosity; all one has to do is measure the amount of solar heat that can be received per unit time on a square metre of black absorbing material, and then multiply by the number of square metres on the surface of a sphere, centred on the sun and passing through the earth. This can be done quite directly with a water calorimeter and a thermometer and gives the surprisingly large value of $1 \cdot 4$ kilowatts per square metre or 2 horsepower per square metre. The amount of solar power falling on a square mile of land roughly equals the total production of electrical power in the whole of Australia.

The source of this fantastic energy generation lies deep in the sun where we cannot make direct observations, because the overlying gas is opaque, but remarkable lines of reasoning have been developed that give rather a good idea of what is happening. As a matter of fact, our state of knowledge of the sun's interior has for many years kept ahead of our state of knowledge about the earth's interior.

Probably everyone knows now that inside the sun thermonuclear reactions take place which, in a sense, are like those occurring in a hydrogen bomb explosion. A closer analogy, however, exists with the controlled hydrogen fusion reactors which are being developed to replace the nuclear fission reactors containing specially treated uranium or plutonium. The fusion process will not require the mining and processing of uranium but will use only hydrogen, which is universally abundant, and is therefore greatly favoured by nuclear power authorities. The only trouble is that it has not been possible to make this process work yet, in spite of the fact that many very able groups of engineers and physicists have been working on it for more than a decade, so it has turned out to be inherently a much harder thing to do than to make a uranium pile work, which proved to be feasible the first time it was tried.

The difference between the sun and the experimental fusion machine in the laboratory is that much higher temperatures are available in the sun, and the reaction volume is surrounded by gas at very high pressures to contain the reaction. Without going into the details of the fusion reaction, which has several stages and requires more than cursory attention, it is enough here to say that protons are consumed, helium nuclei are left over, and energy is emitted. Also, the reaction rate is higher where the temperature is higher. We know that the sun's temperature must increase inwards, reaching a maximum at the centre, because the direction of energy flow (radially outwards) is determined by the direction of the temperature gradient. Consequently, the production of fusion energy is somewhat concentrated towards the centre of the sun.

From its interior origin the solar power makes its way out, through an essentially opaque medium, being absorbed and re-emitted over and over until it emerges through the photosphere surface to continue out into space. Most of it will carry on forever, only a small part being intercepted by the earth and other bodies.

CHAPTER 2

The Character of Sunlight

The sun's temperature

As with other simple questions that can be asked about the sun, the question as to its temperature led in earlier times to a division of opinion, and yet it is one whose answer comes practically within the realm of everyday experience. We know how the colour of a hot object gives a clue to its temperature and from this alone we can conjecture that the sun is perhaps not too much hotter than a white-hot poker. Of course it is more complicated to say how many degrees that temperature measures, but it is only quite recently that tables could be set up such as Table 2, which translates the practical data of the blacksmith into numerical temperature readings. Before ideas of temperature scales for high temperatures developed, it is obvious that the blacksmith could weld, forge, temper and anneal, all operations requiring the metal to be brought to definite temperatures, by using colour as his indicator.

If in the laboratory we were to measure the total power emitted by a white-hot sphere of a given size and compare that with the power known to be emitted by the sun we could begin to discuss the sun's temperature quantitatively. From Table 1 we can calculate that the sun emits 7×10^{10} ergs per square centimetre per second and I can state here that a solid object heated to 3000°C, a temperature which can be reached in a kiln, will be found to emit about 7×10^9 ergs per square centimetre per second, i.e., about ten times less. Does this mean that the sun's temperature is 30,000°? It would, if the amount of heat radiated was proportional to the temperature, but it is not. To discuss this it is more convenient to talk in terms of temperatures measured above absolute zero, which is at —273°C, and to indicate that we are referring to absolute temperatures by using the abbreviation K (for Kelvin). Thus the freezing point of water is 0°C or 273°K, and its boiling point is 100°C or 373°K. More than 100 years ago John Tyndall (1820-1893) made the laboratory measurements using a hot wire and judging the temperatures from the colour. Then Josef Stefan (1835-1893) showed that the very rapid increase of radiation with temperature could be described concisely by saying that the radiated power was proportional to the fourth power of the absolute temperature. So, in order to radiate ten times more power than a body at 3000°C (or 3273°K) the sun needs only to have a temperature that is higher than this by a factor $1 \cdot 78$ (since $1 \cdot 78^4 = 10$). This gives us a solar temperature of $1 \cdot 78 \times 3273 = 5800°K$, which is not so very much higher than the temperatures we work with in kilns. It is, however, a very difficult matter indeed to reach the temperature of the sun in the laboratory, except for short times and in small volumes, because available materials vaporize. There are other ways of arriving at the sun's temperature and the values obtained vary somewhat according to the method. For this reason we generally use the round value of 6000°K in referring to it.

It should be added here that the rough basis for Stefan's law was subsequently supplemented by very elegant theoretical considerations propounded by L. Boltzmann according to which the fourth power law is expected to describe precisely the dependence of radiated power on temperature under certain ideal conditions. We generally summarize these conditions by saying that the emitting body shall be "black" — a technical term meaning that the body absorbs all incident radiation at all wavelengths. If it does not, the law has to be corrected. It turns out that the sun is "black" to a good degree of approximation, but not perfectly, and this is a reason why the temperatures deduced will vary by a few per cent when different methods are used, if the small corrections are ignored. To calculate the radiated power by Stefan's law we need to know the constant of proportionality, which is generally known as the Stefan-Boltzmann constant and is represented by σ. Its numerical value is given by

$$\sigma = 5 \cdot 7 \times 10^{-5} \text{ erg cm}^{-2} \text{ deg}^{-4} \text{ sec}^{-1}$$

Then, for a black body at absolute temperature T, the radiated power per unit area of surface is given by

$$\sigma T^4 \text{ erg cm}^{-2} \text{ sec}^{-1}$$

As an exercise in the use of this powerful principle, calculate the approximate number of watts radiated by the whole of your body, allowing 10^7 ergs per second to one watt. As to whether the black body approximation is good in this case is a question that depends mainly on whether the skin is a good absorber of infrared radiation, because that is the part of the spectrum in which we emit our radiation; for the sake of simplicity, assume you are wearing an absorbing suit. You may be surprised by the result of this calculation.

The spectrum of sunlight

Our information on this subject commences with Newton (1642-1727) whose famous experiments with sunlight that he let in through a hole in the shutter of his window have been widely reported, though I am afraid that the reports are generally too brief to do justice to the experiments. Newton's *Opticks* was not written in Latin, as were his other scientific works, and as it is still in print and readily available, I can strongly recommend that you peruse it. One of the interesting things that Newton reported when he examined the solar spectrum, as spread out by his prism, was that he could see seven colours: red, orange, yellow, green, blue, indigo and violet. Since these early observations of the solar spectrum have long been superseded by accurate quantitative measurements made with reliable instruments of several kinds, all of which show that sunlight is continuously distributed over all visible wavelengths, it sometimes puzzles people that Newton could see seven distinct colours. Of course we have all seen colour pictures of the spectrum as reproduced in printer's ink, and the gradations of hue are generally a little uneven, but very few of us, to tell the truth, have taken the opportunity to have a good look at the actual light from the sun, as dispersed by a prism, in a dark room. You should do this as soon as you can. It is not as easy an experiment to arrange as you might think, and takes a little care, one of the problems being that the sun keeps changing its position. But it is well worth doing as a personal check on the qualities of Isaac Newton as an observer. If you use a slit, instead of a circular hole, you may just possibly be able to see one of the Fraunhofer lines, dark features in the spectrum which were first seen by Wollaston (1766-1828) in 1802.

TABLE 2
Colour Temperature Chart

Colour	Temperature degrees C.
Dark red	700
Dull cherry red	800
Cherry red	900
Bright cherry red	1000
Orange-red	1100
Orange-yellow	1200
Yellow-white	1300
White	1400
Brilliant white	1500
Iron melts	1535
Dazzling white	1600

The spectrum is brightest in the middle and becomes faint at the red and violet extremes. It is rather strange that the violet end of the spectrum seems to have a little red in it, but indeed it is so, as you can soon discover if you try to mix up water colour paints to match the spectral colours. This odd fact is really a question of physiology and cannot be pursued here. If you succeed in producing a good solar spectrum in a dark room you will wish to repeat a striking experiment first carried out by Herschel of placing the bulb of a thermometer in the dispersed sunlight. The highest reading is obtained well to the dark side of the red end, clearly showing the presence of the invisible infrared rays.

With these introductory remarks, we may present a graph of the solar spectrum (*Figure 4*). To a good approximation, this spectrum curve follows the course expected of an ideal black body, especially in the visible region where the intensities are highest. In the visible range the most striking departure from the ideal lies in the presence of numerous points in the spectrum where the

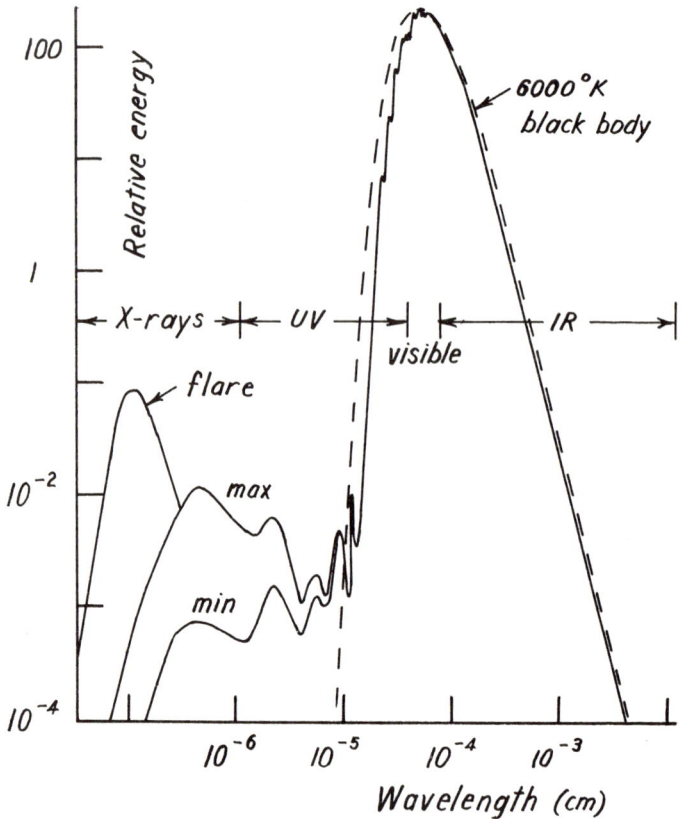

Figure 4. The strong part of the solar spectrum comes from the photosphere and follows the spectrum of a 6000° black body, but the ultraviolet and X-ray emissions, which come from the chromosphere, behave differently and vary with solar activity.

intensity is low. When the spectrum is viewed or photographed through a slit, dark lines are seen at these wavelengths. There are no bright lines visible. Another feature of interest is that the spectrum continues on well beyond the limits of sensitivity of the eye. Heat measuring devices, or bolometers, are used to study the infrared side, but the ultraviolet side may be studied by photography.

The Fraunhofer lines

When the dark lines in the solar spectrum were first noticed they were not understood. Indeed they are a manifestation of the quantum behaviour of the atom that was not elucidated for another century. However, a lot was soon found out about these lines by Kirchhoff and Bunsen (1811-1899) who showed that similar dark lines were produced when vaporized substances were introduced into a beam of white light. Different chemical atoms produced different patterns of lines and it was possible to show without any doubt that certain well-known atoms such as sodium, calcium and magnesium must intervene somewhere in the path between the photosphere and ground level. There proves to be a good way of deciding where the absorbing atoms are situated, which makes use of the Doppler effect, according to which the wavelength of radiation shifts if the source is in motion relative to the observer. If the relative speed is 1 per cent of the speed of light (300,000 kilometres per second) then the wavelength is shifted by 1 per cent and so on (except at very high speeds where relativity effects set in). Now the sun's distance changes by plus and minus 2,500,000 kilometres in the course of the year, because the earth's orbit is slightly elliptical, so the sun is always either approaching or receding from us and therefore any dark absorption lines due to absorbing atoms in the solar atmosphere are always offset a little, towards the violet or towards the red, as compared with an absorption line due to atoms of the same element in the earth's atmosphere. Examination of the Fraunhofer lines at different times of the year when the earth is in different parts of its orbit, reveals that some of the lines follow the expected Doppler shift and some do not, so there is no difficulty in separating the solar from the terrestrial lines. Of the many thousands of lines, the majority are solar.

Let us make a check on this matter to see for ourselves whether it would be easy. Let the earth be precisely at its mean distance of 149·6 million kilometres from the sun (this happens on April 4 and October 4 each year). Then it will take 91 days to change its distance by 2·5 million kilometres, and so its component of velocity along the line to the sun will be 0·5 kilometres per second, which is 1/600,000 of the speed of light. Consequently, in April and October we would expect the numerous lines around a wavelength of 6870 Angström units, which are due to absorption by oxygen molecules in the earth's atmosphere, to be shifted to the red or to the violet by approximately one hundredth of an Angström unit. This well-known unit was used by Anders Angström (1814-1874) in a table of wavelengths of Fraunhofer lines that he produced in 1868; it is 10^{-8} centimetres. One hundredth of an Angström unit may seem extremely small but a shift of this amount is not difficult to detect, if one is able to compare two spectra taken some months apart, because the Fraunhofer lines are so numerous, averaging about five per band one Angström wide. So the actual scaling that has to be carried out on the spectrum itself to determine whether a given line has shifted towards its nearest neighbour only needs to be done with an accuracy of 1 per cent or so.

Not only do vaporized atoms absorb light at very definite wavelengths peculiar to each atom but if we can view the vapour against a dark background, bright lines can be seen at precisely the same wavelengths, showing that while some of the atoms may be absorbing, others are emitting. These laboratory discoveries going back over a century are precisely what we need to understand the red light of the chromosphere, for when the chromosphere was first looked at through a prism, as was done at the eclipse of 18th August, 1868, bright lines were seen against a dark background at wavelengths where dark Fraunhofer lines were known in the photospheric spectrum. The strong red line which gives the chromosphere its characteristic colour was identified as belonging to hydrogen gas — not molecular hydrogen in the form that we know it in the chemistry laboratory, but a gas of single hydrogen atoms.

The negative hydrogen ion

If the sun is a mixture of gases, why does its spectrum not consist of masses of emission lines? Or is it conceivable that the more or less smooth spectrum is just the sum total of line emissions from many different elements. The answer to this question only came painfully and is still not widely known among physicists. Although the solar spectrum immediately reveals what elements are present it does not directly give quantitative information as to how much of each is present. The composition of the solar material is still being refined but we can say now that it consists mostly of hydrogen, some helium, and a sprinkling of everything else. Not so very long ago it was seriously assumed that calcium was a principal constituent, on the very reasonable basis that the strongest Fraunhofer lines are calcium lines. (Being rather far to the violet end of the spectrum, the light that they contribute to the chromosphere does not make as much impression on the eye as the red line of hydrogen.)

At the temperature of 6000°K the hydrogen and helium are of course far above their boiling points and so are in gaseous form, but with the sprinkling of metal atoms it is different. Not only are they vaporized at 6000°K but many of them are ionized, i.e., they have been stripped of their valence electrons which are thus free to move in the gas until they recombine with some other positive ion which has previously lost an electron. But another thing can happen, the electron can attach to a neutral hydrogen atom, thus forming a negative hydrogen ion. Unlikely as this may seem as compared with the attraction offered by a positive metal ion, calculation confirms that it is a frequent event and in fact is the microscopic process by which the ordinary quantum of sunlight is generated. Instead of possessing always the same energy, as happens when an atom undergoes an internal transition, and thus always coming off with the same wavelength, the quantum of radiant energy released when an electron drops into the grasp of a neutral hydrogen atom will have a wide range of possible wavelengths open to it according to the kinetic energy that the particles happened to possess just before attachment.

The light of the corona

We have seen how the white light of the photosphere originates together with its dark Fraunhofer lines and we have seen that the chromospheric light is light redirected by atoms contributing to Fraunhofer line absorption. The light of the corona, however, was for long a mystery. So was the coronal material, because it was not possible to understand how the immense gravitational attraction of the sun failed to pull the corona down.

To convince you of this difficulty, I shall calculate the height to which the corona ought to extend. In the photospheric layer, where the temperature is about 6000°K, the gas atoms are flying about in all directions with different velocities, and, as we know, the more the kinetic energy of the particles, the higher the temperature. The average kinetic energy of a particle is the same regardless of whether it is a hydrogen atom or a heavy metal ion, or an electron. This means, of course, that the average velocity of a particle depends on how massive it is, so heavy metal ions will be moving slower than, and electrons faster than, the hydrogen atoms. Now my plan is to calculate the speed of an atom and then see how high above the photosphere I would expect it to be able to climb. Since hydrogen atoms are the most numerous, we start with them. Two basic principles are needed. First, the kinetic energy of a mass m moving at velocity v is $\frac{1}{2} mv^2$; also, the mean kinetic energy of a particle in a gas at temperature T is $(3/2) kT$, where k is Boltzmann's constant. We also need two numerical facts:

$$m = 1 \cdot 67 \times 10^{-24} \text{ grams}$$
$$k = 1 \cdot 38 \times 10^{-16} \text{ ergs per degree.}$$

By putting

$$\frac{1}{2} mv^2 = \frac{3}{2} kT$$

and solving for v we obtain

$$v = \frac{3 kT}{m}$$

$$= 10^6 \text{ cm sec}^{-1}.$$

This is quite a high speed, but without spending time to find something to compare it with, we may compare it immediately with the velocity of escape listed in Table 1. We see that it is far below the value needed for escape from the sun, so even allowing for the fact that a certain fraction of the atoms are going at speeds above average, it is quite clear that 6000°K is not a high enough temperature to boil off gas against the attraction of gravity. Hence we return to the calculation of how high above the photosphere the particles could climb. As it passes through this to the kinetic energy $(3/2)\,kT$ it possessed in the photosphere a height h it has gained potential energy mgh and if we equate we shall arrive at the maximum height to which the particle can be projected. Of course, many of the particles will not reach this height because they are moving mainly sideways, and furthermore collisions with other particles will impede the motion. However, writing

$$mgh = \frac{3}{2}\,kT$$

we find, for the hydrogen atom,

$$h = \frac{3}{2}\,\frac{kT}{mg}$$

$$= 10^9 \text{ cm.}$$

This distance is only about 1 per cent of the sun's radius. Therefore, it is negligible compared with the extent of the corona as we observe it and it is clear now why difficulties were encountered in explaining the corona. You might care to repeat these calculations to see how far out from the moon a lunar atmosphere would extend, using a temperature of about 250°K, and to express the result relative to the radius of the moon $(1\cdot7 \times 10^8$ cm).

The clue to the nature of the corona first came from observations of its spectrum. First of all it is white light, and since it does not exhibit Fraunhofer absorption lines, it seemed that it must be self-luminous. Otherwise, if it was merely a tenuous exhalation shining by reflected light from the photosphere it would have deficiencies in light at those Fraunhofer wavelengths in the spectrum where the photosphere has deficiencies. There was an immediate cause for concern here, however, because Kirchhoff had discovered in the laboratory that solids and liquids had continuous spectra whereas gases gave out bright emission lines. At the eclipse of 1868 the very innermost part of the corona was found to be emitting a faint green line, which came to be ascribed to "coronium" because no known substance had such a spectrum. In an interesting parallel, an unidentifiable yellow line was seen coming from the chromosphere, also in 1868, and it was said to be due to "helium". In due course, about 20

years later, helium was found leaking out of the earth, and the name has stuck. However, in the case of coronium some 70 years were to elapse before its nature was elucidated. Remarkable to relate, the green coronal line comes from atoms of iron from which thirteen of the orbital electrons have been stripped (iron only has 30). Such a state of affairs is reached if the iron atoms form part of a gas at a temperature of about 1,000,000°K and this in fact is the situation in the corona. If we go back to the height calculation, using this high temperature, we find a height of about one solar radius, so the difficulty about the corona resisting the pull of gravity is overcome. Similarly, we can now explain the white coronal light as scattered photospheric light, for the motions of the particles are so fast that phospheric light, at wavelengths adjacent to Fraunhofer lines, is shifted n wavelength by the Doppler effect, so as to fill in the dark lines.

Shape of the corona

In *Figure 5* we see some pictures of the corona as witnessed at various eclipses. What we are seeing here is a great outstreaming of gas, mostly the protons and electrons into which the high temperature breaks the neutral hydrogen atoms of the photosphere. The motions are quite complex, the strange shapes revealing a control exerted by magnetic fields emanating from the photosphere. The whole phenomenon is made visible by sideways scattering of the intense blaze of photospheric light that passes through the corona.

The solar wind, as this outstreaming has come to be called, has been the subject of much investigation by means of satellites out in the interplanetary space between the earth and the sun, for the corona extends from the sun to distances far beyond those covered in the illustration.

On one of the eclipse pictures the famous "anteater" prominence can be seen, a great wall of dense chromospheric material projecting into the corona.

CHAPTER 3

Sunspots

The solar oscillator

Inside the sun there is something that oscillates with a period of about 22 years and it is a major goal of those studying the sun to find out what this something is. There is nothing in the theory of the interior of the sun, as it has been developed to reconcile the internal pressure, temperature, composition and power generation, to indicate that oscillatory behaviour would be expected. Also, the perimeter of the sun does not pulsate. Yet the fact that something mysterious is going on inside is evidenced by the sunspots. In some years they are numerous and in others not, as shown by *Figure 6*. The precise dates of maxima and minima are shown in Table 2.

Anyone can see sunspots with the aid of a mirror and a lens. A rather weak converging lens is needed, one with a focal length of 3 metres or more. The longer the focal length the larger the image you will get — with a 3 metre focal length you will get a 3 centimetre image. With the mirror you shine the sunlight horizontally through the lens and then allow the rays to go through a window into a room where it is not as bright as outdoors, and catch the image on a piece of white paper. *Figure 7* is a photograph that I took with a 35 mm camera of just such an image on the wall of my office. Another method is to point a tripod-mounted telescope at the sun and catch the light that comes out the eyepiece on a white card. By adjusting the focusing you can obtain an image well outside the telescope where it can be conveniently inspected. *Figure 8* shows enlargements of sunspot photographs made by R. J. Bray and R. E. Loughhead using a 5-inch photo-heliograph installed by Dr. R. G. Giovanelli. You will not be able to see as much detail as in these pictures but you may well be able to discern the sharp division of the larger spots unto an umbra and penumbra.

SUNSPOTS   33

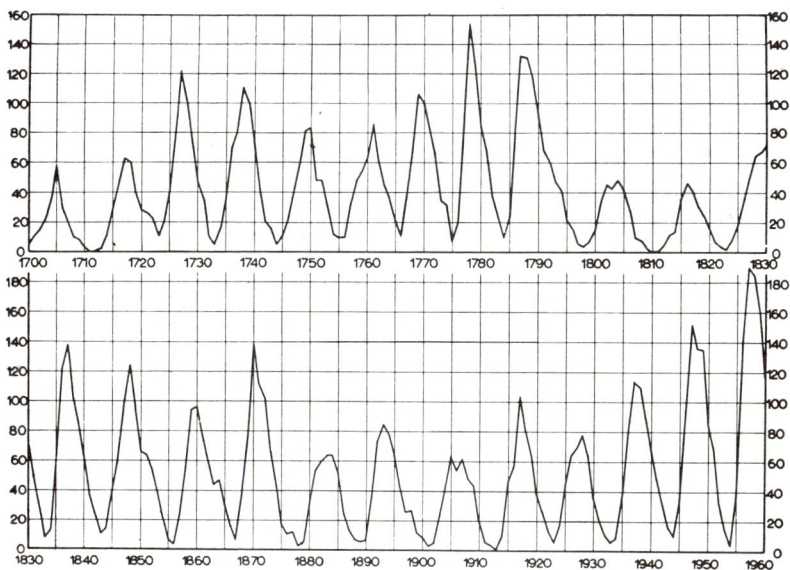

Figure 6. A graph of sunspot numbers since 1700 as compiled at Zürich by M. Waldmeier.

From an examination of *Figure 6* it can be seen that the peaks of activity occur at intervals of about 11 years, but closer examination also shows that the periodicity is not strict. The interval between one minimum and the next has been as little as 10 or even 9 years and as long as 13 years. The period of activity starting from one minimum and finishing at the next minimum is referred to as one cycle, and numbers have been given to the cycles. The one that started in 1965 is cycle number 20. Although the duration of the cycle varies rather widely, there may be some underlying regularity which one could seek for by careful study of the graph. It is worth pausing for a moment to contemplate the staggering amount of human effort that has gone into counting the sunspots, day by day unremittingly, to place us in a position today to survey the record of the coming and going of the sunspots over the past two-and-a-half centuries.

TABLE 2
Dates of Sunspot Maxima and Minima

Minima	Maxima	Minima	Maxima
1610·8	1615·5	1784·7	1788·1
1619·0	1625·0	1798·3	1805·2
1634·0	1639·5	1810·6	1816·4
1645·0	1649·0	1823·3	1829·9
1655·0	1660·0	1833·9	1837·2
1666·0	1675·0	1843·5	1848·1
1679·5	1685·0	1856·0	1860·1
1689·5	1693·0	1867·2	1870·6
1698·0	1705·5	1878·9	1883·9
1712·0	1718·2	1889·6	1894·1
1723·5	1727·5	1901·7	1907·0
1734·0	1738·7	1913·6	1917·6
1745·0	1750·3	1923·6	1928·4
1755·2	1761·5	1933·8	1937·4
1766·5	1769·7	1944·2	1947·5
1775·5	1778·4	1954·3	1959·0
		1964·8	1968

However, the record has been under scrutiny for a long time by many able people and I can say immediately that not very much underlying regularity has been found, certainly not enough to permit prediction ahead of time by a substantial fraction of a cycle. World-wide activity is devoted to the matter of prediction, for much the same reasons as for weather forecasting, namely the impact on human affairs. In this case the main impact is on overseas radio communication. Another interesting development concerns the safety of astronauts, who must not be launched into space on dates when the probability is high that dangerous flare radiation may be emitted. The flares take place where there are sunspots.

Meanwhile, a better understanding of what causes sunspots would no doubt help. One of the main problems of research of this kind, which is aimed at discovering the unknown, is what to do next. I am going to take sunspot research as a basis for a

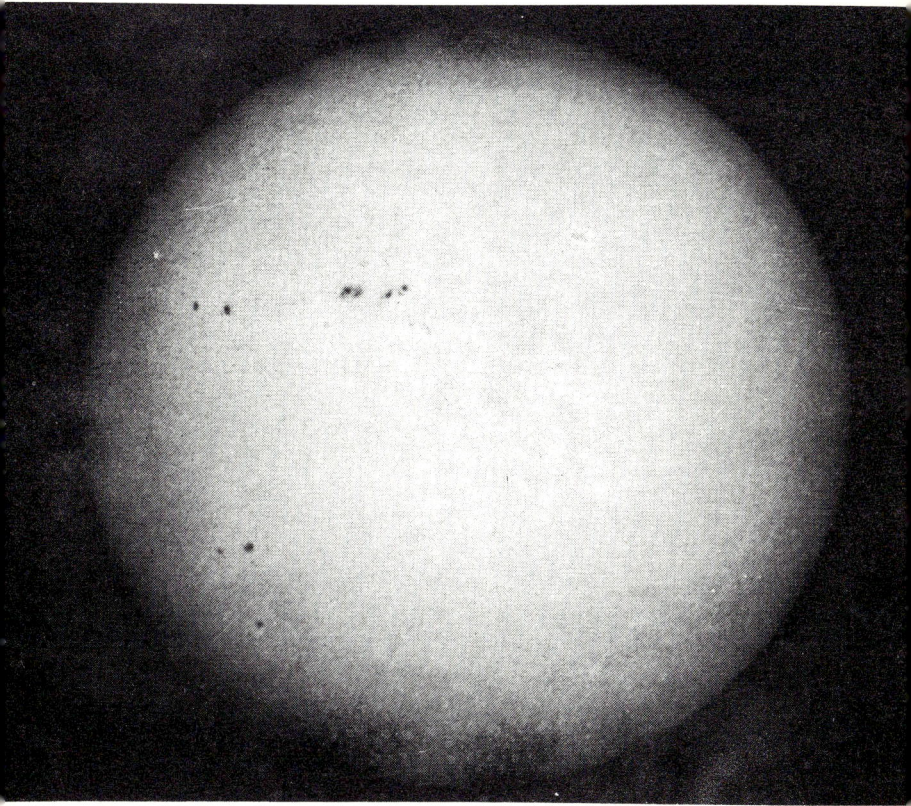

Figure 7. Photograph of a projected solar image showing sunspots.

commentary on this problem, not because you are likely to work in this field yourself, but because you may very well become involved with scientific research in one capacity or another.

Therefore, let us return to the year 1700 when the magnificent sunspot number data that we have today did not exist. At that time, the telescope had been available for nearly 100 years; Newton had already introduced the reflecting telescope, and great improvements continued to be made through the 1700's. As we

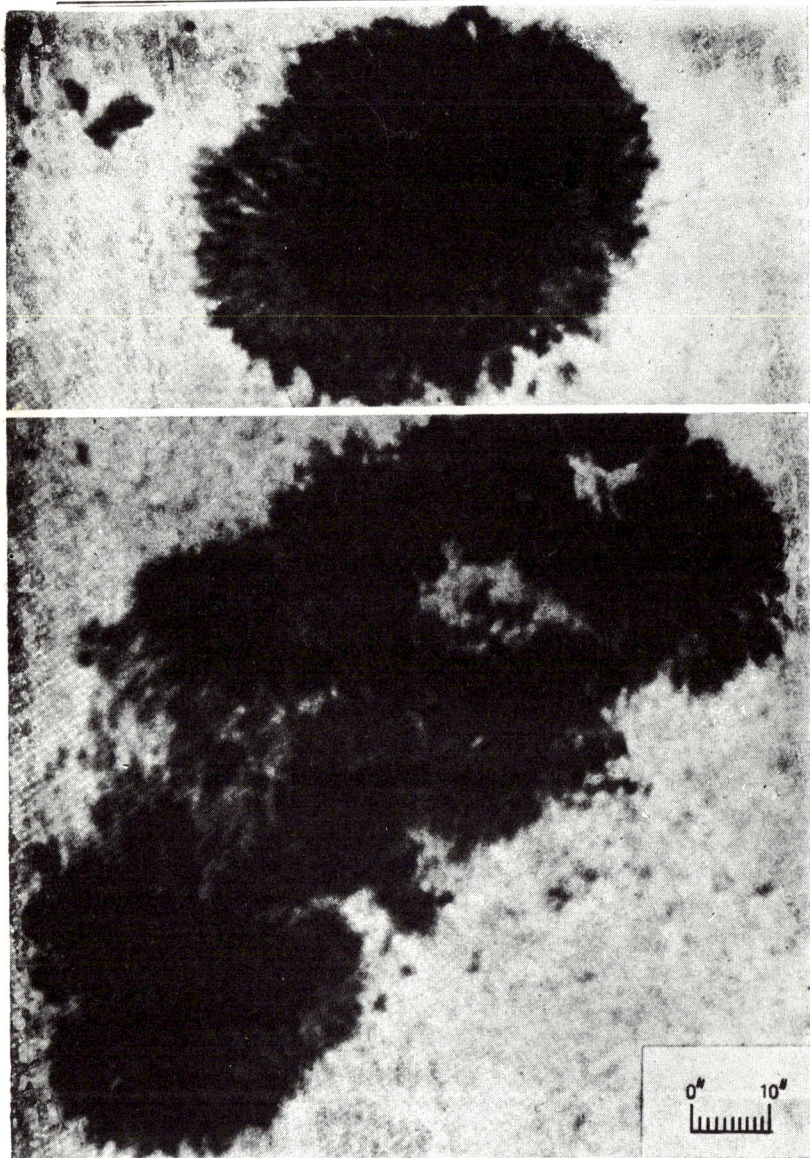

Figure 8. Close-ups of sunspots taken by Bray and Loughhead.

know from *Figure 6*, it is absolutely obvious that the sunspots exhibit pronounced 11-year cycles. Bearing in mind that the effect is so clear-cut, when would you say that the cyclical behaviour was first announced? Would you believe 1843? That is the fact. Please return to the sunspot number graph and locate the date 1843. The lesson to be drawn from this is, that the absolutely obvious can be completely invisible. In conducting scientific research, therefore, I think it is justifiable to assume that some of the answers are staring us in the face. Often, however, the questions we may be asking are not the ones to which the answers are waiting; part of the delay of 232 years between the first telescopic studies of sunspots in 1611 and the announcement of their striking cyclical variation in 1843 is no doubt attributable to this. A technique for allowing the available answers to speak is to accumulate reliable facts carefully, and this will be illustrated many times in the story that follows. Only occasionally has it proved possible to raise a question first and then obtain the answer. So now let us go back to 1611 when the telescope was first turned on the sun.

In *Figure 9* are copies of two drawings made by Galileo (1564-1642) showing the appearance of the sun as seen in the telescope on two different dates. From these drawings you can deduce immediately that the sun has rotated, and with a little care you can probably deduce the direction of the axis of rotation and the number of days taken for one complete rotation. I suggest that you copy one of the drawings on tracing paper and lay it over the other for this purpose. (How would you prove to your own satisfaction that the appearance of spots was not due to the transit of objects in orbit around the sun?)

Galileo concluded that the rotation period of the sun was about a lunar month, that the spots lay on the surface of the sun, and that they came and went with various lifetimes. He also saw that the sun's axis was not exactly perpendicular to the sun-earth line, by noting that the path of a sunspot across the sun was not exactly straight. The angle of inclination of the solar axis to the plane of the earth's orbit is close to 7 degrees.

The discovery of the sunspot cycle was made by S. H. Schwabe (1789-1875), an apothecary who for many years made sunspot

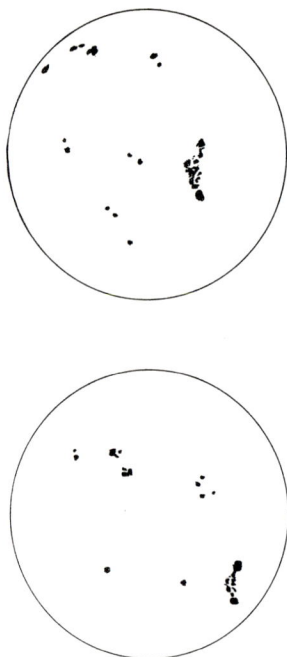

Figure 9. Copies of sunspot drawings made by Galileo on the 3rd and 6th May, 1612, immediately following the invention of the telescope.

sketches from an observatory on the roof of his house at Dessau. Starting in 1826, he made observations daily, or as often as weather permitted, until at the conclusion of the cycle ending in 1843, he could point to the evidence of a cycle in the number of spots. Later his observations were incorporated with those of other observers by R. Wolf in Switzerland, where most of the careful collation has been done on which the sunspot number graph now rests.

Even without knowledge of the true nature of a sunspot, it is apparent that the sun contains something that is oscillating. Our best plan for pursuing this problem is to accumulate some further facts by observation.

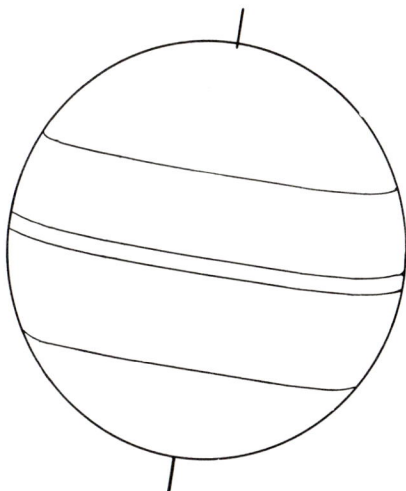

Figure 10. The zones where sunspots occur.

Strangely enough, Schwabe did not report a very odd thing about the sunspot locations. The sunspots are confined to two zones of the solar surface, one in the northern and one in the southern hemisphere, extending between latitudes of about 5 to about 40 degrees, as shown in *Figure 10*. As can be seen from *Figure 9,* however, when several spots are present at the same time, they are not all on the same circle of latitude but are spread out. It is therefore not easy to see, even from many, many months of observations, that the *mean* latitude of the spots changes progressively as the cycle wears on. When the data are suitably presented, as in *Figure 11,* this progression is quite evident. R. C. Carrington (1826-1875) discovered this.

If you think about this and suppose that whatever it is that generates the sunspots wells up from below and subsides again, every 11 years, then you see that it must also be situated at a latitude of perhaps 20 degrees at the beginning of the cycle and be approaching the equator at the end. So it is not back where it began 11 years ago. Now, since spots are beginning to form

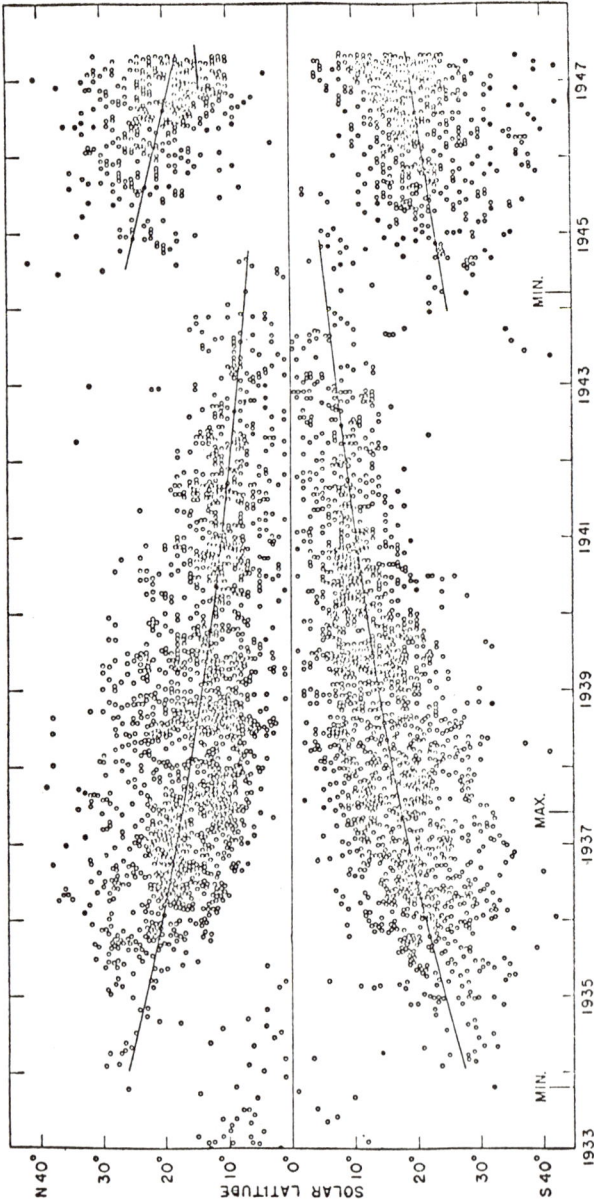

Figure 11. Each small circle shows the latitude of a sunspot group and the date when it was present (from G. Abetti, "The Sun", Faber and Faber, London).

again in latitude 20 degrees while the last equatorial spots of the old cycle are still occurring, we are forced to suppose that there are two of these subsurface things that generate sunspots (plus two more in the other hemisphere). Let me assume that in 1935 Thing A was lying well below the surface along the parallel of latitude at 20 degrees North, that it moved towards the equator over the next 11 years, rising nearer to the surface in 1937, the year of maximum sunspot occurrence, and was subsiding below latitude 5 degrees in 1944, and that the spots seen in latitude 20 degrees in 1944 were due to Thing B, which was beginning to rise. After the passage of ten years, in the middle of 1954, Thing B was sinking near the equator, but new high-latitude spots were appearing. Could they possibly be due to Thing A reappearing? If so, then 22 years would be the time taken for the situation to repeat itself. No matter what you suppose, *Figure 11*, or rather its extended form which goes back to 1874, makes it very difficult, though perhaps not impossible to see how the basic period could be 11 years.

Indeed the period is 22 years, and for the proof of this we now jump to 1908 in California where G. E. Hale (1868-1938) carried out an investigation that succeeded in answering a specific question. He asked, is it possible that the sunspots have a magnetic field associated with them. If they did, then certain of the Fraunhofer lines, which in the light from sunspots are double, could owe this property to a magnetic field, and the clinching observation would be to show that the light was circularly polarized. Hale therefore made a suitable instrument, found the polarization, and thus established the vital fact that sunspots in some sense were magnetic phenomena. It is easy to understand how the problem of the nature of the sunspots had remained insoluble to the observers of the previous three centuries, for an essential piece of the puzzle was missing. And yet in these three centuries some extraordinary puzzles were solved, such as the motions of the planets, and remarkable intellectual achievements were registered; for example, the development of the theories of electricity, magnetism, and relativity. One can only conclude that the intellectual giants could, nevertheless, not have coped with the problem of sunspots in those days. Some problems are just harder than others, and when the time is ripe, they are solved (sometimes after a delay). We still do not know what causes sunspots, and I cannot tell you today whether the time is ripe, or whether there is a piece of the puzzle missing that will not turn up for another 100 years.

To return to the discovery of magnetism in sunspots, Hale found that the groups of sunspots were frequently bipolar, i.e., some of the spots had one magnetic polarity and some had the other. The spot groups are generally elongated in the east-west direction, and in 1908 the situation revealed was that the easterly spots were south poles and the westerly ones of the same group were north poles — in the northern hemisphere; but in the southern hemisphere it was the other way around. This situation remained rather definite, bearing in mind the irregular character of spot groups, until the minimum in 1913. When spot groups of sufficient size appeared with the advent of the new cycle, a somewhat puzzling thing was found — the whole polarity picture appeared to have reversed! Now I hope you can appreciate the humorous side of this story. Can you imagine the photo lab technician explaining to the astronomer, "Now, as I hold this photographic plate up to the light, with the emulsion side *towards* me and the reference mark on *top*, the line on the left has left-hand polarization." Astronomer: "Was the plate put in the plateholder back-to-front or upside down, and what about the polarizer, was it reversed by chance, and are you sure I didn't do the left-hand setting first this time? Perhaps the electrician crossed the wires on the button when he repaired that trouble last month." Apparently the answer was —"Well — no." Still this poses a dilemma. Would you announce this new discovery, and risk your reputation, when all the time maybe the electrician did get the wiring crossed last month. The really funny part of this incident is that it would be another 11 years before a conclusive test could be made. So it was with much relief when it was found in 1924 that the polarity picture again reversed, without any shadow of doubt. Since then, it has reversed in 1934, 1944, 1954 and 1965, so we all firmly expect now that it will reverse with every new future cycle. This observation confirms definitely that the true period is 22 years. (How do we know some further fact may not show the true period to be 44 years?)

This is all I have to say about the internal solar oscillator with its 22-year period. Much speculation has developed along physical lines, mainly inspired by the discovery of the magnetic field, and invoking internal motions such as torsional oscillations of an interior doughnut, which is brought to rest and reverses itself every 11 years as it winds up magnetic field lines like a spiral clock-spring. Some mathematical developments have also been pursued: the one I like best is a theory proposed by G. U. Yule, which aims to explain how an oscillatory system can have a period that varies widely from cycle to cycle. His picture was of a pendulum which, if left to itself would swing perfectly regularly, but which was driven in an irregular manner by small boys firing at the bob with pea-shooters.

Terrestrial effects

Now when a body as important as the sun is for earthly existence, has a pulsation, then the earth pulsates with it. Subtle as this oscillation may be, making no perceptible difference to the diameter of the sun or to its power output as we measure it, and being apparently buried below the photosphere, still it has to be admitted that the terrestrial consequences are very striking.

Figure 12 shows the critical frequency of the F2 region of the ionosphere as observed in England every year since 1932. Just let it be stated here that the ionosphere is that region of the upper

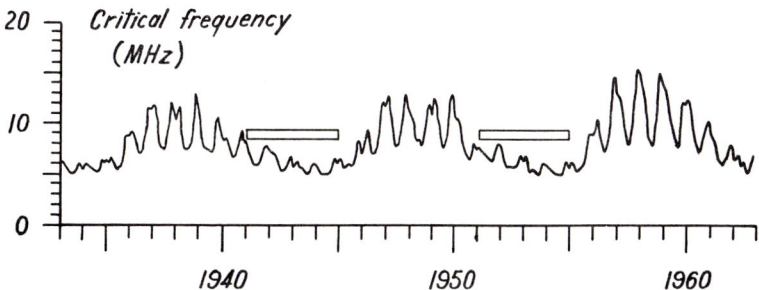

Figure 12. The critical frequency of the ionosphere at noon in England.

atmosphere, at a height of roughly 300 kilometres, where the ultra-violet light from the sun spends itself. Instead of reaching to ground level, where it would be instantly lethal to life, the ultra-violet rays do their damage in the rare upper atmosphere by ionizing the air molecules. The electrons so chipped off give electrical conducting properties to the air, which can then turn radio waves back to the earth and permit long-distance propagation on short waves. If the frequency of an electromagnetic wave emitted from the ground is too high, however, it will go through the ionosphere without benefit of reflection. The dividing line generally falls in the short wave band and is called the critical frequency. From *Figure 12* it is apparent that the critical frequency varies a lot from summer to winter, a fact that is connected with the obliquity of the sun's rays (although it may puzzle you to note that the critical frequency is lowest in summer when the sun is beating down most directly). The seasonal variability is so marked that some bands, such as the 30 metre band (frequencies around 10 MHz) cannot be used at all in some months. Now on top of the seasonal variation, if you will compare with the sunspot graph, you will see that the solar cycle is evidently impressing itself on our ability to communicate by radio. Thus the years 1941-1945 and 1951-1954 were lean years for long-distance communication on 10 MHz *(Figure 13)*, because the ionosphere had thinned out drastically. These years were years when one solar cycle was ending and a new one beginning. So for an abstruse phenomenon buried beneath the solar surface, the impact on human affairs was considerable. If you knew how many dollars a minute it costs to send international press messages by radio you could almost say how much expense the communications companies are put to per sunspot. This is a real cost and shows up in terms of the money that has to be raised to lay new transoceanic telephone cables and to launch communications satellites to carry the traffic that the ionosphere cannot sustain.

A second phenomenon connected with sunspots, and which historically was noticed much earlier, is the behaviour of fluctuations in the earth's magnetic field. Everyone knows that the compass points to the north, not exactly, but not too far off, but you may not have heard that if you sit and watch the needle

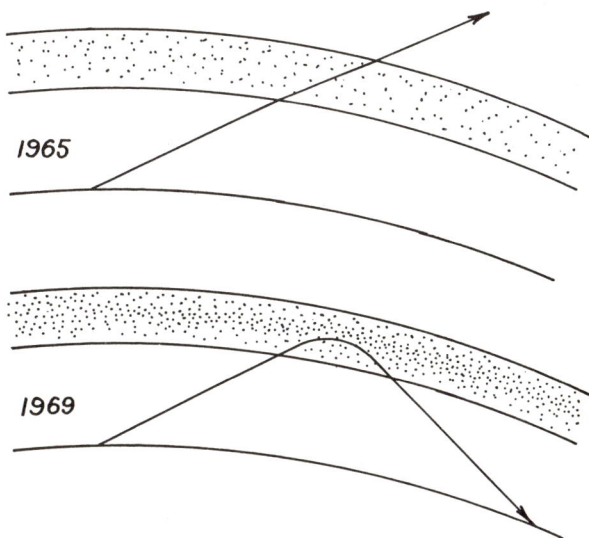

Figure 13. Conditions for radio propagation around the curve of the earth on a frequency of 10 MHz were bad in 1965, a year of few sunspots, but will be good in 1969.

closely for 24 hours you will see it wave back and forth over a small arc. This is due to the fact that the compass needle is acted on largely by the interior magnetism of the earth but also partly by electric currents flowing overhead in the conducting regions of the atmosphere. The strength of the current system varies according as it is night or day. Within a few years of the discovery of the solar cycle by Schwabe, a similar cycle was found in the amplitude of the diurnal variation of the compass, and the two cycles were found to be in agreement. In addition to controlling the diurnal swing of the needle, the sun produces wild disturbances of the compass, on occasion. This is one of the most fascinating stories, and it is worth recounting how it came to be understood. The trail, as we shall see, was rather indirect, and it begins with the work of an English instrument maker, G. Graham (1675-1751), who is well known for his contributions to the mechanism of the clock and who made the telescope with which Bradley discovered the aberration of light. It had been known for a long time that the direction of magnetic north was drifting and this was not a trifling matter because over the period of available records at London north had shifted by 20 degrees. However, the annual shift was only nine minutes of arc, which is a rather small angle, but not particularly difficult to measure. By taking reasonable precautions over the mounting of a needle a foot long and providing good reading arrangements, Graham immediately discovered, in 1722, the diurnal variation of direction, which was fully as large as the expected annual shift. With a needle of similar sensitivity, A. Celsius (1701-1744), who is well known for introducing the centigrade thermometer, and O. P. Hiorter, made a remarkable discovery in Upsala. The following quotation is taken from S. Chapman and J. Bartels, "Geomagnetism", Oxford, 1940.

" '. . . in Hiorter's words: A motion of the magnetic needle has been found which deserves the attention and wonder of every one. Who could have thought that the *northern lights* would have a connection and a sympathy with the magnet, and that these northern lights, when they draw southwards across our zenith or descend unequally towards the eastern and western horizons could within a few minutes cause considerable oscillations of the magnetic needle through whole degrees? The first time that I saw an aurora to the south and noted simultaneously a great movement of the magnetic needle was on March 1, 1741, in the evening, although at various other times I had noted a displacement of the needle, but on account of overclouded skies had not known whether this was accompanied by northern lights. When I announced this to the professor (Celsius), he said he too had noticed such a disturbance of the needle in similar circumstances, but had not wished to mention it, in order to see (this was his expression) whether I too would light on the same speculation. . . . Subsequently this was always observed, especially on the following April 5, 1741, when the needle at 2 p.m. began to be disturbed, so that at 5 p.m. it was $1°$ 40' to the west of its direction at 10 a.m. The most remarkable feature of this motion of the magnetic needle was the following. The professor (Celsius), some weeks earlier, had by letter requested Mr. Graham at London to observe his own needle for some days, so that, should our needles be disturbed, one might know whether this occurred also at many widely separated places, in which case our observations could not be ascribed to any special properties of our observing room and of the iron in it. What happened? The magnetic needle at London had just such an unusual motion at London at the same time as here at Upsala.' "

"Graham described his observations of Sunday, April 5, 1741, as follows:

" 'The alterations that day were greater than I have ever met with before. Tho' no alteration of any thing in the Room could occasion it. ·It was a fine day. I was alone all the time, and observed the Needle with all the care possible, sometimes in 2 or 3 minutes of time a sunder, when I perceived the Needle changing. Whether the cause of the change proceeds from one place, where the Needle was placed, or from any other, I know not; the only thing, in which I am certain, is, that there was no change of position of any thing in the Room, that could cause it, being alone all the day (I mean Sunday the 5th) when the greatest alteration happened. The observations upon the other days were made with the same care, but they are much less, and more regular etc.' "

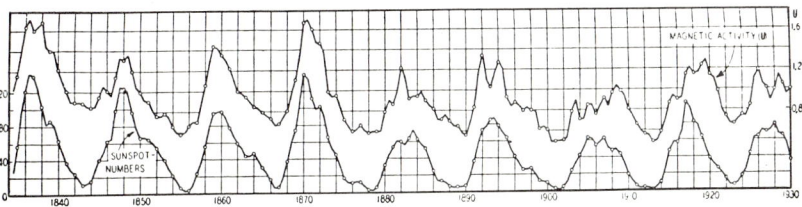

Figure 14.   Disturbances of the magnetic compass are strongest when sunspots are numerous (after S. Chapman).

The taking of regular records of the earth's magnetic field was organized in due course, so we can now present information such as that in *Figure 14* which shows that magnetic disturbances are strongest and most numerous in the years when sunspots are numerous. It turns out that the aurora itself does not act directly on the magnet; the polar aurorae are a separate example of a terrestrial phenomenon evoked by the sun. In years of high sunspot activity they are frequent and at sunspot minimum they become rare.

Weather effects have been examined, and other less likely phenomena, and no doubt many solar cycle influences of a subtle kind will show up. Professor Gamow once wrote that the American Revolution, the French Revolution, the Paris Commune, both Russian Revolutions and also some others fell fairly close to the years of maximum solar activity, and you may wish to compare the table of dates of maxima and minima with some chronological table of world events to see for yourself whether sunspots have an influence on them.

CHAPTER 4

Active Regions on the Sun

The first solar flare

Between 1853 and 1861, R. C. Carrington (1826-1875) made daily observations of the sun, and while I am not going to refer to the several important advances that he made, I can strongly recommend that you glance at his "Observations of the Spots on the Sun", published in 1863, if you are ever in the neighbourhood of a library that has a copy. It is a remarkable record of the work that a single individual can do. The story to be told here is of a unique occasion that, as it turned out, was not to be repeated for many decades. On 1st September, 1859, at 11.18 a.m., Carrington was making a drawing of a very large sunspot, one that had already been accompanied by great auroral displays and magnetic disturbances, and here is what he saw.

"The image of the Sun's disk was, as usual with me, projected on to a plate of coated glass . . . a picture of about 11 inches. I had secured diagrams of all the groups . . . when within the area of the great north group . . . , two patches of intensely bright and white light broke out . . . . My first impression was that by some chance a ray of light had penetrated a hole in the (shading) screen attached to the object-glass, for the brilliancy was fully equal to that of direct sunlight; but . . . by causing the image to move I saw I was an unprepared witness of a very different affair. I thereupon noted down the time by the chronometer, and, seeing the outburst to be very rapidly on the increase, and being somewhat flurried by the surprise, I hastily ran to call some one to witness the exhibition with me, and on returning within 60 seconds, was mortified to find that it was already much changed and enfeebled. Very shortly afterwards, at 11^h 23^m, the last trace was gone."

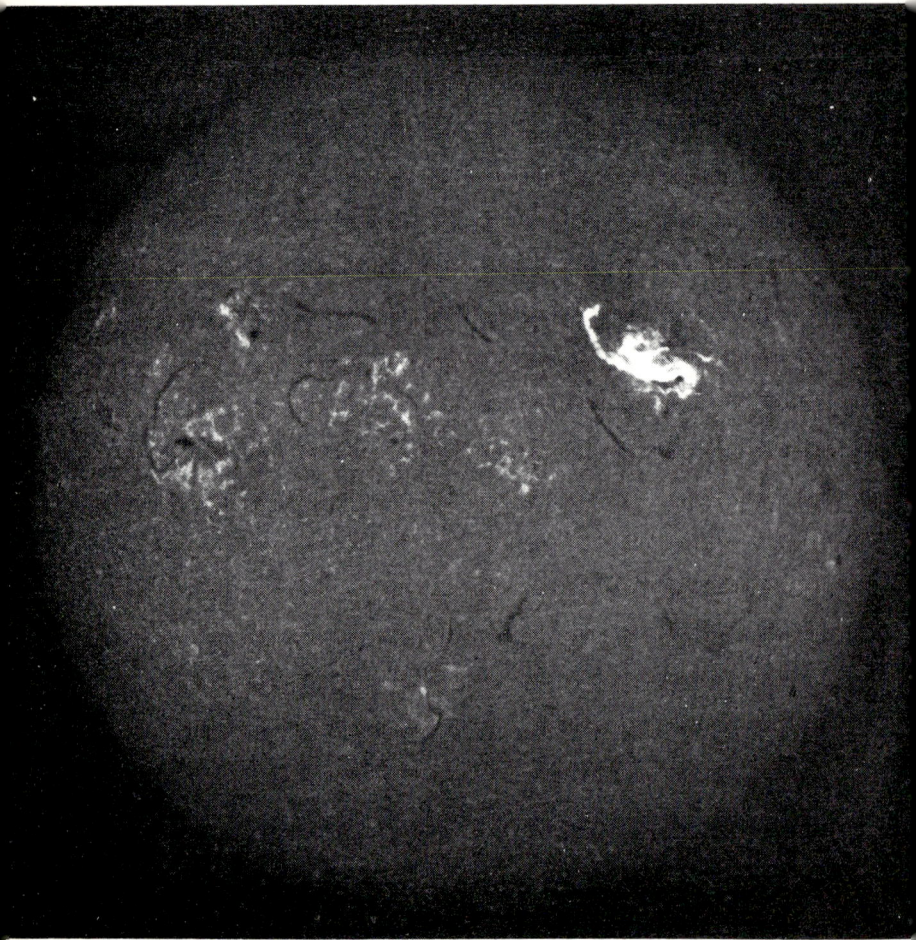

Figure 15. Filtergram of a solar flare taken by G. Morton at Lockheed Solar Observatory through a filter passing the red line of hydrogen (the H∝ line).

Only a handful of observers have since had this experience, but the solar flare, as this phenomenon has come to be called, only becomes visible in white light on rare occasions when the flare is of the greatest intensity. (Surprising to relate, the flare of 1859 was also seen by another solar observer, Hodgson.) More often, a flare emits the line spectrum of hydrogen and other elements and therefore a very sensitive mode of observation is to select a narrow band of light from the spectrum at a wavelength where the flare effect is concentrated. *Figure 15* is just such a photograph, or spectroheliogram, taken with the reduced light of the hydrogen Fraunhofer line at a wavelength of 6563 Angströms.

The beginnings of flare research

At precisely the same time as the flare of 1859 was seen, a small but distinct kink was recorded in the three automatic records of the elements of the earth's magnetic field at Kew Observatory. This very special kind of magnetic disturbance is now known as a magnetic crochet and is known to be caused by a burst of X-rays emitted by the sun at the time of a flare. These X-rays, which are absorbed in the earth's atmosphere towards the bottom of the ionosphere, produce extra ionization which, urged along by dynamo voltages induced by tidal atmospheric motion across the earth's general magnetic field, constitutes an extra overhead electric current for the time being, and it is the magnetic field of this current that shows as a kink in the record. Although this is a long chain of events, the fact that the X-rays and the light travel with the same speed from the sun (taking about eight minutes) means that the visual observation and the magnetic effect take place together. Many years later at the time of the sunspot maximum in 1928, H. Mögel in Berlin identified a special class of short-wave radio fadeout and found that the magnetic crochet accompanied them also. It is now very well understood how the extra ionization, produced at the base of the ionosphere by the X-ray burst from the flare, causes short waves to be absorbed. This may seem to you to be in conflict with the fact that the extra ionization produced during the years of sunspot maximum improves radio communication. The clue to the difference lies in the greater penetrating power of flare X-rays, which are of rather short wavelength (around 10 Angström units) and go through most of the ionosphere without affecting it much.   As they penetrate deep into the atmosphere, ultimately reaching levels below 100 kilometres where the air is dense and of a suitable composition to absorb them, the electrons released by ionization find the atmosphere too dense for them to move as freely as in the upper ionospheric regions.   When an incident radio wave acts on electrons in such an environment its energy is quickly converted to heat.

The clear connection of the solar flare with the magnetic-radio phenomenon was established within the following few years and initiated an era of close attention to flares, which, with the invention of the spectrohelioscope by Hale in 1926, were now conveniently observable.   The discovery of the X-ray emission, however, had to await the appearance of high-altitude rockets.

Rockets and flares

When rockets became available there was plenty of incentive for firing X-ray and ultraviolet detectors into the upper atmosphere. For years it had been supposed that a variety of magnetic phenomena depended upon the presence of solar ionizing radiation in the upper atmosphere and of course the very existence of the ionosphere depended on the same idea.   However, it was a complete mystery what wavelength of electromagnetic radiation would be responsible and opinions ranged over three or four decades from the short X-rays or gamma rays to the near ultraviolet.   One theoretical possibility, that of reasoning back from radio observations of the behaviour of the ionization ran into difficulties for a number of unforeseen reasons, one of which was that the composition of the upper atmosphere is quite different from that at sea level, and in ways that were to prove difficult to disentangle.   If you knew that the familiar mixture of nitrogen and oxygen carried on up to great heights, then you could calculate, from experimental data obtained in the laboratory, what wavelengths would penetrate how far and compare your expectation with the ionization produced, as measured by radio.

Now there were some problems of practical rocketry connected with observing flares; the first success was obtained by H. Friedman and his collaborators, of the U.S. Naval Research Laboratory, using a rockoon.   This is a rocket on a balloon.   Each day a small rocket was lifted to 80,000 feet by a helium balloon, where it would float all day until word of a flare came. Then the rocket would be fired by radio control up into the lower ionosphere. One of these shots succeeded and revealed the presence of X-rays in the range from 3 to 8 Angström units down to a height of 77 kilometres above sea level. These are the penetrating rays previously referred to.

Since then many similar rocket flights from ground level have greatly extended the information and here is the picture as now understood. First of all, the normal corona at its temperature of $10^{6°}$K, emits X-rays over a broad band around 100 Angströms, and traces of this radiation had been picked up as early as 1949 with the aid of V-2 rockets that were brought from Germany and launched in the United States. If we heat a volume of the corona to a temperature of $10^{7°}$K, then we can quickly calculate the approximate wavelength of the radiation that will be emitted, as follows. In a gas at temperature T the kinetic energy of the particles is $(3/2) \, kT$, where Boltzmann's constant k is given by

$$k = 1 \cdot 38 \times 10^{-16} \text{ ergs per degree}$$

and the energy of a quantum of radiation at frequency v is hv, where Planck's constant h is given by

$$h = 6 \cdot 62 \times 10^{-27} \text{ ergs per Hertz.}$$

Equating these two energies, we have

$$(3/2) \, kT = hv$$

and we can solve this equation for the frequency, or better still for the wavelength λ, remembering that $\lambda = c/v$ to obtain

$$\lambda = 10^{-7} \text{ centimetres} = 10 \text{ Angströms,}$$

which is about what is observed. On the basis of this calculation it is therefore thought that a flare heats up a volume of the lower corona to the immense temperature of 10,000,000°. Though easy to say, this temperature is far beyond our experience. To appreciate its significance it is necessary to know that practically any atom in the solar atmosphere, if hit by such a particle or a photon, can be not only ionized but completely denuded of all its orbital electrons. The electron which recombines with a nucleus under these conditions does so with such an impact as to emit an X-ray photon. Oddly enough it is not at all difficult for us to generate such X-rays ourselves, but needless to say it is not done by heating. The necessary energy can be delivered to an electron by attracting it to an anode held at a potential of about 1000 volts. By the time it reaches the anode the electron is going at the same speed as it would have in a 10,000,000° gas and whatever it hits next is likely to give out an X-ray. We can also generate tiny volumes of highly stripped atoms by putting neutral atoms in a vacuum tube and allowing a capacitor charged to 10 or 20 kilovolts to pass a spark through. This is sufficient to allow the spectra to be determined.

Theories of flares

Just how flares are caused is still under discussion and it is hardly possible to do justice to this subject here because a good deal of observational material would have to be referred to that I have not had time to mention. Later we shall encounter further flare phenomena but the whole body of knowledge about flares, accumulated with the spectrohelioscope and other optical instruments, must be passed over.

It is clear that the flare problem is closely tied in with the problem of the sunspot. For one thing, flares are only observed in sunspot groups. Just as the sunspot is a basically magnetic phenomenon, so also probably is the flare. Here is one descriptive account that is widely entertained in one form or another. As a sunspot group grows, the magnetic field above it grows and becomes the seat of much stored magnetic energy. With the passage of time, a limit to the storage is reached, possibly because of twisting of the lines of force; then suddenly the stored energy is dumped. There is a rapid rate of change of the magnetic field, which is a way of accelerating charged particles. Some of these particles escape, and some heat the surroundings, the various radiations from which later strike the atmosphere and surface of the earth. This catastrophic event relaxes the magnetic field but next day it may be all tied up again and ready to produce another flare. The energy for this might come from churning of the solar interior below the level where the magnetic field lines dip into the photosphere, and you can understand how the kinetic energy of rotation is transported above the photosphere by analogy with the way we wind up model aeroplanes by turning the propeller with the finger.

The mechanical energy of rotation is fed through the fuselage in the form of elastic energy that travels along the rubber band. This analogy represents the flare theory quite well, for if you slowly keep feeding more and more energy in, a certain limit will be reached and the stored energy will be suddenly released with a sharp crack. This kind of theory has much in common with the theory of earthquakes where again a steady conversion of convective kinetic energy of the earth's crust into elastic strain energy leads, at irregular intervals, to sudden releases.

Another picture is presented by Professor Alfven, who copes with the feature that so much energy appears so suddenly all in the one place by likening it to the behaviour of an electric circuit when the switch is suddenly opened. If the circuit is carrying current, there will be a spark. Now just before the switch was opened, the energy of the circuit was stored in the magnetic fields surrounding the circuit, which could be quite extensive. When the spark takes place, and has finished, there is no more current in the circuit and so no more stored energy. We see that the whole of the energy which was originally widely distributed, manages to concentrate itself suddenly in the one small spark discharge. Alfven points out that it is a perfectly common property of current-carrying plasmas to become suddenly non-conducting, a fact that causes technical difficulties in connection with the application of mercury arc rectifiers to high voltage power distribution and also with the machines intended for nuclear energy production by controlled fusion of hydrogen. There are certainly large volumes of the solar atmosphere carrying electric currents and it may be that the flare is like a spark trying to burn its way through a temporary blockage in the circuit.

Aspects of solar activity

The coming and going of sunspot groups and the occurrence of flares are among the manifestations of solar activity. A variety of other phenomena occur too. If *Figure 15* is looked at carefully, bright patches will be seen in addition to the very bright flare. Later the flare would subside to a similar level of brightness. These bright patches, or plages as they are now called, are a normal accompaniment of sunspot groups. Another example is given in *Figure 16*, which is unusual in that the moon's limb was obscuring

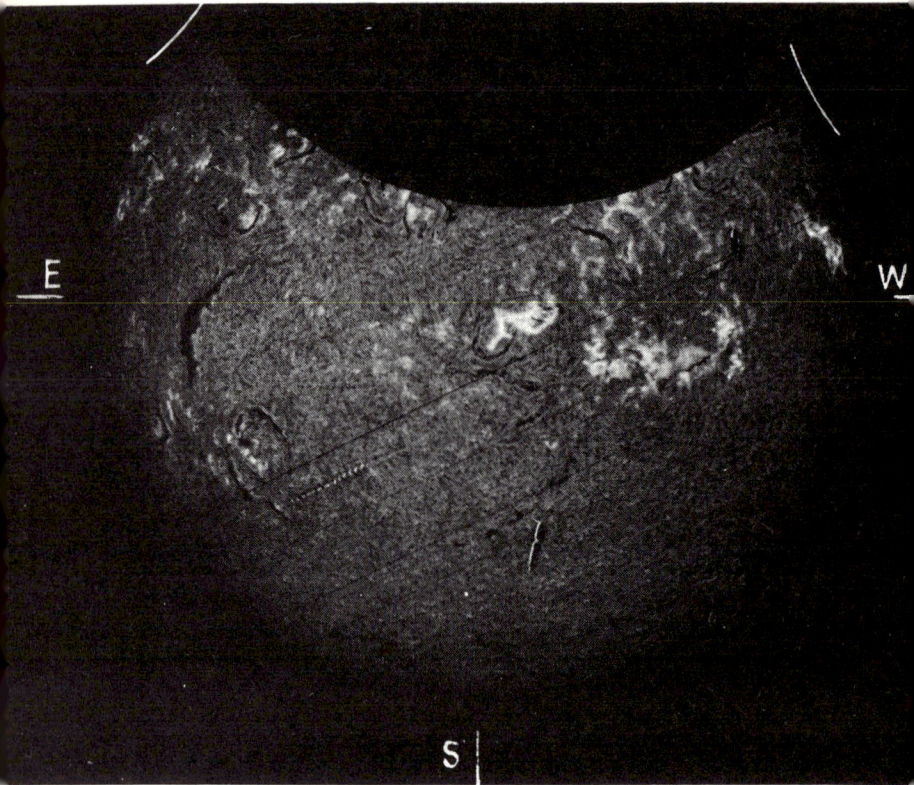

Figure 16. Spectroheliogram made by L. d'Azambuja at Meudon Observatory during the partial eclipse of 28th April, 1949, using H$\propto$ light.

part of the sun at the time. At the time of this eclipse my colleague, T. Straker, and I were observing the bottom of the ionosphere and we saw it go up and then down by about one kilometre. From this one could make some deductions as to where on the sun the ionizing radiation was coming from and prove quite clearly that the radiation was not the same as that producing the rest of the ionosphere.

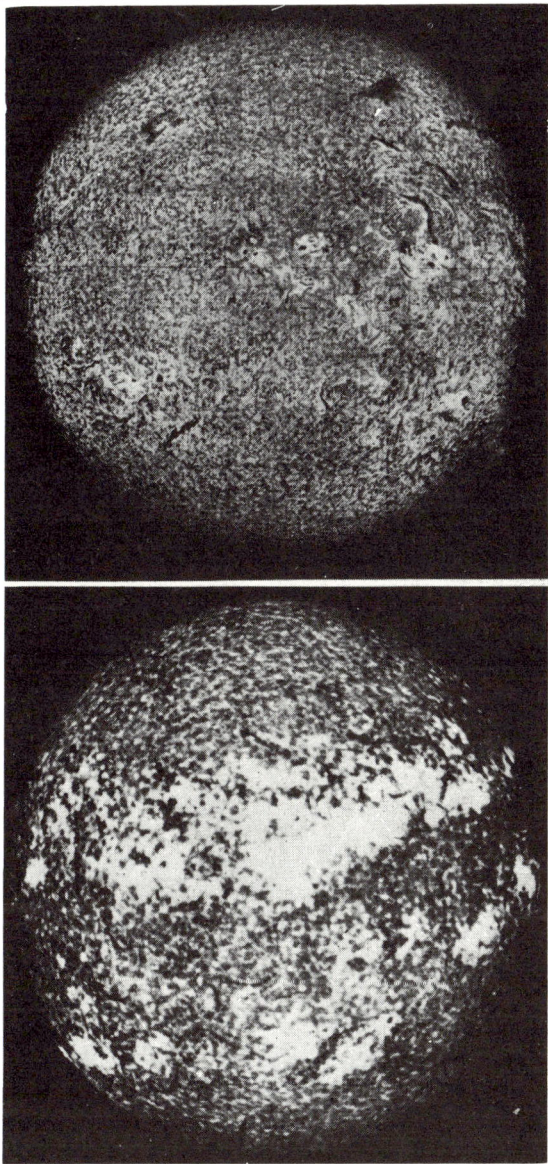

Figure 17. Spectroheliograms made at Meudon Observatory on the same day, one in Hα light and one in the blue line of calcium.

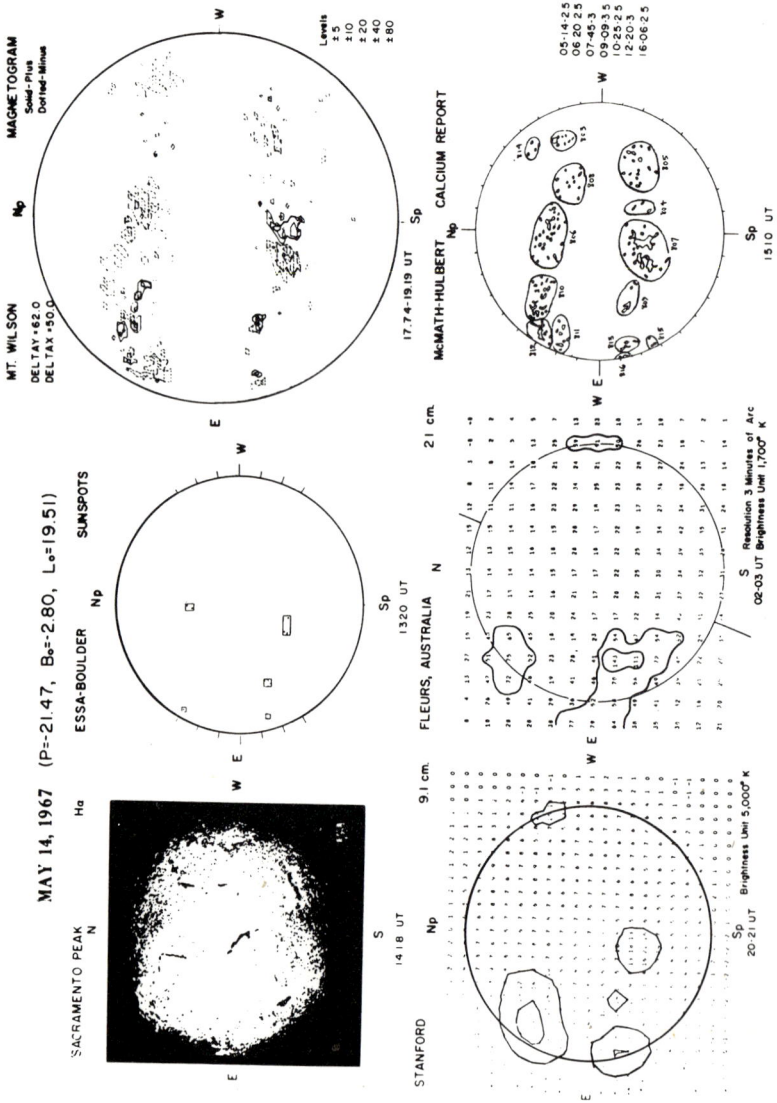

Figure 18. A page from "Solar-Geophysical Data", published monthly in the United States by the Environmental Science Services Administration, giving solar data of several kinds for each day.

In the spectroheliograms the sunspots may or may not be visible. In some cases they are covered by overlying chromospheric material which does not give out much light, but in the narrow band under observation can obscure the spot.

Other things that can be seen are numerous dark markings, or filaments, such as the one near the east limb in *Figure 16*. This one, if it could be seen on the limb, would present an appearance like that of the anteater prominence in *Figure 5*, except that the latter is very large. Other dark filaments can be seen in the neighbourhood of plages. They take their origin there, and may become very long lived, drifting away towards the poles over the course of months.

Figure 17 is a pair of spectroheliograms taken at the same time but one is made in the blue line of calcium. With difficulty, many of the features can be identified in both pictures, but owing to the different range of heights in which the lines are formed the details are quite different. The message conveyed by the calcium spectroheliogram is that the whole of the zones occupied by sunspots (compare *Figure 10*) is strongly affected by solar activity. Clearly it is the radiation from the active regions that is responsible for the main part of the ionosphere and for the solar cycle dependence revealed in *Figure 12*.

Because of the importance of these various solar phenomena in our physical environment, efforts are made to make and distribute synoptic data to interested parties. One such effort is the monthly bulletin of the Environmental Science Services Administration, Boulder, Colorado, a page of which is shown as *Figure 18*. The information includes a filtergram from the Sacramento Peak Observatory, New Mexico, a calcium plage report from the McMath-Hulbert Observatory, Michigan, a magnetic map from Mt. Wilson, California, and a sunspot drawing from Boulder, Colorado. In addition you will notice that there is a radio map of the sun from Fleurs, N.S.W., and it is to radio phenomena that we turn next.

CHAPTER 5

Radio Phenomena Caused by the Sun

Noise storms

One of the well-known stories of solar activity refers to the great stream of radio noise that poured out of the sun in 1945 and 1946, raising the question whether the sun as a whole was responsible or whether it was the sunspots. This poses a problem of some technical difficulty because the sun has an angular diameter of only half a degree and therefore one requires a radio antenna of high discrimination. Now if records could be made merely of the occurrence of the noise it would be possible to say from comparison with sunspot observations whether or not the sunspots were responsible. No regular records were then available, but fragmentary reports dating back to 1942 did point to sunspots. At the time a coastal radar installation suitable for height finding on aircraft existed at Collaroy, N.S.W., and by analysis of records of the radio noise received as the sun rose over the Pacific Ocean, Pawsey, Payne-Scott and McCready were able to show that the source of the noise was above a great sunspot group. At Mt. Stromlo, Dr. D. F. Martyn found that the radiation was circularly polarized, an indication that the particles responsible for the emission are circling under the influence of a magnetic field. In the following 20 years a good deal was learned about the solar radio noise, the most notable point being that there are many different kinds. Indeed the study of radio emissions from the sun has become a very rich field of investigation. The kind of noise that was studied in 1946 came to be referred to as a noise storm. With the aid of very ingenious receiving equipment R. Payne-Scott and A. G. Little found that the source of the radiation was confined to a relatively small volume of space located at a very substantial height indeed above the associated spot group, in fact the height can be as much as one solar radius. Since the duration of a noise storm can run into days, it seems that the magnetic field that imposes the circular polarization also contains the radiating particles, confining them to a certain volume of space until they are spent. Considerable mystery remains as to just what the noise storms are; one of the outstanding questions concerns the short bursts which occur with the noise storm and were first reported on in detail by J. P. Wild. Some noise storms consist of nothing but bursts and this is not yet understood.

Microwave emission from plages

In the general neighbourhood of a sunspot group, as can be seen from the spectroheliograms, there is the disturbed region of the chromosphere that we call a plage, and evidently this appearance is but a thin horizontal slice through a whole volume that is activated in some way by the same thing that causes the spots themselves. To start with, similar bright regions can be seen in white light on the photosphere, regions that we call faculae following the name originally introduced by Galileo when he first saw them. Also, above the chromospheric plage in the lower corona, we have the X-ray emitting regions that contribute to the ionization of the parts of the ionosphere that exhibit the solar cycle variation *(Figure 12)*. So there must be a cylindrical volume surrounding the sunspot group, no doubt including sub-photospheric levels that are invisible by the observational techniques we dispose of at present, and rising into the corona, and in this volume energetic phenomena of different kinds take place that do not take place over other parts of the sun. The effects of this active volume no doubt extend well out into the corona and you may examine the pictures of the corona with a view to discerning such effects. At each level the manifestations of activity are different and we are now going to talk about microwaves, which comprise a band of wavelengths extending from about 1 to 100 centimetres, but, for the sake of definiteness I shall concentrate on 9 centimetres. At this wavelength the corona is mostly transparent, but the chromosphere is absorbent. Consequently, if we look at the sun using the wavelength of 9 centimetres, we receive the radiation from the chromosphere, nothing from the photosphere, although it must contain such radiation, and a contribution from the lowest and densest part of the corona. The amount of radiation received is what we

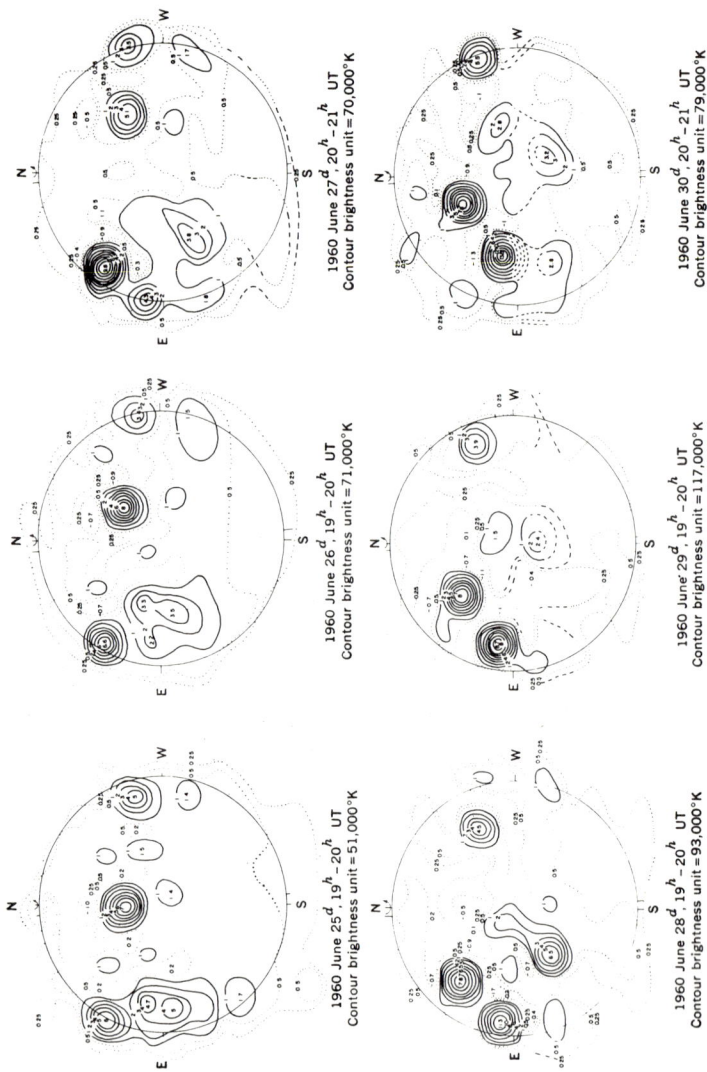

Figure 19. A sequence of microwave sun maps made at a wavelength of 9 centimetres.

1960 June 25^d, 19^h – 20^h UT
Contour brightness unit = 51,000°K

1960 June 26^d, 19^h – 20^h UT
Contour brightness unit = 71,000°K

1960 June 27^d, 20^h – 21^h UT
Contour brightness unit = 70,000°K

1960 June 28^d, 19^h – 20^h UT
Contour brightness unit = 93,000°K

1960 June 29^d, 19^h – 20^h UT
Contour brightness unit = 117,000°K

1960 June 30^d, 20^h – 21^h UT
Contour brightness unit = 79,000°K

would get from a black body at about 30,000°K, which is in reasonable agreement with expectation, and the diameter of the disc from which it comes is about 10 per cent larger than the diameter of the photosphere.

This suggests an interesting digression. We can read in an encyclopaedia that the sun has a diameter of 866,000 miles and a temperature of 6000°K. But if we had evolved with eyes that are sensitive to microwaves instead of to light then we would have seen that the sun was bigger and hotter. Which is the real sun? You have to admit that the brief statements in the reference books about facts of the astronomical universe are heavily biased towards optical means of perception, and necessarily so until recent times when it became possible to receive the radio waves that are sent to us by the celestial objects. The picture revealed by radio astronomy is usually different and supplements the optical view. Even so, the full picture often remains obscure, and we are now drawing on the spectral bands that can only be observed from outside the earth's atmosphere. On the subject of the human eye, have you wondered why its range of sensitivity is so limited in wavelength range, and why it evolved to operate in the vicinity of 6000 Angström units?

Now while we see a solar disc at 30,000°K, things are different over a plage. There we see temperatures of hundreds of thousands of degrees. Maps of the sun made at 9 centimetres are shown on *Figure 19*. Maps of this kind were first made by W. N. Christiansen at Fleurs, N.S.W., with the instrument shown in *Figure 20*, which is still in regular use, and produces maps that are published monthly in the United States (see *Figure 18*) for world-wide distribution; and since the coming into use of the Mills Cross, built by Professor B. Y. Mills, of the Sydney University School of Physics, similar maps have been obtained at a wavelength of 75 centimetres by A. G. Little, also of this School. The revolutionary new solar radio telescope at Culgoora, N.S.W., built by J. P. Wild, not only makes solar maps, but makes them at a rate of one or two per second so that it will be possible to play them through a projector like movies. The instrument at Stanford, California, which is shown in *Figure 21*, consists of 32 parabolic reflectors which follow the sun in unison by means of equatorial mounts.

Figure 20.   Some of an array of 64 antennae at Fleurs, N.S.W., used for mapping the sun.

Figure 21. A microwave antenna array at Stanford, California, with which the 9 centimetre sun maps are made daily.

You can see from the shadows that the antennae are not pointing at the sun in the illustration. The day has ended and the antennae have gone back to the eastern horizon, where they will start up automatically shortly after sunrise. The design of instruments that can point at one small area of the sun while ignoring the radiation coming at almost the same angle from adjacent areas of the sun is a very interesting subject in itself that you may wish to look into on some occasion.

Solar bursts

Mention has been made of the noise storm bursts that can be observed on metre wavelengths, but in addition a whole variety of irregular radio emission phenomena have been reported. To put it briefly, the occurrence of a solar flare is such a catastrophic event, that everything within sight, including the earth, gets jolted.

When the flare occurs, a volume of the chromosphere is suddenly heated, and for the time being compressed, and so the microwave emission from the chromosphere increases, taking a considerable time, perhaps an hour, to increase and fall off. Known as a "gradual rise and fall", this phenomenon may set in just before the flare starts.

A brief impulsive burst of microwaves, lasting only minutes, is received also at the start of the flare, and is probably due to fast protons ejected downwards from the flare volume spending their energy partly in very high frequency gyrations in the chromosphere.

The particles that are ejected upwards force their way through the coronal plasma exciting it into oscillation at radio frequencies over a wide range, then the energy radiated by these oscillations travels on over to the earth where it is received on metre wavelengths as a burst, or group of bursts, of a few seconds' duration.

A great shock wave also runs out from the flare explosion, with a much lower speed than the ejected particles, but nevertheless fast. As this shock travels upwards it again excites the coronal plasma, but for longer times, and the metre wave radiation received at the earth is generally referred to as an outburst. It has the very interesting characteristic that high frequencies are received minutes earlier than lower frequencies, which exhibits the property of the corona that the upper levels have the lower natural frequencies of oscillation. The same shock wave carries on out into interplanetary space, where it can be detected by the jump in magnetic field it produces as it passes by space probes instrumented with magnetometers, and in a day or two collides with the earth's magnetic field and sets in train a whole series of terrestrial disturbances including magnetic and ionospheric storms and auroral displays.

Many other consequences of the flare, both in the form of a radio emission and of other events, could be dealt with here, but they really merit more detailed attention.

The variety of phenomena has certainly proved to be complex, but it seems that the unfolding of these numerous optical, radio, corpuscular, magnetic, and X-ray manifestations has been a prerequisite to the explanation of the simple flare observation first reported by Carrington.

Science and Technology of the Environment

(Four Chapters)

by

GORDON J. F. MacDONALD

Dr. Gordon J. F. MacDonald,
Executive Vice-President,
Institute for Defense Analyses, Washington, D.C.

Science and Technology of the Environment

Introduction

The last few years have seen an explosive growth in knowledge about the earth, its oceans, and its atmosphere. Part of this growth has resulted from the application of new technologies to investigations of the earth. Satellites have added an entirely new dimension by enabling us to watch the detailed movements of cloud patterns and to explore conditions in the electrically-conducting part of the atmosphere, the ionosphere. What may be somewhat more surprising is that satellites have also aided our attempts to understand the inside of the earth. Because of satellites, we now know much more about the way matter is distributed within the earth and satellites have even made possible new estimates of mechanical properties of the earth's deep interior.

Technologies other than space have also contributed to knowledge about the earth. Nuclear explosions have released vast amounts of energy at exactly known times and positions and, combined with new arrays of seismic recorders, have led to a more accurate description of the physical properties of materials within the earth. Electronic computers have played an especially important role. The earth is so large, its properties so uncertain, and the time scale of the earth's processes so long that it is still rare to carry out direct experiments on the earth. I will discuss later the limited way we are experimenting with the environment. Using large computers we can conduct model experiments. These models, while incomplete and in some ways unrealistic, allow study of the environment without actually perturbing it.

Development and application of new technologies that add to our knowledge of the earth is not the only reason for the current excitement and interest in the environmental sciences. As knowledge has developed, rather precise questions about the earth can be asked; there is less emphasis on pure description and more on understanding. The challenges posed in answering questions about the earth possess the same order of difficulty and are as profound as fundamental questions in other fields of science. The intellectual excitement in the earth sciences has in turn drawn to the field some of the most brilliant scientists of our day.

There is another characteristic about the current state of the earth sciences which is worth emphasis. We have come sufficiently far in our basic understanding of the earth and its fluid envelope that it is possible to contemplate realistically the alteration of the environment for the benefit of society. We now perturb the environment, that is the atmosphere, the oceans, the solid earth itself, in numerous unplanned ways, and these alterations usually produce disagreeable results — pollution is the most notable example. In addition to the unanticipated modification of the environment, we are taking the first steps to produce beneficial results. I will discuss in broad outlines the beginnings of environmental engineering, a set of technologies which in the next fifty years may have a more profound effect on the world we live in than developments in nuclear technology. Environmental technologies are essentially global in nature and cannot be confined to a single country, and thus require the participation of many nations in their applications. These complexities, coupled with deep legal, economic, and sociological problems presented by these technologies, imply that they will require major institutional adjustments if they are to be effectively used.

I will begin by considering the atmosphere and explore briefly with you what present abilities are to modify the atmosphere. The oceans form one of the last great unexplored regions on earth. I will discuss what we know of the oceans and how application of this knowledge can alleviate certain pressing problems in the world today. While we know a little about the deep oceans, we know even less about the interior of the earth, for the simple reason that no one or no instrument has penetrated more than a very small distance towards the interior of the planet. Most of the knowledge we have about the earth comes from a study of the waves generated by earthquakes. Earthquakes themselves can be enormously destructive, but in this field the optimistic view is that within a few years we may be able to predict the occurrence of earthquakes and perhaps in the more distant future even prevent earthquakes. Finally, in the last lecture, I will return to the central problem about the earth: how did it originate and what is its place in our solar system?

CHAPTER 1

Weather Modification

Introduction

Considering the science of the atmosphere, it is important to ask whether or not it is or will soon be possible to alter the weather in a useful way. Serious intuition reveals that progress in weather modification must continue. For many centuries, societies have tried to modify their weather. The primitive rainmaker tried to imitate nature. For example, the American Indians blew water from special pipes in imitation of rainfall and in addition religious rites were conducted for the purpose of increasing precipitation. Ghastly practices resulted when the idea grew that rendering gifts or payment to the Rain Gods might be beneficial. In the beginning, the chief rainmaker would be slain for failure to bring rain; then a substitute was found by the chiefs to save their own lives. Often the sacrifice was a captured enemy. The North American Indians, for example, roasted young women from enemy tribes over a slow fire then killed them with arrows before eating their hearts and burying the rest in the fields they wanted irrigated with rainfall. More fortunate tribes used a jug of beer rather than a barbecued maiden as reward to the Rain God.

Yet for all this interest and activity, civilization can point to one major success. Today we still modify weather by constructing shelters. We might also consider clothing as a form of weather modification must continue. For many centuries, societies have tried attribute clothing to psychological rather than climatological pressures. In recent times, weather within the shelter has been modified and we use heating units and air conditioners to control our microclimate. This microclimate can be extended to the scale of a baseball stadium, but clearly there are limits to the use of shelters as a means of modifying weather and climate.

As I will discuss, successes in positive weather modification are still limited. However, it is important to realize that means for obtaining major successes appear to be at hand. While the positive effects of weather modification have had limited success, negative aspects continue at a frightening pace. The citizens of Los Angeles may be the most publicized dwellers of a polluted atmosphere, but they are far from alone.

There are two other reasons why I believe it is useful to examine weather modification. In weather modification, we see illustrated some of the characteristics of modern science. Science and technology and the world are changing. The essence of this change lies not so much in the expanding scale of discovery and application; it lies rather in the complete penetration of science and technology into the domain of human affairs. In its most primitive form, weather modification poses legal, economic, social and aesthetic problems to which we should pay attention. A solution of these problems is as essential to the health of our society as an understanding of the scientific aspects. A second reason that the discussion of weather modification is timely is that, of the many problems linking government and science, one of the thorniest is weather modification.

In the United States over the past twenty years, Congress has devoted substantial attention to weather modification. Indeed, it has frequently displayed considerable impatience with the scientific community's slowness to accept weather modification as an important area for research and development. Congressional reaction is well illustrated by a comment made by Senator Clinton Anderson after my presentation of the National Academy's Panel report to the Senate Subcommittee on Water and Power Resources in March, 1966. He stated: "I am pleased to see that the National Academy has finally concluded that the Senate Interior Committee has been on the right track for more than fifteen years in stimulating scientific and engineering research in weather modification. It is good to have you folks on the team, too."

The present close relationship between weather modification and government is not new. In fact, this relationship has had a long and somewhat chequered history. In the U.S., technological developments accompanying the Civil War led to a number of proposals for rainmaking. It was noted then by a number of retired generals that there was an apparent correlation between battles and rain. The more intensive the battle, the greater the rainfall. This apparent correlation led to the suggestion that the concussion of heavy cannon fire produced rain. In the years following the Civil War, there was limited demand for heavy cannons. The guerilla activities of the Plains Indians required only the attention of conventional cavalry, and besides the Western Plains were already largely defoliated. The suggestion was repeatedly made by consultants to the gunpowder industry that experiments be carried out with surplus powder. I am not certain but I believe at this time there was a lengthy discussion on the peaceful uses of cannon.

The apparent correlation between battles and rainfall actually has had a much longer history, extending at least as far back as Plutarch. One early explanation was religious and stated that the gods were offended by the carnage of battle and sent rains to purge the blood and gore that mortals made. But materialists had another answer. Rain was simply the condensation of blood, sweat, and tears of the warriors as they slaughtered each other. This last explanation however was discounted by scientists some time later when it was pointed out that even if 10,000 men were all liquid and all of them were completely vaporized and later condensed as rain, the result would only be about half an inch of rain on a very small field.

In 1890, the matter of rainmaking by cannonading was brought to the United States Congress, and a total of $9,000 — a very substantial sum for those days — was appropriated for use by the Department of Agriculture. In accordance with more modern practice, the $9,000 was divided into two sums: $2,000 for the investigation of rainmaking possibilities and $7,000 for practical experiments. Mr. George Dryenforth, later to be called George Dry-henceforth, was placed in charge of the Department's activities in this field. Although Dryenforth possessed many of the characteristics which we now associate with the modern entrepreneur of science, his qualifications in the weather modification field were somewhat nebulous. Dryenforth did, however, undertake extensive cannonading in various cities of the southwest United

States. He found this pastime rewarding in that he was always able to supplement federal funds with local funds. For example, at El Paso, Texas, the city fathers appropriated $800 for local experiments. The best Dryenforth could do though was to raise one small cloud and that, unfortunately, happened to be over Mexico.

Today, there are some who would claim that weather modification has not advanced much beyond the state in which Dryenforth left it. However, renewed interest in the subject has developed in the last twenty years. This time a much better definition of problems in various areas of weather and climate modification has been achieved. The subject of weather and climate modification is concerned with any artificially produced changes in the nature and behaviour of the atmosphere. The changes may or may not be predictable and modification may be deliberate or inadvertent. Duration and size are relevant, for weather turns gradually into climate and weather phenomena span the range from local to global. The objective within weather modification centres on control: to produce deliberate beneficial changes in the environment within man's limited ability to do so and to bring under control those changes which are damaging to society.

The modern history of weather modification began in 1946 when two General Electric scientists, Irving Langmuir and Vincent Schaeffer, modified clouds by seeding them with dry ice pellets. Soon afterwards, Bernard Vonnegut demonstrated that silver iodide crystals could serve the same purpose. The General Electric scientists were able to verify experimentally the theory advanced in 1933 by Tor Bergeron, a Swedish meteorologist, and Walter Findiesen, a German physicist, that clouds would precipitate if they contained the right mixture of ice crystals and water droplets, if the water droplets were cooled well below their freezing temperature, $0°C$. Such drops are called supercooled. In the years following the discoveries of Langmuir, Schaeffer and Vonnegut, increased interest in rainmaking led to a number of substantial investigations, particularly in Australia, but also in America, Argentina, France, and the Soviet Union. Besides the serious investigators and operators, there also appeared a number of rather unscrupulous enterpreneurs whose activities in the 1950s cast an uncertain air over the scientific basis of rainmaking.

The historical emphasis on rainmaking, however, should not obstruct our view of weather modification. Weather modification encompasses far more. One can view the atmosphere as an envelope of air which rotates for the most part with the same speed as the underlying continents and oceans. The relative motion between the atmosphere and the earth arises from sources and sinks of energy which vary in location and strength, but which have as their ultimate source the sun's radiation. Because we can neither modify this energy at its source nor intervene between the sun and the earth's atmosphere, we must interfere with either the atmosphere or its lower boundary if we are to alter any aspect of the weather.

The quantities of energy involved in weather systems exceed by substantial margins the quantity of energy under man's direct control. The energy typically expended in a single tornado tunnel is equivalent to about 50,000 tons of explosives. A single thunderstorm tower exchanges ten times this much energy during its lifetime; a moderate hurricane may draw more than 1,000 megatons of energy from the sea. These vast quantities of energy make it unlikely that brute force techniques will lead to sensible weather modification.

The key to weather modification lies in the identification of the atmospheric instabilities to which the addition of a small amount of energy will release vastly greater amounts of energy. An atmospheric instability is a situation in which nature has stored energy in some part of the earth far in excess of that which is usual. To trigger this instability, the required energy might be introduced violently by explosions or gently by small bits of material acting as catalysts for nucleating agents to induce rapid changes. The mechanism for energy storage might be the difference in the seasonal rate of heat input between equatorial or polar regions or the supercooling of water vapour by updraughts of a few tens of minutes. The effects of triggering such instabilities could be worldwide as in climatic alteration, or regional, as in locally enhanced precipitation.

Formation of precipitation

We are beginning to understand several types of instabilities within the atmosphere. For example, water droplets cooled below $0°$ are unstable unless supplied with crystallization nuclei and will remain in their supercooled condition over substantial periods of time. The nucleation of supercooled water producing ice can trigger another instability. The water-to-ice phase change produces heat which locally raises the air temperature. Warm air possesses a greater buoyancy than its surroundings and will tend to rise forming an updraught. Supersaturation of water in the air making up the cloud is an instability similar to supercooled water. Without condensation nuclei, water will remain in the vapour phase rather than condensing to water droplets. A large scale instability is one that develops because of the imbalance of heat between the equators and poles; the atmosphere relieves this instability through the creation of extratropical cyclone systems. Local heat introduced at the earth's surface may produce convective instability. The heat may be added by alteration of the ground's reflectivity to sunlight or directly through a fire.

The first step towards precipitation at ground level results when air becomes supersaturated as a result of cooling or by the addition of water vapour. Air is usually saturated by cooling which results when it is gradually lifted during its horizontal transport. The lifting can either be convective, produced by local heat sources on the ground or in the air, or orographic, which is a forced lifting of the air over elevated terrain. A sufficient number of nuclei must be present to produce the phase transition either from vapour to water or vapour to ice. Without crystallization nuclei, water drops can be cooled to as low as $40°$ below zero before they will convert to ice. Once ice or water drops are formed, they must aggregate sufficiently before they can fall out of the cloud. The number of nuclei available determines the concentration of cloud droplets. This number of nuclei varies. There may be as few as two or three nuclei per cubic centimetre over mountain tops or the surface of Greenland or more than 100 cm³ in the air over continents.

Less well understood than the formation of water drops by condensation on nuclei is the formation of small ice particles in clouds. The process of crystallization is thought to be dominant in many clouds. Ice crystals can be formed either through the crystallization of supercooled water droplets stimulated by the freezing nucleus or by the direct sublimation of water vapour to ice stimulated perhaps by sublimation nucleus. The effectiveness of a particular freezing nuclei is ascertained by determining the temperature at which the crystallization is induced in the super-cooled vapour. Ice crystals acting as nuclei are effective at $0°C$ in initiating crystallization. When silver iodide smoke is introduced into a supercooled cloud, some ice crystals begin to appear when the temperature falls below —$4°C$. Their number increases until at —$15°C$ most of the ice silver iodide particles serve as ice nuclei.

How do particles aggregate sufficiently to become heavy enough to fall from a cloud? Although we are not sure, several hypotheses have been advanced. The Bergeron-Findeisen mechanism mentioned earlier is one which may be applicable to many clouds in temperate regions. It states that ice crystals grow at the expense of supercooled water droplets because vapour pressure is higher over the supercooled water than over ice. Condensation on the ice crystals removes water vapour from the air which in turn becomes saturated relative to the remaining water droplets. The droplets begin to evaporate and in so doing replenish the vapour supply demanded by the ice crystals. The process continues until all the liquid droplets vanish. Particularly in tropical regions, however, many clouds produce rain without reaching a freezing point. In these clouds water droplets must grow by collision and coalescence. The amount of energy stored in the surface of small droplets is greater than that stored in large ones so that there is a tendency for the smaller particles to aggregate. To do this the water droplets must collide and these collisions are thought to be caused by small-scale turbulence which constantly stirs the drops within the clouds.

Fog clearing

Operationally, seeding by dry ice and other agents is used in clearing airports of supercooled fog. In this kind of operation there are still a number of problems. The use of aircraft to implant silver iodide or dry ice is one of the disadvantages because it is necessary for the aircraft to take off in fog and to land in the same fog if the seeding fails. Because the fog is highly stable, ground burners are ineffective in dispersing the silver iodide smoke. In the future, small combustible rockets may be used in place of aeroplanes.

The situation with respect to the dispersal of warm fog, that is fog made up of water droplets well above the freezing temperature, is quite different. In this case, we must work directly on the instability of small drops as compared with large drops. As yet, no efficient way of stirring up fog so that collisions occur has been discovered, though inefficient means of burning off the fog have been tested from time to time.

Orographic precipitation

Some of the world's largest amounts of mean precipitation are found where the moist winds flow against mountain tops. It is not surprising therefore that early in the history of weather modification attention was drawn to the possibility of using orographic cloud systems for cloud seeding. The late 1950's saw an extensive programme of randomized seeding trials conducted in the mountainous regions near Santa Barbara, California, where wintertime storms were responsible for the clouds. Results of these trials showed that in the first year seeding produced perhaps as much as 100 per cent increase in precipitation. The second year, however, showed no such enhancement. There then followed a long and acrimonious argument among statisticians and meteorologists on whether or not the experiments had been properly designed and if any substantial conclusions could be drawn from the trial. The most conclusive results have been obtained in Australia, where a long series of experiments in the orographic regime show quite clearly that under certain conditions precipitation can be enhanced by seeding.

Tropical cumuli

Since the conversion of supercooled water vapour to ice releases latent heat of crystallization, it has been suggested that seeding could markedly influence cloud dynamics. During some early seeding trials in Australia, Krauss and Squires observed that some of the cumulus clouds seeded with dry ice grew upwards in a spectacular manner. These effects were attributed to the increased buoyancy induced by the latest heat of fusion. Robert and Joanne Simpson conducted two extensive studies on tropical cumuli in the Caribbean to test the hypothesis. The cumuli were seeded by dropping pyrotechnic devices loaded with silver iodide from aircraft flying at high altitudes. In both 1963 and 1965 trials, cloud growth was observed shortly after treatment. The Caribbean experiments are among the most exciting developments in weather modification because they point to the possibility of not only increasing precipitation, but also of altering the large-scale motions of cloud systems in tropical regions. Cloud systems in the Caribbean are similar to monsoon systems whose occasional failure to deliver rain in India and other countries have led to disastrous droughts. The work of the Simpsons points to possible application of their technique in regions dominated by the monsoonal circulation.

Commercial rainmaking

The cloud seeding studies I have reviewed so far were designed specifically as experiments. Other good sources of data on the effectiveness of seeding are the records of the many operational programmes conducted by commercial cloud seeders for clients seeking additional precipitation on their agricultural lands, forests and watersheds. Unfortunately, the open scientific literature contains little information about these operational programmes, even though a large number have been conducted. Evaluation of the operational experiments was undertaken by Professor James McDonald of the University of Arizona during the course of the National Academy's study on weather modification which I mentioned earlier. From the eastern United States, McDonald selected 14 projects running from 19 days to 5 months. From the mountainous regions of the western U.S. he selected four winter orographic projects of longer duration. Using original sources, he tried to recover relevant data from rain gauge stations, weather maps and other meteorological information. Since the operational experiments were not randomized, McDonald's analysis compared historical rainfall records both in the target areas and those surrounding regions used as control areas. McDonald defined a good control region as one in which, on an historical basis, the rainfall was correlated with that in the target area. In all these

operations, the nucleating agent (silver iodide smoke) was introduced at ground level some kilometres upwind from the target
region, which in the eastern U.S. experiment had characteristic
dimensions of about 50 kms. The hypothesis was that horizontal
transport of air, combined with convective motion associated with
clouds, would place the nuclei into the ice which forms part of
the cloud.

Among the 14 eastern experiments carried out in summer,
autumn and winter the percentage of increase of precipitation
ranged from 0 to 57. While the statistical significance of any
particular experiment was small the combined operations indicate
that the precipitation had been increased in the target regions in a
significant way. The four orographic projects in the western U.S.
lasted from 8 to 14 years and the percentage of increase ranged
from 6 to 18 per cent. The National Academy's Panel undertook
a number of additional studies to determine if bias of any sort had
entered into the experiments. Their final conclusion, based in large
part on McDonald's analysis of operational rainmaking, was that
there was accumulating but as yet somewhat controversial evidence
that under certain meteorological circumstances precipitation can
be increased significantly.

The most important practical consideration in cloud seeding
is whether or not the removal of water from a cloud will reduce
precipitation downwind. At first it might appear obvious that
because of the conservation of matter, an increase of precipitation
in one place must be balanced by a decrease in another. Clouds,
however, are dynamic entities with strong moisture from the
surrounding air and perhaps also from the ground over which
they pass. Analysis by U.S. Weather Bureau scientists of
McDonald's 14 eastern cases disclosed no perceptible downwind
effects, at least to distances of 250 kms or so. However, the
results are of low statistical reliability and further studies are
urgently needed to settle this most important question.

Hail suppression

Hail suppression is another controversial area of weather modification. Considerable damage is done by summertime hailstorms
in several areas throughout the world and attention has been devoted
to methods by which this harm might be mitigated. What dis-

tinguishes a hailstorm from a normal thunderstorm? Apparently three conditions must prevail for a thunderstorm to produce hail. First, the clouds must have sufficiently high updraught velocities to support the growing ice stones. Second, liquid water must be supercooled in the upper regions of the storm cloud so that as the ice stones pass through these regions, the supercooled water crystallizes on them causing the hail stones to grow. Third, these updraughts must persist long enough to enable the stones to grow large enough to fall through the cloud and reach the earth as ice rather than water.

Various schemes for hail reduction have been suggested. One such suggestion is to create explosions within clouds and use the resulting shock waves to break up the hailstorm. Although this proposal has received favourable attention in several countries, an analysis of the energy required to produce a strong enough shock wave shows it to be an unpromising line of research. Other suggestions are based on increasing the number of hailstones or reducing the amount of supercooled water or air, or both. The more hailstones are formed, the less likely it is that they will grow large enough to fall through the cloud. A reduction in the amount of supercooled water brought about by introducing silver iodide into the cloud's supercooled regions would remove the icy stones' growth source.

The greatest hail suppression activities are now conducted primarily in the Caucasus and Trans-Caucasus regions of the Soviet Union. Here, summertime hailstorms do very substantial damage to orchards and vineyards. Although no complete statistical study of these investigations has been released, it is quite clear that reductions have been achieved. The Caucasus experiments are also interesting in that they are supported by rather complex logistic systems reminiscent of Dryenforth's operations. Radar sets, otherwise used in air defence operations, are used to determine when there is a buildup of ice in a cloud. The radar set then locates the best place in the cloud in which to introduce silver iodide. Artillery then fires projectiles carrying silver iodide into the cloud. Attempts at hail suppression in Switzerland, France and Argentina have not been successful; the techniques differ greatly from those employed by the Soviets.

Lightning prevention

Widespread damage resulting from the effects of lightning have triggered substantial effort, particularly by the United States, to study methods for lightning prevention. Over the world approximately 100 lightning strokes occur every second. Although most of these strokes are concentrated in the equatorial regions of the world, in temperate regions they are recognized as the outstanding cause of forest fires. In the western U.S. more than 10,000 are caused annually by lightning. These fires damage forest resources, such as timber, watershed, wild life, and recreational areas.. In addition, lightning is becoming increasingly hazardous to aircraft.

Development of lightning suppression techniques is hampered by lack of an adequate theory of thunderstorm electricity. Two methods of lightning suppression, however, are currently being investigated. In Project Skyfire of the U.S. Forest Service, silver iodide seeding is employed to introduce an abundance of ice crystals in the supercooled part of a thunder cloud, where most ground lightning strokes are thought to originate. The Project's hypothesis is that artificial ice crystals become additional corona points which increase the leakage current between the charged centres, thus suppressing the formation of single discharges. The basic concept, however, is open to question since an alternative theory of thunderstorm charge generation suggests that an increase in the quantity of ice particles would be expected to increase the effectiveness of charge generation. In the latter theory, the charge generation is thought to result from a collision between ice crystals and soft hail particles and from the freezing of water droplets. Despite the unsatisfactory theoretical basis, results obtained thus far in Project Skyfire appear to be quite favourable. In one pilot project heavy seeding led to a 40 per cent decrease of ground strokes but the support statistics are poor.

Hurricanes

In addition to those aspects of weather modification which I have reviewed, there have been numerous suggestions, some highly speculative, concerning climate alteration and the modification of severe storms such as hurricanes and tornadoes. Little progress, theoretical or experimental, has been made in these directions, although a substantial effort has been directed towards the modification of hurricane dynamics. In Project Stormfury, jointly sponsored by the U.S. Navy and the Weather Bureau, pyrotechnic silver iodide sources have been dropped into portions of the eyewall of hurricanes to convert supercooled water to ice thus releasing heat. The resultant vertical motion, it is argued, upsets the balance of forces within the hurricane and could lead to an outward migration of the clouds forming the wall which surrounds the eye. This consequent spreading of energy would reduce the hurricane's force. The statistical variability of hurricane development is so large that the few experiments conducted so far to test the hypothesis behind Project Stormfury have not led to a definitive conclusion. This work on hurricanes is closely allied to the studies made by the Simpsons on the dynamics of tropical cumulus clouds. Progress in such studies is certain to further the understanding of hurricane dynamics.

Future prospects for weather modification

I have presented perhaps a more optimistic view of the future of weather modification than would be put forward by others in the field. I justify my view in the light of three recent developments. First, increasingly sophisticated models of atmospheric processes have been created by atmospheric physicists. The atmosphere is no longer as mysterious as it once was. Secondly, the advent of high-speed computers will enable these atmospheric models to be studied in great detail. These computers are particularly important to weather modification because they enable scientists to carry out experiments on a computer having the same scale and time characteristics as actual operations in the atmosphere. Thus, it will be possible to test the viability of various schemes for manipulating the atmosphere. Thirdly, in recent years a new array of instruments has been developed to observe and detect atmospheric changes. Most dramatic and perhaps most powerful of the new tools is the meteorological satellite which provides a platform from which the atmosphere can be observed not only in geographically inaccessible regions but also with entirely new physical parameters. Thus meteorological satellites of the future will determine humidity, temperature, and pressure averaged over substantial volumes of the atmosphere. These quantities are needed to explore the numerical models mentioned previously. As satellite observations of the atmosphere have developed, so has the sophisticated instrumentation necessary to look at the detailed processes within the smaller parts of the atmosphere. We thus have a powerful means of studying clouds and their interaction with the surrounding air as well as the atmosphere's interaction with its own lower boundary.

I am not as optimistic about solutions to weather modification's non-scientific aspects as I am about its future technological studies. The political, legal, economic, and sociological consequences of deliberate weather modification can be so far-reaching that our present involvement in nuclear affairs will seem simple by comparison. Our present understanding of the basic environmental science and technology of weather modification is primitive. Even more primitive, however, are current notions of the proper political forms and procedures thought necessary to deal with the consequences of weather modification. Past experience demonstrates that much smaller technological changes than environmental control have transformed political and social relationships. Experience also shows that such transformations are not entirely predictable and that guesses based on precedent, which we might make now, are likely to be quite wrong in relation to future activities or needs. These non-scientific, non-technological problems, however, are of such magnitude that they deserve consideration by serious students throughout the world if we are to live comfortably in a controlled environment.

CHAPTER 2

How Can We Effectively Use the Oceans?

Introduction

Our knowledge of the oceans is nowhere nearly as complete as is our understanding of the atmosphere. Yet the oceans can contribute to the welfare of the world since they serve as a source of food and minerals and are a place for a great variety of recreational activities. Before examining how we might use the oceans more effectively than we do at present, I will briefly review two of the outstanding problems in understanding the oceans and the processes within the oceans.

Origin of ocean waves

Looking out over the sea, one gets the impression of endless irregularities. If the winds are light the irregularities are small. When there is a heavy wind, giant waves can be truly awesome. Even when the surface is glassy calm, the ocean may move up and down very slowly as a result of a long, smooth swell which may have originated thousands of miles away.

Description of wave motion has progressed very rapidly in recent years because of the recognition that it is so highly irregular and that the methods of modern statistics can be applied with great benefit to the study of ocean waves. The understanding of how waves originate is another matter. About five years ago, the distinguished English mathematician, Professor Michael Lighthill, in a review paper on the origin of ocean waves concluded that the subject was closed and that the mechanism by which waves are generated is clearly understood. Today, it is no longer clear how waves arise, although two fundamental processes are clearly at work.

The source of energy for ocean waves lies in the atmosphere. The question is: how is the energy transferred from the wind to the water? One mechanism is that of tangential stress as the wind blows across the water tending to push it along. This is similar to blowing across a cup and making the coffee slosh about. The idea that tangential stress can indeed induce surface motion is a quite natural one. However, there are reasons to suspect that this is not the sole process since, when one does the necessary arithmetic, the numbers do not come out.

An alternative mechanism involves the fact that the pressure of the atmosphere on the sea surface is not uniform everywhere but varies over short distances because of the turbulence in the lower part of the atmosphere. In some regions the pressure is greater than in surrounding regions. The irregular pressure distribution can disturb the uppermost surface in an irregular way. Pressure irregularities in the atmosphere are not stationary but move with the wind so one can create waves much in the same way as if one put a great number of fingers irregularly distributed into a bathtub of water and moved them along. The wind is always turbulent even over a glassy calm sea surface, and there will always be pressure differences between parcels of air to initiate and feed the wave motion. Once the sea begins to grow waves, there will be a back reaction of the waves on the wind which also causes pressure differences. This process of transferring energy by moving pressure irregularities is particularly efficient if the velocity at which the ocean wave propagates bears some fixed relationship to the velocity propagation of the pressure fluctuations. If the relations were random, then waves grow linearly with the pressure fluctuation; that is, the same percentage increase in the pressure differences gives rise to a similar percentage increase in the height of the wave. On the other hand, if the pressure fluctuation maintains some fixed relation to the surface wave, the sea may be expected to grow not linearly but exponentially.

Many attempts have been made to incorporate both the shearing tangential stresses and the pressure fluctuations into a single theory. The subject is complicated by the fact that as soon as the waves begin to grow to any appreciable extent, quantities do not add, rather non-linear effects become important. Thus, as waves begin to grow, doubling the shearing stress does not necessarily double the rate at which energy is fed into the motion. In order to treat these non-linear effects, highly sophisticated methods of mathematical analysis must be employed, comparable in difficulty to the kind of analysis necessary in highly theoretical parts of modern physics. Because of these mathematical difficulties, much of the apparent physical simplicity of the theories is lost.

While the origin of ocean waves still remains very much a problem, semi-empirical methods of predicting wave heights at a distance from the storm have been developed. These methods are based on experience and observation of a large number of storms and waves. Using these techniques, it is now possible in many cases satisfactorily to predict wave conditions following a major storm over the open ocean. The real difficulty in the subject arises from the fact that storms of moderate size, such as squalls, develop with great rapidity in the oceans bordering the continents and the usual techniques for predicting storm heights and wind velocities do not apply. These small-scale disturbances can be most damaging to structures near the shore. The petroleum industry, operating on the shallow waters bordering continents, for example, identifies as one of its major problems the forecasting of local and short-term storms, such as squalls, that might affect the profitability of drilling operations. Recent disasters in the North Sea and the Gulf of Mexico have adversely affected insurance rates. Lloyds of London, for instance, recently doubled its premiums on big oil rigs to 10 per cent of the liability because in the span of six months, storms had sent 5 of 80 rigs insured by Lloyds to the bottom. Such problems, along with markedly higher operating costs, decrease the profitability of off-shore operations. It is doubtful that semi-empirical methods can be successfully applied to the prediction of such short-term storms. What is needed is the further development of the theory of the origin of waves so that this knowledge can be applied to the short-term forecast problem.

Ocean weather

The weather within the oceans is very incompletely known. By weather, I mean variations from conditions that on the average prevail. In general, we now have some idea of what climate within the ocean is like; that is, the conditions averaged over long periods. For example, long oceanographic voyages have established the existence of great current systems that move vast masses of water within the ocean. What is not known is how conditions vary locally and over what time scale. Preliminary observations indicate that the range of phenomena may be as complicated as the weather in the atmosphere.

The weather in the ocean is simpler due to the fact that no condensation associated with precipitation takes place. But it is far more complicated through its variable chemical composition. For example, currents can result both from variation in the amount of salt and from the fact that the ocean is strongly influenced by biological activity. The description of the weather within the ocean and the understanding of how weather is created is one of the great scientific adventures of the coming years. The weather stations within the oceans will be for the most part buoys containing a sophisticated array of instruments. The buoy system will have to be worldwide because of the very large-scale character of the major circulation patterns. Such buoy systems that will monitor the oceans on a continuing basis are now possible because of satellites. The buoys will communicate to the central processing centres via a satellite, the satellite acting as a communication relay.

Understanding of the weather in the ocean and prediction of it will become of increasing importance as man develops more complex structures in deeper and deeper parts of the ocean. The engineering design of such structures should take into account not only the mean currents but also their variability. The ocean-wide buoy systems of the future will also be of great importance to the atmospheric sciences since they will monitor the exchange of energy between the oceans and the atmosphere.

Weather in the atmosphere is thought to be very greatly influenced by the oceans. Similarly, the great ocean current systems are driven both by prevailing winds in the atmosphere and the thermal gradients resulting from differential solar heating. Understanding of the interactions of the oceans with the atmosphere is essential to the problems of long-term weather forecasts as well as to the forecasting of weather within the oceans.

I have briefly touched on two of the key areas of scientific investigation. I would now like to discuss why we will in the future devote considerable efforts to the study of how to work in the oceans. In the following, I will restrict my considerations to three major resources: food, minerals, and recreation. To indicate the current magnitude of activities in these areas, the world's fishery catch of 1964 was valued at 5 billion dollars. During the same year, sport fishermen in the United States spent over 1·5 billion dollars, while in the decade preceding 1964, the continental shelves of the U.S. yielded about 3 billion dollars in bonuses, rentals, and royalties.

Food from the sea

Adequate nutrition is prerequisite to all other human activities. For most of humanity, life is supported by a diet which is largely, if not exclusively, of vegetable origin. Of the more than three billion world population, at present only 600 million consume the major share of animal protein. Approximately, 1·5 billion persons largely in tropical and sub-tropical areas, live on diets which are dominated by one staple crop, although mixtures of vegetables and cereals are available. Many of the vegetable diets fail to provide protein of either the quantity or quality needed for adequate human nutrition. The quality of protein depends on its composition of amino acids. Vegetable proteins are frequently absolutely or relatively deficient in one or another of the ten amino acids essential for human nutrition. For example, corn is deficient in tryptophane and is not adequate in lysine content.

The problems of today will be magnified as we go into the next thirty years. It took thirty years for the world population to grow from 2 to 3 billion. The next billion will take perhaps only 15 years and by some time near 1990, the earth's population will be roughly 6 billion. The largest growth in population will be in those tropical and sub-tropical regions where present protein deficiencies are the greatest.

Chronic protein deficiency is a consequence of inadequate amino acid in the diet and is a most serious health problem. With the continuing growth of population in regions where proteins are lacking in the basic diet, protein deficiency will become the greatest health problem of man. Protein deficiency lowers the resistance to infectious diseases such as diarrhoea, pneumonia, and other respiratory infections, and magnifies their effect. These diseases have not yielded to the miracles of modern medicine and require supportive treatment. Such treatment requires adequate medical personnel and these are just not available in the numbers required in the lesser developed countries.

Malnutrition is the leading cause of death between weaning and five years of age in all countries in the equatorial zone. Some estimates state that 50 per cent of the deaths in this age range are due to malnutrition and in particular to protein deficiency. In addition to the direct deaths caused by protein deficiency, there are indirect effects. Studies on the height and weight of groups having protein deficiency, and similar groups without, show marked differences in height and weight. Protein deficiency can cause irreversible mental damage during the critical periods of growth, and such diseases as kwashiorkor can limit not only the lifespan but the productive capacity of the adult.

There are many ways in which the food and population problem could be attacked. Control of population growth is one complex way. Another approach might be enrichment of the diet through a redistribution of agricultural products to ensure that instead of a single staple, a mixture of vegetables and vegetable products with a balance of amino acid composition is consumed regularly. Such a task is one of enormous complexity, involving changes in agricultural traditions and means of distribution, and will require major research efforts.

An alternative is to provide a nutritional supplement of 10 to 20 grams of animal protein per day to a predominantly vegetable diet, particularly during the critical period between weaning and school age. The specific animal protein is of little consequence. Beef, pork, chicken, rabbit, fish, molluscs, and crustaceans — all will serve. There is an argument, however, that the relative inefficiency of converting agriculturally produced grains and grasses into animal protein make it increasingly difficult to use these proteins to supply the needs of a world of rapidly increasing population.

The oceans can provide a major help in alleviating the protein shortages of the world. In 1964, the world fish catch contained $17 \cdot 1$ billion pounds of protein, an amount which would have supplied slightly more than 10 grams of protein per day to 2 billion individuals, and would have been effective in either eliminating or alleviating chronic protein deficiency for the people of the lesser developed countries. This opportunity for upgrading nutrition has not been adequately exploited because of cultural and economic barriers, as well as great problems of processing and distribution.

What are the potentials for protein production in the sea? It is estimated that at least 40 billion tons of organic material are produced annually in the sea, only a tiny fraction of which is harvested by man. In the sea, as on land, food is produced by plants that utilize energy in sunlight to synthesize organic materials from inorganic substances. The grass of the sea, composed of microscopic plants, phytoplankton, is eaten by the grazers, zooplankton, which in turn are consumed by larger animals such as fish. There are unresolved questions relating to this food chain, some of which I will turn to later. Nutrients essential for the phytoplankton are replenished by natural processes such as regeneration due to microbial activities and inflow of fresh water which contains nutrients from the land, including agricultural fertilizers and sewage. With the death of animal and plant life in the sea, the organisms sink and are decomposed releasing nutrients. These nutrients are concentrated in bottom waters, where, due to the absence of life, they cannot be used for photosynthesis. In areas of upwelling the nutrient-rich bottom waters are brought to the surface where they sustain large populations of phytoplankton. Wherever this occurs, such as in the Humboldt current off the coast of Peru, phytoplankton flourish and a vigorous food chain is sustained, leading to the production of large quantities of fish.

The present world fish catch is about 114 billion pounds. It has almost doubled in the last ten years. The increased catch in recent years has resulted largely from finding new fishery stocks through the use of fishing fleets with factory ships. Helicopters and sonars have helped both the Soviet and Japanese fishing industries. The most dramatic new find has been off the coast of Peru, where a catch of 20 billion pounds of anchovy was taken in 1964, where ten years previously the catch had been 2 per cent of that amount. That the waters were rich in fish was known to the local inhabitants from the great activity of the guano-producing birds, but an industry did not flourish. Development of the industry in Peru did not result from any major advance in technology but from the activities of local entrepreneurs who were willing to invest both in primitive fishing vessels and processing plants. The local infusion of capital quickly led to the establishment of a major industry. The Peruvian example once again illustrates that science and technology are not by themselves the answer to the problems of the development of the sea.

Certain fish are brought to market directly for human consumption. A large fraction of the total fish catch is not utilized directly by man. This is particularly true of fish of relatively moderate and small size; for example, anchovy, menhaden, and hake, which are caught in great numbers by simple trawling and seining procedures. These fish are processed for industrial oil and fishmeal. Fishmeal is used as a high protein source, particularly for poultry but also for livestock feed. From the standpoint of human nutrition, this use is particularly wasteful because some of the protein in the fish is lost in its conversion to poultry and livestock protein.

Nevertheless, the problems of storage and transportation, rapid spoilage, cost of processing small fish, and cultural habits of many people, make it apparent that at present only a small fraction of the catch of the small fish can be used directly as food by man. The major portion of the catch, such as the small-sized fish which abound in the Humboldt current off Peru or off the California coast, must be processed into a form which is readily stored and transported and accepted as food by people of many countries. The Bureau of Commercial Fisheries in the United States has developed a solvent extraction process for the preparation of fish protein concentrate (FPC) from various species of hake. Unlike the coarse meal, FPC is intended for human consumption. In powder or flour form, FPC has almost no taste and no trace of odour. The FPC is approximately 85 per cent protein and it is estimated that this material can be produced commercially, but not distributed, for about 25c per pound. At this cost, $2 a year would provide the daily protein requirement of 10 grams.

The major problems in the use of FPC are developing techniques for incorporating it into acceptable foods, such as breakfast cereal, wheat flour, noodles, etc., and the ever-present problem of distribution. Very recently the Food and Drug Administration of the U.S. approved FPC for human consumption. Now that this legal hurdle has been passed, great opportunities exist for innovation in the area of processing FPC and incorporating it into food forms acceptable to a wide variety of cultures.

The opportunities to enrich and amplify man's food supply by fishing in the open oceans are highly significant, but nonetheless limited. An entirely different set of opportunities is offered, however, by the potential crop that might be obtained by systematic and scientific farming of restricted areas of the sea — aquiculture. Aquiculture could take the form of farming more restricted areas or ranching, in which fish would be kept in fenced areas, fed sufficient nutrients, and later harvested.

The yield of fish, whether in the natural environment or in aquiculture, will depend on the nutrients supplied by upwelling. Opportunities exist for the development of schemes to utilize natural hydrodynamic or atmospheric energy sources to bring to the surface nutrient-rich deep water to fertilize selected marine habitats, such as bays, coral lagoons, and fenced-in areas. But the problems involved are biological as well as technological and their solution requires the marriage of engineering and marine biology on a scale not previously attempted. Natural energy must be harnessed to drive the upwelling. The amount of nutrients delivered must be controlled so that desirable phytoplankton are produced in limited numbers. Production exceeding the carrying capacity of the environment, particularly for oxygen, would lead to mass mortality of marine life.

Current attempts at aquiculture are still primitive but they do demonstrate long-term possibilities. Japan is a current world leader in marine aquiculture. Its efforts have been directed to production of organisms such as fish, shrimp, and shell fish, including oysters for pearl culture. Forty years ago, the Japanese began growing oysters on long ropes hanging from floating rafts or ropes sustained by buoys. With this method, oysters were grown throughout the water column not only on the bottom. Furthermore, oysters are free from bottom predators. The further exploitation of oysters in aquiculture seems most promising, since oysters feed directly on phytoplankton and thus are extremely efficient in transferring nutrients into protein. Oysters are not alone among the species susceptible to aquiculture. Shrimp and crab have been grown artificially in Japan, as have squid. Further, there is some experience in Formosa on the use of fertilized seawater in the growing of milkfish.

The review of the problems of obtaining food from the sea raised major technological problems ranging from methods of food processing and distribution to the driving of upwelling currents. There are also substantial unanswered questions of a scientific character which must be dealt with if the sea is to be used in an effective and efficient way to produce food.

The conversion of photosynthetic plants to animal protein on land is relatively well understood. In the sea, however, photosynthetic plants are restricted largely to microscopic planktonic algae, phytoplankton. Conversion to animals large enough to serve as food for man usually involves many intermediate steps, though as mentioned above, oysters directly feed off phytoplankton. The knowledge of the complex and diverse food chains and food webs of the sea is very sparse. The natural food of even the best known marine animal species are unknown except in general terms. Central and prerequisite to scientific control and ultimate management of marine food resources is further knowledge of the essential nutritional requirements, of feeding habits and food preferences, and of efficiency in converting planktonic algae into animal protein.

Photosynthetic plants in the sea use solar energy to synthesize organic material. In agriculture, solar energy is channelled into production of plants that are useful to mankind either directly as plant products or indirectly as animal food. The growth of plants in the sea on the other hand is a process over which we have no control and little knowledge. Some species of planktonic algae are recognized as food organisms for marine animals. Others are weed species of little or no nutritional value. Still others, such as red tide organisms, are noxious or lethal to marine life. To increase significantly the amount of food obtained from the sea we must learn to control the kinds of phytoplankton produced as primary food sources. There is also lack of insight into the complex relationship of organisms to the environment. The general correlation of strong centres of phytoplankton activity with up-welling has been noted. It is not known what determines the capacity of the environment for carrying the phytoplankton. A delicate balance exists since as the phytoplankton increase in number, their use of oxygen in the water depletes this gas to such an extent that marine life is no longer possible. To understand these changes, long-term studies on the fluctuations of the meaningful parameters and resultant changes are needed.

A good example of our lack of understanding is provided by the case of the sardine and anchovy off the coast of California. The two fish stocks are strong competitors. Over the past few years, selective harvesting of sardines has been accompanied by a marked reduction in the population. At the same time, the untouched anchovy population has increased markedly. This observation led to the suggestion that harvesting sardines was responsible both for the decrease in sardine population and increase in the anchovy population. The decrease in sardines proved disastrous to the California fishing industry, since house-wives buy sardines but not anchovies. A further study, however, has raised questions of whether or not harvesting had anything to do with the sardine-anchovy problem. Fish scales are deposited in sediments off the coast from both anchovies and sardines. Population studies from the deposited and preserved fish scales determine the past fluctuation in the relative population of the two fish. These studies show that there have been other years of marked fluctuation of relative abundance of the two fish stocks. Thus, it is not at all clear whether the current decrease in sardines and increase in anchovies is due to harvesting or due to a natural fluctuation accompanying some major change in the oceanic circulation.

In addition to an understanding of the biological and physical environment, there is a further need in understanding interactions among various members of the biological community. One new species may account for most food production but rare ones often provide essential services, such as parasite removal from other species. Eliminating these services may be catastrophic. In addition, cryptic species may be present which, while not differing appreciably in morphology, have quite different behavioral, physiological, and population characteristics in the environment.

It is quite surprising that in an age where technology has grown closer to physics and chemistry, there is still a great need for comprehensive study of the systematic, taxonomic biology of marine organisms involving morphological bio-chemical and behavioral differences among the species. Such studies can provide a basis for selecting races or strains within a single species with characteristics which render them particularly appropriate for exploitation and cultivation by man.

Minerals from the sea

The possibility of mining the sea floor has caught the popular imagination because of numerous articles about the potential riches of the sea. These speculative notions perhaps hide the fact that the sea and the sea floor do provide substantial quantities of many essential materials. For example, about 12 billion pounds of salt are produced each year from sea water. Of these, 2·4 billion come from 80 square miles of salt ponds at the south end of San Francisco Bay. Similarly, substantial quantities of bromine are obtained from sea water and essentially all the world's manganese is extracted from the sea. In the U.S., from continental shelf wells, petroleum, natural gas, and sulphur produced minerals valued at 820 million dollars in 1964, while beaches and in-shore sea floors yielded about 44 million dollars worth of sand, gravel, and limestone.

Mineral resources certainly exist under the oceans, at least in the continental shelf area, since these are but extensions of the continents. Less certain are the amount of minerals that lie on or beneath the deep ocean floor. In general, we can distinguish three classes of mineral deposits: surface deposits on the shallow continental shelf, bulk deposits within rocks under the shelves, and deposits on or in thin sediment layers of the deep sea floor.

The surface ore deposits of the continental shelf are mainly of two types: placer ores concentrated in submerged river channels or beaches, and blanketing layers of nodules such as phosphorite precipitated from sea water. These types of ores have been mined at various places around the world; for example, diamonds currently mined off the coast of southwest Africa. The beaches of southwest Africa have long been a source of diamonds, and geologists suspected that the alluvial deposits continued off-shore. Development of these resources require new, if primitive, technology for dredging very large quantities of sand. Diamonds have been found in depths up to 200 feet in mixtures of sand, gravel, and boulder. Tin is mined in waters up to depths of 200 feet in southeast Asia. Iron ores are mined off Japan and attempts, largely unsuccessful, at mining gold have been made in various places.

Phosphorite occurs in nodules or as coatings on rocks with sizes ranging from grains to larger pieces, as much as a metre in diameter. Deposits are generally found where there is an upwelling of deep waters rich in tricalcium phosphate. It is estimated that a billion and a half tons of phosphorite are available off the coast of California. Here, however, an attempt to mine the phosphorite has been frustrated by a concentration of unexploded naval shells along with the phosphorite.

The further development of the surface ore deposits on the continental shelf will require an extension of conventional geologic methods of exploration of the continental shelf and the evolution of appropriate technology in securing materials.

The fact that the rocks of the continental shelf are essentially identical to the rocks of the continents themselves has been established in many cases, due to the commercial exploitation of oil and gas. Extension of underground mining techniques on continental shelf deposits, however, appears problematical since there is little doubt that the cost of mining a few hundred feet under water will exceed the cost of mining on land.

The deep sea floor is paved in many places with nodules containing manganese, iron, cobalt, copper, and nickel in concentrations which approach the levels mined on land. The resources are enormous but the economic potential is uncertain. The general distribution of nodules is known in certain areas, but there remain numerous uncertainties both with regard to metallurgical extraction of the important metals and means of harvesting the nodules. Substantial experience in offshore drilling and mining operations has led to the clear definition of technological requirements for the further exploitation of the sea. This situation can be contrasted with that in aquiculture, where the primitive nature of the subject implies rather diffuse requirements.

Recreation

One of the very great resources of the sea is that offered for recreation. A substantial proportion of the world's population lives within a few miles of the coastline of large inland bodies of water. In the U.S., about half the population lives near the margins of the oceans or the Great Lakes. In Australia, the percentage is even greater. Near-shore environment is of critical importance both as a source of food and raw materials, and also as a great recreational area. This near-shore environment is being modified rapidly by human activities in ways that are unknown in detail but often highly undesirable. There is the problem of pollution, which renders beaches unsafe for swimmers, destroys valuable fisheries, and generally degrades the coastline. In addition, man through his ineptness in engineering, has failed to build structures that would preserve portions of the coastline or create additional harbour or shelter facilities.

Indeed, I know of no branch of engineering in which less progress has been made in the last few decades than that of coastal engineering.

Problems of pollution of the shoreline are similar to pollution problems of lakes and rivers. Production of industrial waste and sewage in restricted marshlands stimulate phytoplankton productivity to such an extent as to destroy all other marine life by removing the oxygen from the water. Sewage can destroy valuable areas suitable for clams and oysters. Large-scale pollution, such as dumping of oil waste, can destroy or markedly reduce the recreational value of the near sea and of the beaches.

The recent Torrey Canyon incident, in which large quantities of oil were released, damaging beaches in southern England and France, illustrate the dimensions of the problem. The next few years will see tankers with very much larger capacities, perhaps as large as 500,000 tons, sailing the seas. The accidental release of this much oil would be truly catastrophic, unless ways are devised to remove it safely.

Prevention of erosion problems and of construction along the seashore have not been adequately discussed in recent times. The technology applied is, for the most part, the technology developed in the late 1800's with emphasis on overcoming the force of the sea by brute force structures. Today, many of the problems remain because little that is new in technology has been devoted to beach problems and even the rather substantial advances in the science of the interaction of the sea with the land has not been adequately applied.

Anyone living near the coast is aware of cases where beaches have been destroyed by breakwaters designed to extend the beaches, where harbours and marinas have silted up as a result of construction designed to provide shelter, and the enhancement of wave action by building of jetties supposed to lessen wave erosion. This high failure rests in part on the lack of understanding of basic mechanics of the interaction of the ocean with the beach and also on the inadequate application of new developments in technology.

Erosion is usually accompanied by the movement of sand along the coast as a result of wave-caused currents. This littoral transport is responsible for many of the shoreline problems. Sand is being removed from some places where people want it and being deposited some place else where it is not wanted. A typical, though not isolated example, is the New Jersey shore near Barnegat Inlet. Atlantic Ocean swells strike the coast near here at its most prominent point. The point is eroding rapidly as the littoral currents move the sand away in both directions. In the past 50 years, 50 million dollars have been spent on shore work in an attempt to stabilize the shoreline. The present rate of annual expenditure is of the order of 2 million dollars and the results are far from satisfactory. The most recent suggestion is to provide new sand near the vicinity of Barnegat Inlet and let the currents carry the sand to adjoining beaches and thus enhance other recreational beaches. Whether or not the project will be successful is not known, nor has an adequate estimate been made of the quantity of sand required.

An example of a different sort is found in the experience in Santa Barbara, California. A breakwater was constructed in order to provide an area of deep, quiet water to harbour a recreational area. The littoral currents carrying sand along the beach reach the breakwater and push the sand around the tip of the breakwater.   Here the sand reaches quiet waters and the sand grains are no longer suspended by the wave-created turbulence.   As a result, the harbour fills in.   In addition, further along the beach sand is not being supplied since it is trapped in the harbour and the down-current beach is stripped by the ongoing erosion.   The solution that has been applied in this case is a continual dredging of the harbour.

As can be seen from the above examples, the current solution appears to involve the transportation of sand and dredging. It would appear that the solution to many coastal problems lies not in the direction of better dredges and sand transporting equipment but rather in the better understanding of the seas with the coast and the use of this better understanding in developing new construction techniques for coastal structures.

Future of ocean technology

As is the case for weather modification, many of the problems hampering the proper development of the oceans are not wholly scientific or technological in character. But they do involve complicated questions of economics, law and sociology. A most basic problem arises out of ambiguities in the law regarding the ownership of the oceans and its resources. There have been a great number of international conferences among nations over the past 35 years, and endless litigation in the U.S. between the individual states and the Federal Government with regard to how rules of law could be established governing activities within the oceans. As things now stand the boundary between a nation's territorial sea and the high seas (which are open to all) ies somewhere along a line between 12 and 13 miles from the low tide line. Nations may choose any distance between these limits and many are understandably tending to assert the 12 miles limit, but there is no clear obligation as yet on the part of other nations to recognize the sovereignty of any nation beyond three nautical miles. Within the 12 mile region, however, in the so-called contiguous zone, the coastal state may exercise such controls as are necessary to prevent infringement of fiscal, customs, or sanitary regulations.

Nobody owns the floor of the open ocean, with the exception of the economically important continental shelves. In addition to the minerals on them, their shallow plant-rich waters are also the preferred habitat of most of the world's presently desirable fish. But the ownership rights are ambiguous.

Under an international convention on the continental shelf, the right to explore its natural resources belongs to the adjoining coastal state. The continental shelf is defined as an area adjacent to the coast up to a depth of 200 metres or to whatever depth the exploitation of the resources can be carried out. The resources are defined as mineral and other non-living materials on the sea bottom and subsoil together with living organisms belonging to sedimentary species.

International conventions adopted at Geneva in 1958 go far towards clarifying certain legal questions, but obviously many remain. The Convention on Fishing and Conservation of Living Resources of the High Seas states, for example, that each nation has a "dominant" interest in the fishing resources of its "home waters" but there are problems about how "home waters" are defined. The problems of migrant species of fish remains unsolved. Under present regulations, it is not at all clear that it would be profitable to carry out fish farming in other than territorial bays or estuaries, or to use the ocean as a pasture, since the grazers might be harvested by other fishermen.

Despite these legal problems and the very great uncertainty as to the economic profitability of ocean-going operations, it is quite clear that the next few decades are going to see a very much enhanced effort in the exploration of the oceans. From these efforts will develop the technologies to use the ocean in helping to feed the world; to supply scarce minerals and to provide abundant opportunities for recreation.

CHAPTER 3

Earthquakes and the Interior of the Earth

Introduction

In the past ten years man has begun a new phase in his exploration of the solar system. Probes designed to seek out information of a primitive kind have been sent towards the moon, Venus and Mars. In the next few years, we can expect a wealth of information resulting from this kind of exploration. As we begin this new phase in the exploration of the solar system, it is worth while to consider what we know about the most familiar planet, earth. The earth is but a single member of a complex solar system. However, it is the planet that we know most about. Questions that we will ask about the earth are questions that we will ask about other planets. The answers that we obtain from other planets will in turn guide us in understanding our own planet.

The surface above the sea on the earth has been explored extensively, and there remains few portions of the earth's surface on which man has yet to set foot. Manned exploration of the ocean bottom is just beginning. Scientific probes have been dropped to the bottom for many years, and in the past few years small submersibles have carried man to the deepest depths. We have, however, only begun to probe directly the earth beneath its surface. Tens of thousands of holes have been drilled into the earth, some of them going down as far as 10 kms below the surface. If we were to reach the centre of the earth, we would have to travel 600 times farther than we have drilled so far. Even the boldest technological explorers have yet to think of a way to probe very much deeper into the earth. The recently killed Project Mohole was designed to go only about 10 kms below the ocean surface but at a point where a great deal of information could be secured.

While very great public attention has been focused on the exploration of the moon and the planets, it is not so widely recognized that over the past few years, there have been very significant advances in the understanding of the earth and its interior. Although the exploration is just beginning, we are getting answers to such questions as: What do we know of the matter inside the earth? What is its chemical composition? Is it like the rocks that we see at the earth's surface? Or is the interior of the earth made up of matter very different from the familiar matter at the earth's surface? What are the physical conditions existing within the earth? By conditions, I mean what are the temperatures and pressures? What are the mechanical properties of the material existing at these temperatures and pressures? If we were to stretch the material making up the earth's interior, would it break like a piece of glass or would it flow like toothpaste?

The above questions are just a sampling of the great variety of queries to be posed about the inside of the earth. The answers are important for several reasons. We would like to satisfy our curiosity. We wish to understand the chemistry of our planet as compared with the chemistry of other planets, and in this way perhaps to understand the process by which the planets and their satellites were formed. Besides such purely cosmological considerations, there are many problems of importance to technology. Can we learn about new kinds of materials with desirable chemical or mechanical properties by looking at the natural laboratory, the laboratory consisting of the earth's interior? There is still another reason for attempting to understand our earth. Over the past few years, there have been a number of great earthquakes. Indeed, on the average, we can expect a very large earthquake about once each year. In a densely populated area, the destructiveness of such an event may be measured by a loss which can approach billions of dollars. In the past, severe earthquakes have indeed struck populated areas with losses and damages of the above magnitude. Unlike other natural catastrophes such as storms, floods or volcanic eruptions, earthquakes occur with as yet unrecognizable warning. However, it is the opinion of those who have studied the problem over the years that the development of a system by which earthquakes could be predicted depends in a very clear way on an understanding of the nature of the earth's interior.

In the following, I will discuss three events, two natural and one man-made, all three of which have led to a deeper understanding of the interior of the earth. These events also give rise to the hope that in the future we will be able not only to forecast in some meaningful way the occurrence of earthquakes but also to develop a technology that might prevent earthquakes. The two natural events are two of the great earthquakes of all time: the Chilean earthquake of 1960 and the Alaskan earthquake of 1964. These earthquakes are important not only in the sense that they were extremely destructive and drew society's attention again to the problems of earthquakes but also because they excited very long waves within the earth, the study of which has provided significant new data regarding the makeup of the earth's interior. The man-made event refers to a series of earthquakes in the vicinity of Denver which almost certainly have been due to human intervention.

Two great earthquakes

One of the great earthquakes of all time devastated southern Chile on May 21-22, 1960. The epicentre, or the place on the surface where the motion was most intense, was located in southern Chile but later investigation showed that the earthquake was due to a break in the earth's crust running for more than 1000 kms on a line very nearly parallel to the coastline. The first shock arrived at 6.02 a.m. on a Saturday morning when, in this part of Chile, it was still completely dark. The first shock was not as intense as some of the later shocks, but was sufficient to destroy many buildings, church towers and installations in the large towns on the coast. Many hundreds of people were crushed in their beds by the falling in of the walls. Half an hour later, still well before sunrise, there was a second shock as powerful as the first. More destruction was added to the ruins already there. Although the terror was even greater than during the first shock, there was scarcely any additional victims. The entire population had gathered in the village square and thus removed from the vicinity of structures into open spaces a large percentage of the population. People had moved as a result of long years of tradition — they knew that the greatest danger in an earthquake is the falling in of a house.

A calm lasting some 33 hours followed this double shock. The weather was particularly fine on this Sunday, a most unusual thing for the end of autumn in southern Chile, and a great stroke of fortune for the inhabitants. The largest shock of all took place at 3.00 p.m. in the afternoon on Sunday, May 22. The greater part of the population was out of doors taking advantage of the cheerful sunshine. If the shock had occurred on a weekday or at night, or even if the weather had been indifferent, the casualties, instead of being a few thousand, could have been reckoned by tens or even hundreds of thousands. At night, the people would have been caught by the collapse of their dwellings; of the 353,000 houses which stood in the zone shaken by the earthquake, 59,000 were completely destroyed. On a weekday, the factories, offices, workshops and schools would have been filled with workers and children. The fine weather and the very hour of the day — it was the time of the weekly football game which brought tens of thousands of people into the open air — spared countless numbers from death. During the eight following days, there were three more powerful shocks, and these were not the usual aftershocks but

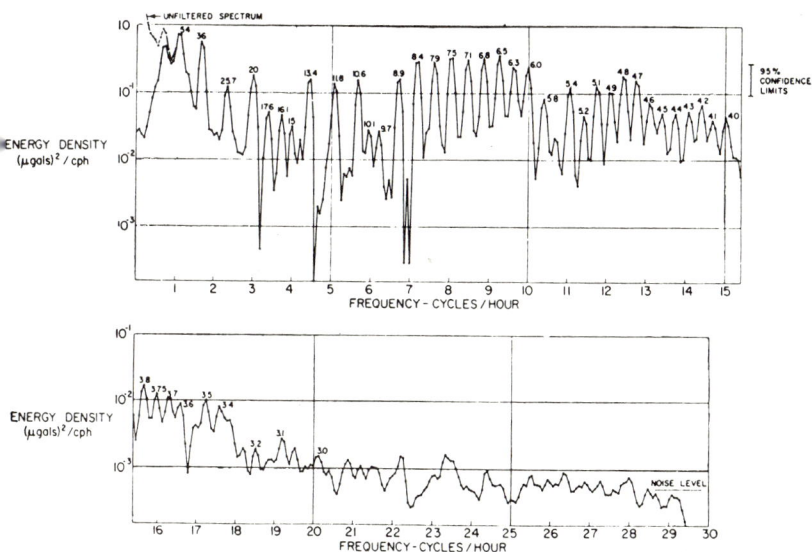

Figure 1. Power spectrum of Chilean earthquake obtained by the LaCoste-Romberg gravimeter at UCLA.

ENERGY DENSITY
(μgals)²/cph

FREQUENCY – CYCLES / HOUR

. ENERGY DENSITY
(μgals)²/cph

FREQUENCY – CYCLES / HOUR

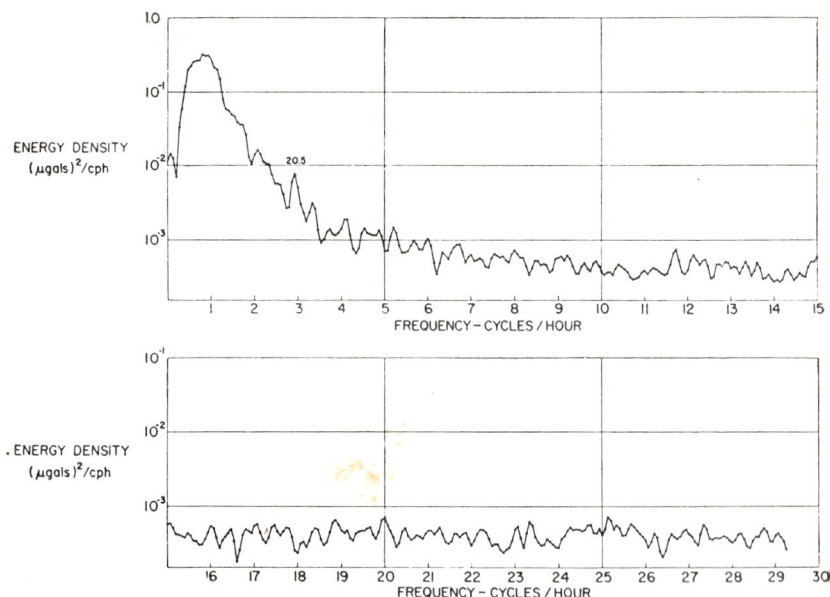

*Figure 2. Power spectrum of quiet period two months after
Chilean earthquake.*

rather earthquakes in their own rights. However, they caused only
slight additional damage.

The damage caused by the earthquake was further intensified
by a large tsunami, or sea wave, associated with the earthquake.
Several coastal villages were completely inundated and massive
damage was done to harbour facilities in coastal towns.

Accompanying the earthquake was a general subsidence of the
land with the accompanying transgression of the sea. What had
been meandering rivers in several places were turned into estuaries;
the earthquake permanently altered the landscape.

The most recent of the giant earthquakes devastated south-central
Alaska. At dusk on Good Friday, March 27, 1964, some 50,000
square miles of south-central Alaska were severely shaken by one
of the most violent earthquakes ever recorded. Only the low
density of the State's population and the hour, 5.30 p.m., when the
schools were empty and business areas uncrowded and tides low,

prevented the death toll from exceeding about 120. The initial estimates of property damage, however, exceeded 400 million dollars. Unlike other large earthquakes, no surface rupture or breakage of the ground was observed. However, the violent ground motion that accompanied the earthquakes triggered numerous rock slides, snow avalanches, and landslides throughout southern Alaska. The fractures developed chiefly in the unconsolidated settlements resulting in much land slippage though no actual deep fault approached the outer surface. Extensive damage in the coastal areas resulted from submarine landslides, which in turn induced waves and destructive tsunami effects. The coastal towns located where outwash and alluvial streams deposited deltas of water-saturated sediments were badly hit, since these water-saturated sediments were set into violent motion by the earthquake.

The waves induced by the earthquake were as high as 30-35 feet and completely devastated harbours on Kodiak Island and along the coast. As in the case of the Chilean earthquake, there were large-scale regional uplifts so that portions of islands were actually lifted out of the water and other portions submerged. While there were no surface breaks, large areas were tilted to a significant degree. The Chilean and Alaskan earthquakes clearly illustrate the devastating power of an earthquake. Many lives are lost and property damage runs from hundreds of millions to billions of dollars. In one case it was only a combination of fortuitous circumstances that prevented the loss of life from being much higher than it actually was. In both earthquakes, loss of life resulted not only from the direct damage that accompanied the earthquake but also from the secondary effects of the earthquakes. Near-coastal regions suffered severely from the tsunami and earthquake-induced waves, while particularly in Alaska, most of the damage resulted in the landslides set off by the earthquake.

Man-made earthquakes

From April, 1962, to November, 1965, Denver experienced more than 700 earthquakes. They were not damaging. The greatest magnitudes were 4·3 on the Richter scale but the community became increasingly concerned. Many people took out earthquake insurance and there was talk in the press that Denver might be removed from the list of possible sites for a large accelerator because it was becoming known as an earthquake area. The occurrence of so many earthquakes caused very great surprise since previous records had indicated that Denver was in an asesmic region. The U.S. Coast and Geodetic Survey reports that on November 7, 1882, an earthquake was recorded in Denver and the nearby counties. No earthquake epicentres were recorded in the Denver area between 1882 and the first of the new series of earthquakes in April, 1962.

Some months after the first earthquake, a number of geologists recognized that the epicentres appeared to be centred at the Rocky Mountain Arsenal. Since 1942, the Rocky Mountain Arsenal had manufactured products on a large scale for chemical warfare and industrial use under the direction of the Chemical Corps of the U.S. Army. One byproduct of this operation is contaminated waste water and until 1961 the water was disposed of by evaporation from the earth and reservoirs. When it was found that the waste water was contaminating the ground water and endangering crops, the Chemical Corps tried evaporating the water from water-tight reservoirs. This failed, so the Corps decided to drill a well and dispose of the waste waters by injecting them into the surrounding rock at a great depth.

The well was drilled to a total depth of 12,045 feet and water could be injected into the rocks at depths between 11,975 and 12,045 feet. The rock at this depth is a pre-Cambrian gneiss with vertical fractures. After drilling had been completed, salt water was pumped from the well and after 11,000 barrels of salt water had been recovered, the well went essentially dry. It was believed that as the fluid was withdrawn from the fractures, they were squeezed shut and this prevented further fluid from entering the well. In March, 1962, the Arsenal disposal programme began and 4·2 million gallons of waste were injected. Denver earthquakes started the next month. The amount of fluid introduced into the well varied during the next four years and for a period between October, 1963, and September, 1964, no fluid was injected. There appeared a statistically significant correlation between periods of high injection and the frequency of earthquakes. For example, during May through July in 1962, an average of about 8 million gallons per month was injected or about an average of 25 earthquakes per month, while during the period when no fluid was injected the average was about five earthquakes.

It appears that movement had taken place in the fractured reservoir as a result of injection of water pressures from 900 to 1950 pounds greater than the reservoir pressure. The mechanism by which movement was initiated is still uncertain. The simplest explanation is that the water under pressure drew apart neighbouring rock blocks and reduced the frictional resistance to sliding; the elastic energy stored in the rocks over very long periods of time was then released in small jerks rather than building up for one large release. Whatever the mechanism, this is the first instance where it is quite clear that man has caused a large number of earthquakes. As an aside, I should note that the Arsenal is no longer pumping water.

It is quite clear from considerations of the Chilean and Alaskan earthquakes and the Denver earthquakes that a major problem for society exists. The challenge to science is to determine what are the basic causes of earthquakes and whether there are methods by which earthquakes can be predicted, and perhaps in the more distant future whether methods of earthquake control can be devised.

Experiences in connection with the Denver earthquakes suggest that seismology may in the not too distant future become an experimental science. Potentialities for prevention of earthquakes are great but so is the danger of setting off large earthquakes unintentionally. What is required is a far deeper understanding of how earthquakes originate.

Earth's free oscillations

The Chilean and Alaskan earthquakes are of interest purely from a scientific standpoint. Both of these earthquakes excited the earth's free modes of vibration, and these were observed for the first time. The detection of the free modes broadens the spectrum over which geophysicists may look into the earth's dark interior. Prior to 1960, almost all the information regarding the earth's interior had been derived from detailed studies of the arrival of elastic body waves as recorded by seismographs. These waves are excited by earthquakes or by artificial explosions. The elastic waves travel in different paths through the earth and thus are able to sample different regions. They contain most of their energy in the high frequency part of the spectrum from about 10 to 0·1 cycle/sec. The interpretation of how these waves travel through the earth is very similar to geometrical objects where one attempts to calculate the path of a light ray through a medium whose properties differ in its various parts. The discovery and use of the low frequency normal modes as a tool for the investigation of the earth is somewhat analogous to the astronomer's use of radio frequencies to supplement observations in the visual range.

The observation and interpretation of the free oscillations following the Chilean and Alaskan earthquakes is the latest chapter in a long series of investigations of the transfer of elastic energy through solids. If an elastic solid is tapped by a hammer, the disturbance is initially carried out by two travelling waves. The fast wave (P wave) carries with it the compression and rarefaction of an ordinary sound wave. The slow wave (S or shear wave) transmits particle motion at right angles to the direction of propagation. If the solid is sufficiently isolated from the surroundings, reflections from the boundaries may set up standing waves. The solid then rings or vibrates at the normal mode frequencies. For an elastic sphere these vibrations can be classified into two groups: the toroidal or torsional oscillations are those in which a particle executes motion on a spherical surface. There is no radial component of the motion. The more familiar horizontal shear waves of classical seismology can be thought of as being made up of combinations of these toroidal oscillations.

The toroidal oscillations are characterized by the fact that the particles initially on the spherical surface remain on that same spherical surface. In the lowest order toroidal oscillation, one hemisphere rotates one way while the other hemisphere oscillates in the opposite direction, out of phase with the other hemispheres. Thus, there is the twisting of the sphere as a whole.

The spheroidal oscillations combine both radial and tangential motion to produce compression and rarefaction. One particular spheroidal oscillation involves only radial motion; the sphere changes volume by alternately expanding and contracting but its spherical shape is maintained throughout. In another low order spheroidal oscillation the sphere alternately assumes an oblate and prolate form. As a result this mode of oscillation is commonly referred to as a football mode.

The classification of the free modes into these two classes was first made by the noted mathematical physicist, Horace Lamb, in 1882. Lamb treated a homogeneous sphere, that is one in which the elastic properties are constant throughout. Later the great astronomer Jeans extended the problem by noting that the gravitational field exerted by the body would influence the spheroidal oscillations since these involve radial motion. Indeed, at low frequencies the gravitational restoring forces are of the same order of magnitude as the forces produced by elastic distortion. A method for mathematically dealing with the gravitational effects was suggested by Rayleigh and later employed by Love and Jeans.

The proper treatment of the gravitational effects did not remove all difficulties in computing the periods of the free oscillations of the earth. The period of free oscillations is determined by the distribution of elasticity and density within the earth; indeed, knowledge of the frequencies aids the determination of the internal properties. Various means for taking into account the inhomogeneities in these parameters were suggested, but only with the development of electronic computers could the variation of these quantities be properly taken into account.

Despite the considerable theoretical efforts, only in the last ten years has an attempt been made to observe the earth's oscillations. An American seismologist, Benioff, constructed a strain-measuring seismometer in the form of a silica glass rod 25 metres long, with the particular purpose of investigating the low-frequency part of the spectrum. The rod was solidly attached to the rock and its changing length provided a measure of the changing strain. Benioff suggested that an apparent 57-minute periodicity visible on the strain records of the Kamchatka earthquake of 1952 was indeed the spheroidal mode of oscillation in which the earth oscillates between the shape of a prolate and oblate spheroid. It was following this suggestion of Benioff that the theorists went to work with the computer to calculate in detail the periods that might be expected for the earth.

Between 1958 and 1960, several instrumental developments

made possible the observation of the free oscillations excited by the Chilean earthquake. Benioff at the Seismological Laboratory of the California Institute of Technology, modified the circuitry associated with the strain seismometer, so that the effects of the finite amplitude earth tides were reduced, and a greater magnification was achieved. A lower noise level was achieved on the LaCoste-Romberg gravimeter operated by the Institute of Geophysics at the University of California at Los Angeles. In addition, the Lamont Geological Observatory at Columbia University installed a strain gauge of the Benioff type in a mine shaft in New Jersey.

The free oscillations were excited by the Chilean earthquake. It is certain that other large earthquakes have also excited the free oscillations, but for the first time they were observed following the Chilean earthquake.

The principal requirement for an earthquake to excite these low-frequency modes of oscillation is that the source must have large dimensions, of the order of 1000 kms. Only the largest earthquakes involve such large sources. Thus, it is expected that only the largest of earthquakes will excite the low-frequency oscillations.

The free oscillations excited by the Chilean earthquake were detected both on the gravimeter and the strain seismometer. These instruments complement each other. The strain seismometer is sensitive to the strain produced by vertical and horizontal motions. It therefore records both the spheroidal and toroidal oscillations. On the other hand, the gravimeter records only vertical accelerations and thus only the spheroidal oscillations. The combination of the observations from the two instruments permits a separation and identification of two classes of motion. *Figure 1* shows a power spectrum of the variations in gravity at Los Angeles for four days following the Chilean earthquake. This should be compared with *Figure 2* which shows a quiet interval some weeks after the earthquake. This spectrum is almost structureless although one should call attention to the fact that the peak at 20·5 minutes is still evident.

The spectrum of seismic disturbances is thus characterized by well-defined peaks for periods that vary between about 1 hour and 8 minutes. At higher frequencies the isolated peaks begin to merge into a continuum as a result of the finite width of the individual peak and the increased number of peaks.

From the observed frequencies of the normal modes of vibration, one can compute the structure as defined by the distribution of density and elastic wave velocity. This has been done both for the spectra observed at the Chilean earthquake and that observed in the Alaskan earthquake.

The observations of the free oscillations permit a determination of the internal structure of the earth. It is in some ways independent and superior to the methods used before. Prior to observations of the free oscillations, the velocity variation with depth was deduced from observations of numerous earthquakes at many stations. These observations were then combined with the determination of the mass and moment of inertia of the earth to yield the density variation. The difficulty with this procedure is that numerous observations of varying quality must be combined. The quality of these observations varies as the quality of the instrumentation is far from uniform. In addition, the readings at any given station depend on the skill and experience of the seismologist, and as a result a large number of uncertainties enter into the construction of the travel time curve by classical methods. The determination of the structure from the free oscillations depends on the identification of the frequencies of the normal modes from observation obtained by a single instrument at a single station or at most by a few instruments at a few stations.

The combination of classical methods with the normal mode observations has thus permitted a far more accurate delineation of the interior constitution. This has been further aided by the availability of observations of body waves generated by nuclear explosions.

In addition to information about the density and elastic constants, observations of the free modes of oscillations provide information about other properties of the earth's interior. If the solid parts of the earth were perfectly elastic and the fluid parts perfectly fluid, then the spectral peaks should show up as individual lines broadened only by data reduction techniques. This is the equivalent to instrumental broadening in spectroscopy. The deviations from perfect elasticity or fluidity result in a natural broadening of the line. The degree to which a line is broadened or alternatively the rate at which a given peak decays provides in time a measure of the anelastic properties of the earth; that is, the extent to which the earth deviates from perfect elasticity or perfect fluidity. The distribution of the anelastic properties can be obtained by studying the decay rate of oscillations at different frequencies, since these oscillations affect different parts of the earth to varying extents. Furthermore, the mechanisms of dissipation will differ in the various oscillations. The spheroidal oscillations, for example, the so-called football mode, involve the entire earth, including the core, and the motion contains components both of compression and shear. Toroidal oscillations are affected only by the outer regions of the earth; that is, the solid mantle.

A study of the rate at which the various oscillations decay point up significant and surprising results. The earth ringing in the football mode with a period of 54 minutes rings like a rather poor bell. For more complicated vibration which sample progressively outer regions of the earth, the dissipation is greater. The most surprising result is that the earth in the purely radial mode of oscillation, that is the one in which the particles initially on a spherical surface remain on that spherical surface but move up and down, the earth rings very much better than a good bell; indeed, following the Chilean earthquake this mode of vibration was still detectable six weeks after the earthquake. The earth thus hummed for a very considerable length of time. The reason for this remains a mystery.

Origin of earthquakes

I have sketched very briefly some of the principal results arising from the study of the free vibrations of the earth excited by large earthquakes. Now I would like to touch on the question of how earthquakes originate. Some earthquake sources are quite well understood. For example, earthquakes can result from the collapse of caverns near the earth's surface or as a result of landslides. In addition, earthquakes are associated with volcanic activity in every volcanic region of the earth. The source of energy is probably associated with the movement of magma upwards and the collapse of the regions from which the melted material is drawn. But the most important and large earthquakes are found in what are termed tectonic regions; that is, regions on the earth's surface which are undergoing or have in the recent past undergone mountain building. These earthquakes are concentrated in two belts. One belt passes around the Pacific Ocean and affects countries with coastlines bordering on this ocean, for instance New Zealand, New Guinea, Japan, the Aleutian Islands, Alaska, and the western regions of North and South America. The second belt passes through the Mediterranean region eastward through Asia and joins the first belt in the East Indies. The majority of the large earthquakes are located on the first belt. Breaks in the earth's outer surface which accompany the largest earthquakes extend over very large regions. I have indicated that in the case of both the Chilean and Alaskan earthquakes, there was disruption at least for a linear distance of 1000 kms.

While the details of the origin of tectonic earthquakes remain a matter of great controversy, it is generally agreed that elastic strain energy is gradually accumulated in certain regions of the earth and then suddenly released in the earthquake. What is the mechanism by which the energy is suddenly released? A much oversimplified view of the matter is that the outer crust is distorted and distortion eventually gives rise to relative movement along a pre-existing break or fault in the earth or if the distortion is sufficiently great, a new fracture is formed. The behaviour of the rocks in Denver suggests that friction between blocks separated by fracture zones plays an important role and if this friction can be reduced, for example, by the injection of water, then the release of elastic strain takes place at a lower level of distortion than would otherwise be the case.

The reason for the buildup of strain within the earth and its apparent concentration along continent and ocean boundaries still remains much of a mystery. I have studied the problem from the point of view of the thermal balance within the earth. Radioactive elements within the earth decay and give off heat so that as one goes deeper within the earth the temperature rises. Distribution of radioactive elements differs between continents and oceans. In continental regions, the radioactivity is concentrated towards the surface. Rocks such as granite have far greater concentrations of the radioactive elements (uranium, thorium and potassium) than do rocks found in the ocean basins. Yet it appears that about the same amount of heat is being produced in the continental regions and oceanic regions. If this is so, radioactivity within the oceans is buried deeper than under the continents. One would expect that under the oceans the rocks are warmer at depths of 50 kms and below than under the continents. These temperature differences can give rise to elastic stresses, the distortion being concentrated along the boundaries of the continents and oceans. The chemical differentiations of the earth in the development of continents may be responsible for earthquakes concentrated along continental borders. In a body with no continents and oceans, earthquakes may be rare. The moon shows no continent-ocean structure. An important observation for the understanding of earthquakes is the determination of whether or not there are moonquakes.

New techniques of instrumentation have led to a major revolution in the old science of seismology. I have referred to two new instruments, the giant strain gauge and the ultra-sensitive gravimeter. There have been a number of other developments. As a result of the need to distinguish between nuclear explosions and earthquakes, great seismic arrays have been constructed in which many sensitive elements are used to form an antenna to pick up and locate weak sources of elastic energy. As this new information develops, we are certain to gain a much better understanding of the origin of earthquakes.

In addition, observations in the earthquake regions clearly indicate that there may be means for predicting earthquakes. For example, it may be possible to accurately measure the slow buildup of elastic energy that is thought to precede an earthquake. Changes in the rate of buildup might be associated with the onset of an earthquake.

There are tantalizing possibilities of actually preventing or altering earthquakes as is suggested by the experience with the Denver earthquakes. These possibilities strongly suggest that seismology will be one of the more interesting and exciting of the earth sciences.

CHAPTER 4

Origin of the Earth

Introduction

Origin of the earth is part of the larger problem of origin of the solar system. The origin of the solar system and the course of its history must be included among the great problems of natural philosophy, comparable in general interest to questions regarding the origin of life and the development of man. Indeed, the study of the origin of life and the development of man cannot be separated from cosmological considerations. An understanding of the history of the sun, the earth, and perhaps of the other planets is required in order to fix both the conditions requisite for the development of primitive life and the changing conditions which stimulated the evolution of those forms of life now present.

Today, the solar system consists of the sun, nine planets (many of which have extensive satellite systems), numerous smaller objects (so-called asteroids), a large family of comets, and a mixture of dust and gas and the space between the major members of the solar system. Study of the origin of planetary systems differs greatly from the study of the origin of stars. Many millions of stars are known to astronomy. Individual stars can be seen at different stages in their development. As we look out we can see young stars in the process of being formed and also very old stars. We can thus reconstruct much of the history of the star just by observations made today, even though a star may live for many billions of years. Compared to the many millions of stars, we can only study in detail one planetary system. Stars with planetary systems like the sun do not appear to be isolated freaks. There is evidence for believing that in fact planetary systems are common. For example, a few planets have been detected by telescopic means. This suggests that our planetary system is not unique.

However, at present, we cannot study these other planetary systems because the light they emit is far too weak to give detailed information. We are thus limited in the investigation of the origin of the solar system to our own system.

Many of the older discussions of the origin of the solar system have been based largely on conventional astronomical observations of the motions of the planet. Today, however, we can add several new dimensions to this age-old consideration of planetary origins and constitutions. We owe our enlarged insight, which consists more of puzzling new questions than of firm answers, first to optical ground-based observations, supplemented in recent years by ground-based observations of the planets in the radar and infrared parts of the spectrum, and spectroscopic analysis of planetary atmospheres.

Stimulating this new ground-based assault is a second factor, a recently-won space capability which has for the first time permitted *in situ* exploration of the moon and Venus and new photography of Mars. The third factor contributing to our enlarged understanding is new knowledge about the earth, some of which I have discussed previously. Progress in this last area has resulted from rapid strides in the science of seismology, and laboratory studies of mineral behaviour under high pressures and high temperatures which characterize the earth's, and presumably other planets', deep interior. This progress also owes much to improvements in our theoretical ability to deal with conditions prevailing in these deep parts of the earth and other planets which we will not be able to study directly, at least in the foreseeable future. A further aid has been new instrumentation which allows us to determine accurately not only chemical composition but also the ratio of isotopes of the elements. Of particular importance has been the study of the chemical and isotopic composition of meteorites, the only non-terrestrial sample of the solar system which we have been able to take to our laboratory, although shortly samples of the moon will be available for laboratory study.

Radio observation of stars in our galaxy suggested to many observers a general picture of how stars are born: a mass of gas and dust in the space between already existing stars is in motion and various forces acting on the gas and dust are in balance. The main forces are those due to the gravitational attraction between the particles and the centrifugal forces due to rotation. At some stage these forces may get out of balance and condensation or aggregation will proceed with the gas and dust locally gathering together. This then is the initial step in the star's formation. While this method of star formation is accepted by many scientists, there is no measure of agreement regarding how a star acquires a planetary system. In order to illustrate some of the difficulties in formulating a satisfactory theory for the development of the solar system, let me review what we know about the solar system. Much of this information is, of course, based on studies of the earth.

Time of origin

While we still know very little about the early stages of the formation of the sun and the planets, we do know quite accurately a critical time, probably the time at which the planets were formed. Beginning of solar system time can be determined by studying the abundance of elements and isotopes that are produced by radioactive decay. Radioactive elements such as uranium, thorium, and potassium decay or break up into lighter elements at a precisely fixed rate. By examining the number of daughter elements produced in the breakup of the parent element, one can obtain an estimate of the time the radioactive element has been in place. Analysis is often complicated in the sense that the material might have contained initially some of the daughter elements.

Studies of the radioactive elements have been carried out on terrestrial rocks and also on meteorites. Various radioactive elements give an age for meteorites of about $4\frac{1}{2}$ billion years, while the oldest rocks found on earth are somewhat greater than 3 billion years. By examining the total abundance of certain isotopes of lead, it is possible to obtain an estimate of the time that has passed since the earth was assembled, regardless of the thermal process mixing of loss of materials that have taken place since. If this method is used the earth is found to be about $4\frac{1}{2}$ billion years old. Thus there is substantial data to indicate that the earth was formed about $4\frac{1}{2}$ to 5 billion years ago. Although we know the time of the event rather precisely, we need much more information about the length of the accumulation or condensation process.

Some estimates would make the time as short as a few million years while others a few hundred million years.

Let me first review the information provided by astronomy on the solar system and survey the conditions imposed and the possibilities revealed by these observations.

Organization of the solar system

The first generalization which can be drawn from conventional astronomy is that the solar system today is highly organized. Every planet follows a monotonous journey about the sun which is predictable, on a short time scale, in every respect. With two exceptions, the innermost planet Mercury and the outermost Pluto, the orbits are very nearly perfect circles; that is, they are ellipses having a very low eccentricity. Furthermore, the orbits lie very nearly in a common plane. But was the system always so highly organized? Are the paths now travelled by the planets the same as the paths travelled 4 or 5 billion years ago when, as indicated by the radioactive studies on terrestrial rocks and meteorites, the planets had already assumed their present form? Is it possible to begin with the present configuration of the members of the solar system and trace changes in their orbits through space and time to the beginning? The answers are that we cannot yet trace the history although fairly elaborate attempts to do this have been made; consequently we do not know why the planetary orbits are as they are.

Any attempt to analyse these problems requires a detailed understanding of how every individual member of the solar system interacts with all the others. The planets and the sun attract each other through gravitational forces which are complicated by the fact that the bodies of the solar system are not rigid or perfectly spherical but are deformable and have shapes that deviate appreciably from sphericity. While we can work out in great detail how two bodies can interact with each other gravitationally, the problem of three bodies is immensely more difficult and the problem of many bodies has not been adequately treated. In order to understand the evolution of the planetary orbits, we require a detailed understanding of how every individual member of the solar system interacts with all the others. We have only a primitive understanding of the long-term consequences of such many-bodied interactions. As in many other fields of physics, the question of many bodies interacting with each other through long-range forces is a central one, and one that we must answer if we are to determine the solar system's history.

Angular momentum in the solar system

The second generalization which can be drawn from classical astronomical observations is that the sun contains almost all the mass of the solar system while the planets possess nearly all of the system's angular momentum. Rotation of the planets on their axes and the sun on its axis contributes very little to the total angular momentum. Indeed, in terms of angular momentum per gram of matter — the angular momentum density — the nine tiny planets by virtue of their orbital motion about the sun contain 100,000 times more angular momentum than the much more massive but slowly rotating sun. This inverse allotment of mass and angular momentum between the sun and the planets has been a major stumbling block in all theories seeking to account for the origin and very early evolution of the solar system. The problem hinges not so much on the inequity in mass distribution, the sun being almost a thousand times more massive than all the planets taken together, as it does on the profound dichotomy in the distribution of angular momentum.

If the general picture of condensation of the sun from a mass of gas and dust is correct, how can this condensation proceed so that all the mass falls to the centre but leaves behind the angular momentum in the thin material which remains? The mystery is further compounded by certain regularities in the distribution of angular momentum among various objects. The major planets, Mars, and certain classes of stars (the so-called early type stars) all possess an angular momentum density which is related to their mass in a simple way and this is illustrated in *Figure 3*.

On the other hand, late type stars such as the sun show a marked deficiency in angular momentum. In the rates of rotation, the late type stars are low for their mass in comparison with the early type stars. The bright massive early type stars rotate at a rate consistent with those observed for the larger planets, while the late type stars are smaller, less bright and rotate at a much smaller rate.

Significantly, the total angular momentum of the solar system, including the orbital angular momentum of the planets and the rotational momentum of the sun, is consistent with the mass angular momentum relationship shown by the bright early type stars.

Consistency of the relationship between angular momentum and mass for many objects, and the inconsistency of the sun to conform to this relationship, raises a number of questions. It would appear that rotating masses having angular momentum characteristic of the major planets and many stars are unstable as a single high density phase if their mass is less than three or four solar masses and

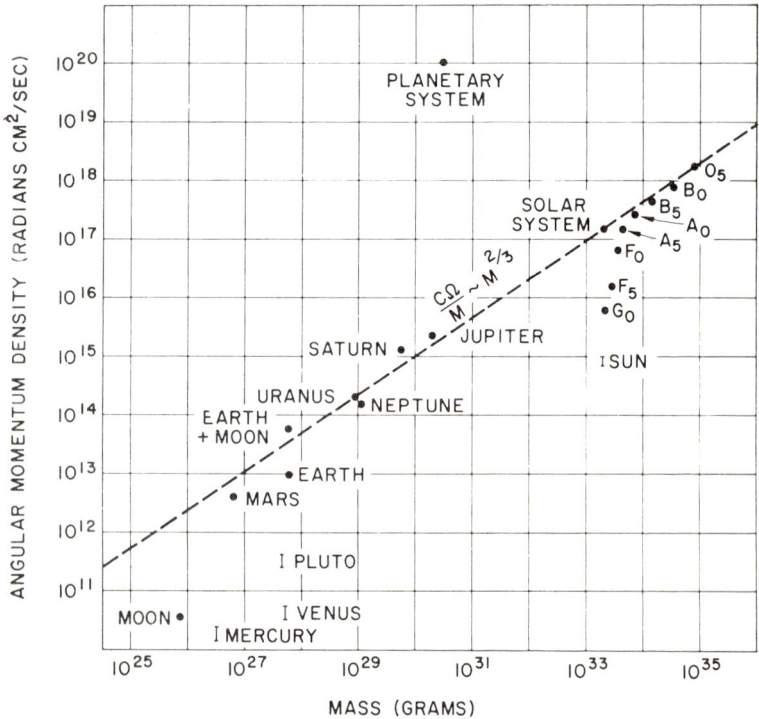

Figure 3.   Variation of angular momentum density with mass for stars and planets.

greater than about one-tenth to one-hundredth of a solar mass. There would appear to be a range of masses for which the object cannot remain as a single body and maintain its angular velocity. If the object has this intermediate range of masses, it breaks up into phases: a centrally condensed, very massive high density phase and one of low density but carrying all the angular momentum. The reasons why such a separation for the intermediate size masses should take place are vaguely understood but details of how this separation comes about certainly are not.

One solution to the angular momentum problem is then to suppose that an object having a mass somewhat in excess of the present sun begins to condense and as it rotates it tends to flatten out into a disc. As the condensation due to the dominance of gravity over centrifugal forces continues, a point is reached where matter separates into a high density angular momentum phase and a low density high angular momentum phase, somewhat analogous to the kind of phase instability one has as one approaches the critical point of a fluid. In such a picture of early solar system evolution, the dominant forces controlling everything are the gravitational forces and the forces due to rotation.

Another possible solution to the problem of angular momentum involves the action of magnetic forces and it is thought by some that such magnetohydrodynamic processes could in fact account for the distribution of mass and angular momentum in the solar system. In this theory, ionized gas thrown from a collapsing sun may, by the action of magnetic forces, slow the sun's rotation and in addition remove gases from the system. The sun would have contracted rapidly to about the dimensions of Mercury's present orbit. At that time, its rotation and increasing temperature produced strong magnetic fields near the surface. These fields produced a wiry rigidity in ionized gas since the magnetic field would be tied to the electrically conducting gas. The rotating sun would drag the outermost gas behind it and wind up its magnetic lines of force and disc about the sun. The angular momentum of the sun is then transferred to this gaseous disc along the magnetic lines of force. Along these, the gases from which the planets were condensed were slowly carried upwards. There are many difficulties with this theory. A major objection is that it provides no easy explanation for the observed dependence of the angular momentum on mass that is found among the planets not affected by other forces later in the development of the solar system and among many stars.

Small and large planets

Whether momentum was transferred to the proto-planets by magnetic or hydrodynamic processes, it is clear from visual and spectroscopic observation that the planets, however formed, fall into two fundamentally different groups. One consists of smaller dense planets near the sun and the other of less dense, very large planets at very great distances from the sun. Pluto is an exception to this statement, but we know very little about this object and it may not have been a planet initially but perhaps a satellite of Neptune. The giant planets possess complex satellite systems while the inner terrestrial planets have either no satellites, one in the case of the earth, or two tiny ones in the case of Mars.

This separation of the giant gaseous planets from the smaller dense ones clearly demonstrates that either chemical homogeneity did not exist during the early stages of solar system evolution or else it did not persist through the late history of the solar system when planets were being born. The giant planets contained abundant quantities of gases, such as ammonia, methane, hydrogen, and helium that are rare in the terrestrial planets. Indeed, it may be that the composition of the giant planets, such as Jupiter, is very similar to the composition of the sun. One may even speculate that Jupiter is a planet that was not quite massive enough to become a star, for the temperatures reached within its interior were insufficient to bring about the nuclear reactions which provide the energy of stars.

Satellite systems

In addition to the mostly classical observations I have outlined above, orthodox discussions of the origin of the solar system usually stress the similarity between the planet satellite systems, and the solar system itself. The analogy is only partly applicable, however, since in addition to the gravitational forces which dominate long-term evolution of planetary systems, there is another force which particularly influences the development of a satellite system.

In discussing the dynamics of planets and their satellite systems, it is useful to distinguish four time scales. As an illustration I will refer to the earth and its satellite, the moon, though these considerations are much more general. In the earth-moon system, the shortest time scale is that of the rotation of the earth. I have discussed earlier that the rotation rate of the planet is very probably determined by a process of deformation, although that rate may have been altered by processes taking place later in the history of the solar system. A somewhat longer time scale is that of the motion of the moon about the earth (about once every 28 days) and the motion of the earth and moon about the sun. The third time scale involves the motion of the orbit of the moon. The orbital plane of the moon is not fixed in space but undergoes a precessional motion. That is, if we draw a line perpendicular to the orbital plane of the moon that line will trace a cone about a direction perpendicular to the plane containing the earth and the sun. This precession now takes place once every $18 \cdot 6$ years. The earth's axis of rotation also precesses. That is, the axis of rotation moves in a cone perpendicular to the plane containing the earth and sun and other planets with a period of once in 26,000 years. This well - known motion is usually termed the precession of the equinoxes. The precession of the moon's orbit and the precession of the earth's axis of rotation have intermediate time scales which are long compared with the rotation of the earth or rotation of the moon about the earth but short compared with changes which may take place over very much longer times.

This additional gravitational interaction which George Darwin discovered and labelled tidal friction depends both on the fact that the planets and satellites are deformable bodies and on the existence of friction. If the planets and satellites possess no friction whatever, then the tide raised by the satellite on the deformable planet and by the planet on the satellite do not influence in any way the long-term orbital motion of the satellite or the rotation of the planet. If friction accompanies deformation as it does in reality then tides will tend to slow the planet's rotational motion, provided the day is short compared to the month, and tend to move the satellite away from the planet. If, on the other hand, the day is long compared with the month, as is the case for one of the small satellites of Mars, then the tidal friction speeds up the planet and moves the satellite towards the planet. A remarkable feature of the tidal friction interaction is its very strong dependence on the distance between the satellite and the planet. When the satellite is close the interaction is very strong. As the satellite moves away the strength of the interaction decreases rapidly.

The sun also raises tides on the planets and the planets raise tides on the sun. However, because of the large distances involved we find the planetary orbits to be stable, in that solar tides produce very small changes over times of the order of billions of years. Even the innermost planet, Mercury, undergoes only minute changes in its path because of the tides raised within it by the sun. In this way, the dynamics of the satellite systems differ greatly from the dynamics of the planetary systems because in the satellite systems the tidal interactions are very important. It is complicated, however, by the motions alluded to earlier, that is the precession of the orbit of the satellite and the precession of the axis of rotation of the planet.

The precession of the axis of rotation of the planet and its intermediate time scale is due to the fact that the planets in general are not spherically symmetrical. Because of its rotation, a planet assumes a spheroidal form with an equatorial bulge. The attraction of the sun and the satellite on this bulge, provided the bulge does not lie in the plane of the planets and satellites, results in a torque; the net effect of which is to cause the axis of rotation to move slowly about its mean position. The precession of the moon's orbit is due to similar torques. If the moon's orbit does not lie in the equatorial plane of the planet, then the planet's equatorial bulge will exert a torque on the satellite as it moves about the planet, the net effect of this torque being a precession of the orbit. Similarly the sun can exert a torque on the satellite orbit, again provided that the orbit does not lie in the common plane of the planets and sun.

The dynamical considerations reviewed above permit a classification of natural satellites by their orbits. Satellites can be classified into two groups. The first group is comprised of satellites moving in equatorial or near-equatorial orbits. With the exception of the moon and Triton, it includes all lunar-sized satellites. This is illustrated in Table 1.

TABLE 1

Planet	Satellites	a/R
EARTH		
MARS	1 PHOBOS	2·76
	2 DEIMOS	6·92
JUPITER	1 Io	5·90
	2 EUROPA	9·40
	3 GANYMEDE	14·99
	4 CALLISTO	26·36
	5	2·54
SATURN	1 MIMAS	3·11
	2 ENCELADUS	3·99
	3 TETHYS	4·94
	4 DIONE	6·33
	5 RHEA	8·84
	6 TITAN	20·48
	7 HYPERION	24·83
URANUS	1 ARIEL	8·08
	2 UMBRIEL	11·25
	3 TITANIA	18·46
	4 OBERON	24·69
	5 MIRANDA	5·49
NEPTUNE		

TABLE 2

Planet	Satellites	a/R
EARTH		
	MOON	60·27
MARS		
JUPITER	6	160·7
	7	164·4
	8	326
	9	332
	10	164
	11	313
	12	290
SATURN	IAPETUS	59·67
	PHOEBE	216·8
URANUS		
NEPTUNE	TRITON	15·85
	NERIED	249·5

TABLE 3

Planet	(a/R) Critical
EARTH	10
MARS	13
JUPITER	32
SATURN	43
URANUS	84
NEPTUNE	70

A similar listing of satellites moving on non-equatorial orbits is given in Table 2. An important point is that the satellites included in Table 1 (that is, those moving in an equatorial orbit) occupy smaller orbits than do the non-equatorial satellites. In these tables, the orbital distance a is given in terms of the radius of the planet R.

The fact that the close-in satellites lie in the equatorial plane seems at first surprising since all the planets are precessing with their axis of rotation moving slowly about a direction perpendicular to the planet's orbital plane. The satellite's position on the equatorial plane implies that the equatorial bulge must be dragging satellites along with it, while the more distant satellites nearly lie in the common plane of the planets and the sun.

Indeed, there is a critical distance such that a satellite orbit lying within this distance will maintain a nearly constant inclination to the planet's equatorial plane. This nearly constant inclination is maintained despite the precessional motion of both the orbit and the equatorial plane. For orbits much larger than the critical orbit, the satellite's orbit plane no longer maintains a constant inclination to the planet's orbit around the sun.

That the satellites do in fact obey this dynamical rule is shown in Table 3. With the single exception of Triton (the satellite Neptune in a retrograde orbit) the equatorial satellites are those lying within the critical radius and the non-equatorial satellites are those lying outside. It is not surprising that the satellites lying outside the critical orbit are not found on the equatorial plane. Indeed, even if one of these satellites were placed on the equatorial orbit, it would soon precess off of it. The surprising result is that, with the single exception of Triton, the 20 satellites found closer than the critical distance from their planets all move on equatorial orbits. This fact certainly calls for an explanation.

Tidal effects of the sort discussed earlier will tend to drive the satellite which is within the critical radius into the equatorial orbit. Once there, it becomes locked and remains. However, in a number of cases, particularly for the satellites of the major planets, the dissipative properties of the planets are sufficiently low so that no substantial change in the orbits could have taken place over the 4 to 5 billion years' history of the solar system. This would suggest that the satellites initially formed in the equatorial plane and have remained there. Furthermore, it is important to note that it is the largest satellites which are today found near or on the equatorial plane of the planets.

In considering the formation of satellites by the aggregation of a disc of materials surrounding the planet, it is important to recognize that the equatorial plane occupies a unique position. A disc of particles placed in any other plane would rapidly disperse into a band as the individual particles precess at different rates. Beyond the critical radius, a band could not form because of the perturbing effects of the sun unless it was placed in the orbital plane of the planet. Thus, in the formation of certain planets, it would appear that a disc of material is associated with the equatorial plane and this band aggregates to form the innermost and more massive satellites. The outermost satellites may have been formed some distance from the planet and later captured by it.

Earth and moon

While the effects of tidal interactions on the orbits of certain of the satellites have been small, this is not the case for the earth-moon system. We know, for example, that the moon is retreating from the earth at a small but measurable rate, a retreat caused by the tidal friction interaction. This rate can be precisely determined by astronomical observations extending over long periods of time and is equivalent to increasing the radius of the moon's orbit by about an inch a year. In this process, angular momentum is transferred from the earth to the moon which enables the moon to enlarge its orbit, but the earth as a consequence slows down in its rotation since angular momentum must be conserved. As the earth slows down, the day gets longer.

Sufficient details about tidal friction interaction are known to permit a calculation backwards in time. While the present rate at which the moon is moving away is small, it may have been substantially larger in the past, since the effectiveness of tidal friction depends greatly on the distance between the planet and the satellite. If we extrapolate backwards in time for the earth-moon system, we find a striking result. Suppose the present rate of tidal dissipation of the earth's rotational energy is representative for all geologic time — and it may well be if the earth's internal structure took its present form quite early in the history of the planet. If that were indeed true then the moon must have been very close to the earth $1\frac{1}{2}$ to 2 billion years ago, a time short in contrast to the $4\frac{1}{2}$ to 5 billion year age of the earth. The indicated result of tidal interaction creates major problems for understanding the development of the solar system. Is the present earth-moon system itself a more recent development, not dating to the earliest stages of earth's history? Has the moon arrived in our vicinity relatively recently? Or did the earth at one time possess a system of moons differing greatly from the present one? Indeed, the analysis of tidal interaction suggests that the moon may have been an interloper which was captured by the earth and that the moon travelled a different path in the very early history of the solar system. Alternatively, the moon may have formed from several smaller objects circling the earth. These objects had been formed in the near vicinity of the earth at these early times. Both of these ideas are much more tenable than the old notion that the moon was torn from the body of the earth.

What about other satellite systems? Are they more stable than ours through tidal interaction? The complex satellite systems of Jupiter and Saturn, especially the presence of massive inner satellites, demonstrates that the frictional properties of the giant planets differ greatly from those of the earth. If the properties had been similar then the large inner satellites of these giant planets would have moved into the planet, perhaps even sweeping up the smaller satellites that lay in their paths. This has not happened and we can thus conclude that the interiors of these planets differ dramatically from the earth, at least insofar as their dissipative properties are concerned. These planets must be much more nearly perfect elastic bodies than is the earth since they dissipate energy and transfer angular momentum to their satellites at a rate only one-hundredth to one-thousandth that of the earth. But even at these low rates of dissipation, major tidally-induced changes in the orbit have almost certainly taken place in certain satellites — particularly Triton, the satellite of Neptune.

Rotation of Venus and Mercury

Tidal forces not only enlarge a satellite's orbit but they reduce a planet's rotational speed. While the angular momentum associated with the planet's orbital motion about the sun is only slightly affected by tidal interaction, the angular momentum due to its rotation on its axis can undergo large decreases.

This effect explains why two of the four terrestrial planets, Mercury and Venus, have much lower rotational speeds than do the other planets and why the earth has a lower angular momentum density than would be expected from its mass. Indeed, if the earth had not lost angular momentum through interaction with the moon and the sun we would be rotating with an angular velocity corresponding to a period of 12 hours. The earth initially had a day that was about 12 hours long if its original angular momentum density was consistent with that observed today among planets and stars. The tidal effect also has altered the rotation of the moon. The earth, by raising tides on the moon, has brought the moon into a rotation so that the same face of the moon is always turned towards the earth. For the moon the length of the day and the length of the month coincide, but this is not the case for either Mercury or Venus. For both these planets tidal friction has not caused the length of the year to equal the length of the planetary day.

At the same time that Mariner IV was making its historic journey to Mars, equally historic ground radar studies showed that Mercury spun on its axis once about every 59 days not once every 88 days as the length of its year had been widely assumed. Since tidal interaction had brought the moon's length and its period of revolution about the earth to coincidence, it had been widely assumed that tidal interactions would similarly affect other members of the solar system. It was thought that tides raised on Mercury by the sun were so great that they brought the rotation of the planet into coincidence with the revolution of the planet about the sun. The discovery that Mercury had a period of rotation of 59 days quite convincingly showed that effects other than pure tidal interaction are at work.

For a circular orbit the sun's tides would seek to equalize Mercury's rotation with its orbital angular velocity. But the orbit of Mercury, as has been noted earlier, differs from that of a perfect circle and is quite eccentric. The strength of the tidal action depends very greatly on the distance between the planet and the sun and this strength varies in an eccentric orbit. Tidal effects are four times as strong when Mercury is closest to the sun as compared to only one-third as strong when the planet is farthest from the sun. In addition to tidally varying torques there are effects due to the fact that Mercury, like other planets, probably does not have a density distribution which is completely symmetrical about its axis of rotation. The earth's equator is slightly elliptical and so is the moon's equator. Mercury is very probably similarly misshapen. There is a tendency for the largest axis of the equator to assume a preferred position with respect to the orbital motion. Indeed, the most probable rotation rate for a planet having deviations from symmetry about the axis of rotation with as large an eccentricity as is the case for Mercury is one in which the year corresponds to $1\frac{1}{2}$ days. This is an important discovery since it implies that no area of Mercury is permanently in shadow; the planet keeps showing a new side to the sun and to the earth. What is surprising is that optical observations had not revealed this faster rate of rotation, since Mercury, unlike Venus, has quite definite surface detail.

Venus has been shown by similar radar studies to have a slight retrograde rotation; that is, its rotation is opposite its orbital motion. The rate of retrograde rotation is very nearly that which would be expected if the planet were asymmetrical about its axis of rotation and if there were tidal interaction with the earth. Thus it would appear that the earth, even though far distant from Venus, and of relatively small mass, is effective in pulling Venus into its current period of rotation.

Early history of solar system

Observations and interpretations discussed above provide a sketchy outline for what may have been the early history of the solar system. A cloud of gas and dust began to accumulate as a result of the imbalance between the gravitational and rotational forces. The aggregation leading to the formation of the sun was such that angular momentum was transferred to a thin disc of material in the sun's vicinity. The mechanism by which this transfer took place is quite uncertain.

An unanswered question is whether some instability was reached in which a sector of the disc suddenly collapsed forming a planet which then swept up the few remaining materials or were there many smaller aggregations which were then slowly collected as the largest aggregations travelled about the sun? For two objects of the solar system, the moon and Mars, we know that at least in the last part of their aggregational history an infall of material contributed to the outer layers. What is not known is whether this outer surface is like many similar surfaces buried at depth or if the impacting objects were leftover material from a major condensation process.

Prior to the flight of Mariner IV many considered Mars to be a planet somewhat similar to the earth. Nothing in the many thousands of visual observations and photographs made from the earth suggested that the planet had a geologically inert surface. Indeed, earth - based observations suggested a contrary set of possibilities. The 22 closeup photographs of Mars taken by Mariner IV revealed a surface rather densely populated with impact craters up to 120 kms in diameter. On Mars we see a barren moon-like visage which implies Mars may be more like the moon than like the earth. This was not the only discovery made by Mariner IV. Even more significant was the failure to discover Martian equivalents of such earth-like features as mountain chains, continents, or depressed basins. Apparently the principal Martian landscape has not been produced by mountain and continental building stresses originating within the planet as has been the case within the earth. Thus Mars is a very different planet from the earth. We expect the same to be true for Venus but our uncertainties here are multiplied by the dense atmosphere which covers the solid surface. The giant planets also differ very greatly from the earth.

Thus, we can learn a great deal by the intensive study of one planet, the earth, but to answer questions about the solar system as a whole, we will have to explore each of the planets and find their unique features and what these features tell us about the earliest part of the solar system's history. Only in this way can we hope to unravel the processes which led to the formation of the sun, earth, eight other planets, and the many other small objects which comprise our solar system.

The Time Scale
Of Creation

(Three Chapters)

by

ROBERT M. MAY

Dr. R. M. May,

Reader in Physics,
School of Physics, University of Sydney.

THE TIME SCALE OF CREATION

Preamble

The earlier chapters in this book have dealt with our earth and its satellite, the moon, and have gone further into space to discuss our personal star — the sun — and our solar system as a whole. Now we are going to venture out into the broader domains of astronomy and cosmology, where stars, or more usually galaxies, are to be thought of as the smallest intelligible units of study.

The central theme of the following chapters is an account of the present status of cosmology. Cosmology deals with man's attempts to understand the universe as a whole, and to answer such questions as what was the universe like in the remote past? What will it be like in the future? In particular, we will address ourselves to current attempts to answer the question — *how old is the universe, what is the time scale of creation?*

Of course, such questions have fascinated man literally from time immemorial, and have in the past produced much thought, some of it profound and some of it crackpot, but essentially *none* of it based on rational, verifiable data. On the other hand, the present discussion will cleave to experimental facts (most of them very recently discovered), and it will be seen that a few things can now sensibly be said in answer to the old questions of cosmology.

In this part of the book, Chapter 1 contains a brief review of the cosmologies, the pictures of the world, held by older civilizations. This is followed by an account of the various pieces that go together to form our current picture of the universe, from which we proceed to a description of the various modern cosmologies.

Since the eventual aim is to single out *the* correct cosmology on the basis of agreement with experiment, we next embark on a discussion of the pertinent experimental facts. This experimental evidence is often oblique and devious (and much of it has simply been omitted because its present footing is too speculative and insecure). In Chapter 2 we discuss the information which can be gleaned from an understanding of the lives of the stars; such an understanding of the birth, evolution and death of stars can lead to a lower limit to the age of the universe. In Chapter 3 we consider the evidence obtained by "looking backwards in time" at far distant galaxies, radio sources and the glamorous and ill-understood quasars: we can indeed look backwards in time simply because light travels at a finite speed, so that light which we see now from very distant objects must have started its journey a long time ago.

It will be seen that our present knowledge is insufficient to dictate a definitive choice of the correct cosmology, but that there are many straws of evidence all blowing in the direction of a universe which exploded from a primeval fireball some ten billion years ago, which is still uniformly expanding, but which will at some future time (about one hundred billion years from now) cease to expand and will contract back to its primeval state, possibly again to explode in a never ending cycle of cosmic expansions and contractions.

Before beginning, it is as well to emphasize again that cosmology in the past has suffered from too much speculation based upon too few experimental facts. In what follows I have endeavoured to discuss those aspects where some foundation is beginning to exist, and avoided other aspects, which are not yet respectable (for example, theories on the genesis of galaxies). The immediate present is a particularly exciting time in cosmology, for right now the subject is in the position that nuclear physics was in in the late 1920's and early 1930's, in that at last critical experiments are beginning to be done, and the subject placed on a firm footing. Even so, cosmology still enjoys a vaguely disreputable reputation among more firmly grounded scientific disciplines; it is the bearded Bohemian among its flannel-suited colleagues. Indeed it is perhaps in the nature of cosmology, which seeks to understand the overall nature of the universe as a whole, that one must say of it "The first man knew him not perfectly, no more shall the last find him out" (Ecclesiastes VIII, 17).

CHAPTER 1

Cosmologies, Past and Present

Earlier cosmologies

Man's speculations upon the nature of the universe around him go right back to the beginning of human thought. Long before there was any astronomy, the prehistoric hunters of the old stone age pondered cosmological questions as they chased the reindeer and the woolly rhinoceros: we have fragmentary records of the answers which they embodied in their primitive religions and folk lore.

Somewhere around the seventh millenium B.C. (somewhat later in the New World) there occurred a revolutionary change in man's habits, as he began to abandon the nomadic life of a hunter and food gatherer and to settle down to an agrarian existence planting and raising crops. This change brought in its wake many momentous consequences — for example, a rapid increase in the world population by a factor of at least 10 and possibly 100 — but the consequence which concerns us here is the development of astronomy.

If one is going to sit quietly and grow crops, it is essential that one have some form of calendar, for Nature makes it quite clear that "to everything there is a season; a time to sow and a time to reap". The task of reliably predicting these "times" was thus a technical exercise of great practical relevance to these early agrarian people, and the solution gave rise in time to the science of astronomy. It was soon noticed that the seasons could be correlated with the motions of the sun.* From this beginning ancient

---

* Unfortunately all the early calendar makers were tempted down the plausible false path of assigning to the moon a role more or less equal to that of the sun in determining the procession of the seasons. Since the moon is in fact quite irrelevant to the seasons, with the period from full moon to full moon bearing no relationship to the length of the solar year, this misapprehension tended to mess up early calendars. Our own peculiar calendar, with its "Thirty days hath September, and so forth", is a legacy of this ancient error, as is the shifting date of Easter Sunday (the first Sunday after the first full moon following the vernal equinox or spring day on which day and night are of equal duration).

astronomers went on to a surprisingly accurate tabulation of the motion of the sun and the planets against the background of the "fixed" stars.

As one of many examples we can notice that to the ancient Egyptians it was a matter of deep concern to be able to predict when the Nile would overflow its banks, supplying the rich mud and water which were vital to their agricultural system. Before written Egyptian history began, it had been observed that when a certain star, Sirius, appeared over the eastern horizon exactly at daybreak, that is when Sirius rose with the sun, then the Nile flooded and a new year would begin. To facilitate these observations great temples were built with long narrow corridors directed at the exact spot where Sirius would appear. Since the tilt of the Earth's axis in fact changes slowly, moving around or "precessing" with a period of 26,000 years, the priestly Egyptian meteorologists noticed as the centuries rolled on that the position of Sirius and the other "fixed" stars moved slowly in the sky, so that in time other stars had to be found which rose with the Sun when the Nile was to flood, and new temples had to be financed out of the public works budget to observe these stars with proper accuracy and respect.

At this point it is possible to make a distinction of sorts between astronomy and cosmology. Astronomy deals with the motion and character of heavenly bodies, originally the sun, moon and planets, and later other stars and even other galaxies. It began as a practical science, and has remained as a branch of physics dealing with measurable quantities. Cosmology, however, deals with the larger questions of the nature and history of the Universe as a whole. Clearly any such distinction is a somewhat fuzzy one, and certainly to the ancients the two were inextricably intertwined. Thus the sun and other celestial objects which served to define the appropriate time for specific agrarian activities came also to be gods and provided the basis for primitive cosmologies which blended sympathetic magic and anthropomorphic religion with hard-headed astronomical measurements. That of course is why the Egyptian "observatories" referred to above were housed in massive temples of religion. Similarly, Stonehenge, which almost certainly began as a species of simple observatory cum analogue computer, ended its life as an elaborate religious edifice.   Ancient Grants Committees were lavishly generous to their cosmologists.

Setting aside for the moment the host of deities and assorted quaint and sinister creatures that are necessary in a detailed discussion of any one ancient concept of the universe, two main themes can be distinguished in these cosmologies.   Most of them envisage a universe which was created at some specific time in the past and which will go on indefinitely into the future.   Such is the cosmos created in the churning of the ocean by Vishnu, or the Universe created out of "the darkness upon the face of the deep" (Genesis I, 2) by the Hebraic Jehova.   In contrast are the cyclic cosmologies which see the cosmic story as one of endless repetition from birth to death to rebirth to death again, perpetually.   Notable in this class are the cosmologies of ancient Central America, and of some eastern peoples.   Particularly interesting is the Mayan civilization, which flourished in the Guatemalan highlands and the Yucatan Peninsula roughly between 1000 B.C. and 900 A.D., and which achieved great sophistication in observational astronomy.*   Their cosmology envisaged cyclic creations and destructions, with each cycle enduring for 5125 years; in their system the next Armageddon is due on 24th December, 2011 A.D., and I, for one, will be happy if we last that long.

It is amusing to notice that the two major alternatives of the modern relativistic cosmology which we shall meet below, namely continual evolution from some fixed point in past time or endless oscillation, are thus foreshadowed.   However, such a remark properly belongs at a cocktail party: in contrast with modern ones, earlier cosmologies were based purely on speculation (or else on some curious accident; for example the Mayan cyclic cosmology may well have derived from the exigencies of their complex numerical and calendric systems).

The essential difference between modern cosmologies and those of older societies is that these days we insist that scientific theories be derived from, and in agreement with, objective facts obtained by experiment and observation.

---

* They figured the average period from full moon to full moon to be 29·53020 days, quite remarkably close to the actual value of 29·53059 days.

In this context it is as well to note that fanciful beliefs about the universe are not exclusive to the ancients. Bertrand Russell recalls that, less than a century ago, one of his aunts was regarded as an eccentric by her Victorian contemporaries because she did not share the current belief that the universe was created in the year 4004 B.C. This date was derived by Bishop Ussher in the early seventeenth century, in what must be one of the earliest theoretical astrophysics calculations, based on the ages of the patriarchs as recorded in the Bible. This chronology was until quite recently included in the Authorised Version of the Bible, although it always had its sceptics, as noted by Samuel Pepys,* (May 23, 1661): "To the Rhenish wine-house, and there came Jonas Moore, the mathematician, to us . . . and spoke from many things not so much to prove the Scripture false, as that the time therein is not well computed nor understood."

Indeed, even today we find astrology columns in most magazines and newspapers. The belief that the planets move for so frivolous a purpose as to warn people to avoid dealing with blondes on Tuesdays is a great deal more silly than the beliefs of the ancient Mesopotamians who constructed the zodiac.

In short, we should not snigger at the speculative efforts of the ancients. They mixed much accurate observation and much sound thought in with their more fanciful flights of imagination. Moreover, as we shall now see, due to technical limitations no really direct evidence on cosmological issues was available until as recently as the 1920's.

The current picture

By now sufficient observational evidence has accumulated for a fairly clear picture of the present composition of the universe to be given: we live on a planet which circles one of many, many stars which make up our own galaxy; beyond our galaxy is an emptiness bestrewn with myriads of other galaxies; and this universe of galaxies is expanding, in the sense that the galaxies are flying apart from one another. This last fact can be explained in two general

---

* This passage suggests that the social habits of cosmologists have not altered much in the past 300 years.

ways, which generate the two main cosmological theories of recent times: the "Big Bang" theories and the "Steady State" theory.

We shall now go on first to elaborate this picture of the expanding universe, and thence to a more detailed exposition of contemporary cosmologies.

Most of the distances and numbers involved in astronomy are so vast as to defy any intuitive grasp of just how big they are. Thus the usual unit of astronomical distance is the light-year, the distance light travels in one year at its constant speed of 186,000 miles per second, or some six million million miles.

Our earth is one of several planets which orbit the sun. The distance from earth to sun is eight light-minutes, and the extent of the entire solar system is about five light-hours.

The sun is "our own star", and, just like the other stars visible in the night sky, it burns with a thermonuclear fire. The stars

Figure 1.1. The constellation of Orion and surrounding region. This figure of a minute part of the Milky Way makes it believable that there are some 100 billion stars in our galaxy. (Mount Wilson and Palomar Observatories.)

derive their energy from nuclear fusion reactions, in a way which will be discussed in detail in Chapter 2.

The star nearest to our sun is Alpha Centauri, which is four light-years away. The distance to our nearest neighbour stars is about 10,000 times as great as the size of the solar system itself.

These neighbouring stars are in turn part of a great congregation of some 100,000,000,000 stars which make up our local galaxy. This galaxy has the shape of a huge spiral disc, which is slowly rotating, and is roughly 100,000 light-years across. Thus the galaxy itself forms a unit which is roughly 10,000 times as big as the separation between neighbouring constituent stars.

Figure 1.2. A typical spiral galaxy. Our own galaxy would look much like this to an outside observer. (Mount Wilson and Palomar Observatories.)

In this galaxy, the sun, with all its planets, is no more than a medium-sized star well out towards the rim of the disc. The "Milky Way" is our view of the galaxy, looking inwards.

As well as stars, our galaxy contains a deal of interstellar gas and dust. From this material new stars are even now being spawned by condensation, while on the other hand gas is being returned to interstellar space both by steady loss and violent explosions from older stars. Sometimes this gas and dust is noticeable when it is dense enough to black out the sky behind it; for example the dark streak across the middle of the Lagoon nebula in *Figure 2.10* (page 202) is due to dust between the bright area and ourselves.

Beyond our own galaxy lie an estimated 10,000,000,000 other galaxies, each of a comparable vastness. Between these galaxies is the unimaginable emptiness of intergalactic space. The typical distance between galaxies is about ten million light-years, so that the galaxies themselves (huge though they be) are small compared to the distances between them.

Figure 1.3. An unusual cluster of galaxies, displaying many different types in a single photograph. (Mount Wilson and Palomar Observatories.)

The speculation that the universe consists of tenuous inter-
galactic space dotted here and there with galaxies similar to our
own has been current for some time.   After the first primitive
telescopes were built in the seventeenth century, it was observed
that the night sky is illuminated not only by stars, which appear
as pin-points of radiance, but also by more diffuse luminous objects,
which were called "nebulae".   Various people, among them the
philosopher Immanuel Kant, guessed correctly that these "nebulae"
were other galaxies like our own Milky Way (in Kant's imaginative
phrase, other "Island Universes"), but so distant from us that the
individual constituent stars could not be distinguished, so that the
whole appeared as an ill-defined luminosity.   One of the alternative
speculations was that the stars were embedded in the opaque
fabric of a fixed heavenly sphere which encased the solar system,
and that these "nebulae" were holes in the fabric through which
could be glimpsed the empyrean, the eternal celestial fire of Greek
astronomy.

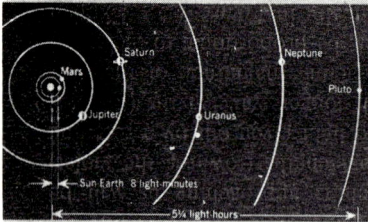

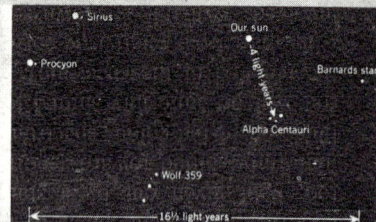

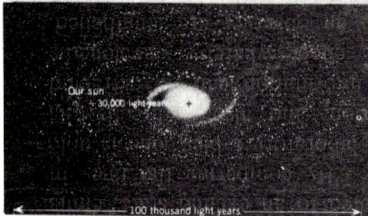

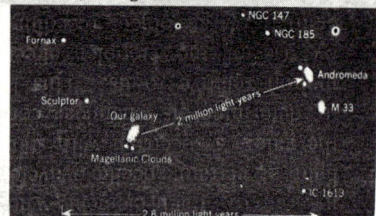

Figure 1.4.

Facts replaced speculation when in the 1920's the new 100-inch telescope at Mt. Wilson, in Southern California, came into operation; this telescope was sufficiently powerful that it could distinguish some individual stars in the nearer galaxies.

This hierarchy of increasingly vast distance scales is illustrated schematically in *Figure 1.4*.

Two Crucial Cosmological Experiments

(1) The Universe is expanding

Following up the abovementioned work on the Mt. Wilson telescope, the late Edwin Hubble, one of the giants of modern astronomy, made an even more dramatic observation of profound cosmological significance: the universe of galaxies is expanding. All the galaxies are speeding away from each other, and the farther they are apart the faster is their speed of recession.

Hubble derived this information from observations on the detailed nature of the light emitted by distant galaxies. When this light is decomposed into its constituent wavelengths, or colour spectrum (as it can be in an ingenious device called a spectrograph), bright and dark lines show up at certain wavelengths depending on the details of the chemical elements which emitted the light at its source. Each element, when heated to an excited state, emits light at certain specific wavelengths. By comparing the lines in the galaxy's spectrum with standard laboratory tables, one is able to determine its chemical composition, and also, indirectly, its motion relative to us. This latter information can be extracted by use of the Doppler effect, the effect whereby wavelengths emitted by an object receding from an observer are lengthened. The effect is familiar in such everyday examples as the increasing and then decreasing pitch of the whistle of a passing train. So too, light given out by a star which is approaching us will be uniformly shifted towards shorter wavelengths (towards the violet-blue end of the colour spectrum), and light given out by a receding star will be uniformly shifted towards longer wavelengths (towards the red end of the spectrum). The actual magnitude of the shift will, apart from relativistic corrections, be proportional to the speed involved. All this is shown schematically in *Figure 1.5*. Explicit photographs of the observed spectra of some stars are to be found on page 188, and actual red-shifted spectra of distant galaxies are displayed in *Figure 3.3* on page 214.

To make the above discussion more quantitative, let λ be the wavelength of a particular spectral line in the laboratory (e.g., the fundamental line of the hydrogen atom spectrum is the so-called Lyman - α line with a wavelength $\lambda = 1 \cdot 216 \times 10^{-5}$ cm).

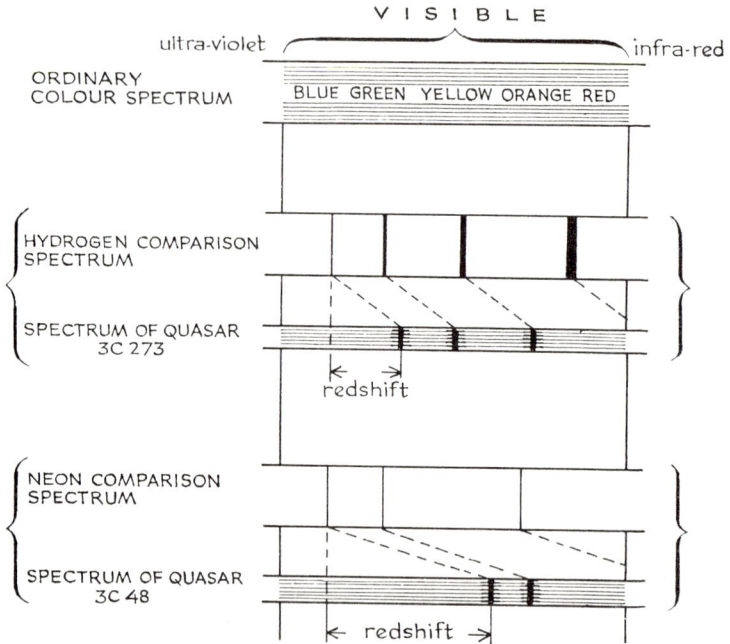

Figure 1.5. The red-shift. The top spectrum represents the ordinary colour spectrum which results when visible light is decomposed into its constituent wavelengths. At both ends the spectrum becomes invisible as the wavelengths move into the infra-red and ultra-violet regions.

The next spectrum is a laboratory spectrum obtained from the chemical element hydrogen. Below it is the spectrum resulting from passing the light from 3C273 through a spectrograph and photographing the result. The spectrum of 3C273 is seen to correspond to a hydrogen spectrum with a considerable shift towards the red end of the spectrum. This red-shift results from the high speed at which 3C273 is receding from us.

Similarly at the bottom of the diagram is a laboratory comparison spectrum for the chemical element neon, and below it the spectrum obtained from 3C48. Again the spectrum of the object 3C48 displays a large red-shift.

Then if this particular spectral line can be identified in an object which is receding from us, and if the observed wavelength is shifted to a value $\lambda' = \lambda + \Delta\lambda$, so that $\Delta\lambda$ is the *change* in the wavelength, then the relative velocity of recession of the object, v, is given by

$$\frac{v}{c} = \frac{\Delta\lambda}{\lambda} \qquad (1.1)$$

when v is significantly smaller than c, the velocity of light. In passing we may mention that it is conventional to define a quantity z, called the "red-shift parameter", which is simply equal to $\Delta\lambda/\lambda$. The fully relativistic relationship between velocity of recession and red-shift (that is the fully relativistic Doppler effect) is in fact

$$z \equiv \frac{\Delta\lambda}{\lambda} = \sqrt{\frac{c+v}{c-v}} - 1 \qquad (1.2)$$

which reduces to *Eq. (1.1)* when v is significantly less than c.

From his observations Hubble found that the colour spectra of the galaxies beyond ours were consistently "red-shifted". Next he went on to compare the magnitude of the red-shift with the galaxy's distance from us. This distance can be estimated by observing the relative brightness: if all galaxies were identical in respect of their intrinsic brightness, then since the energy they radiate falls off in intensity according to the inverse square of the distance, we could say that if galaxy A and galaxy B appear equally bright then they are equally distant, or if galaxy A appears four times brighter than galaxy B then it is twice as close, and so on. Of course all galaxies are *not* identical, but individual differences average out if one deals with a sufficiently large number of examples. The result of all these observations was that the magnitude of the red-shift was directly proportional to the distance; Hubble enunciated this result in 1929.

This law, that the galaxies are receding from each other with a speed proportional to their separation, is called Hubble's law. It is the centrepiece of modern cosmologies.

Two corollaries to Hubble's law are worth mentioning.

First, it might seem that we occupy a unique position in the universe in that all other galaxies are fleeing us. Careful reflection shows that just the opposite is the case; if the whole universe of galaxies is expanding uniformly, then it is necessary that the speed of recession between *any* two galaxies be proportional to their separation. The situation can be visualised by considering a currant pudding, with the currants (representing galaxies) distributed uniformly throughout the pudding. When the pudding is cooked it expands, and as it expands all the currants move away from one another at speeds which are proportional to their separations.

Secondly, and speaking with the wisdom of hindsight, we remark that Hubble's law explains why the night sky is dark!

To justify this statement, suppose the universe were *not* expanding, and divide it up into concentric spherical shells surrounding the earth as the layers of an onion surround its core. Further, let these spherical shells each have a thickness, T, sufficiently great to include many galaxies. Now consider a particular shell which surrounds the earth at a distance R (such that R is much greater than T). As we observed above, the apparent luminosity of a galaxy a distance R away *decreases* as R^{-2}. On the other hand, the number of galaxies in the shell at radius R is proportional to the volume of this shell, and since this volume is $4\pi R^2 T$, the number of galaxies in this shell *increases* as R^2. Hence the total light received on earth from all the galaxies in this shell of radius R is *independent* of R. Thus we can go on adding shell to shell, onion skin to onion skin, and since each contributes the same constant amount of light at the earth (regardless of the radius of the shell), the total amount of such light will increase without limit. Our sun's light becomes an infinitesimal fraction.

This paradox was first propounded by Olbert in 1826, and it is surprising that for a century people did not worry about it much.* They should have.

---

* Below we will meet the modern cosmologies which involve non-Euclidean geometries, that is peculiar geometries where surface areas increase faster or slower than as R^2. One may hope to escape Olbert's paradox in such geometries. However, the nub of the paradox is that while surfaces and thus numbers of galaxies increase as R^2, conversely their apparent brightness decreases at R^{-2}, whence the two effects exactly cancel. Even in funny non-Euclidean geometries, although the surfaces and thence the numbers of galaxies increase more rapidly or less rapidly than R^2, the diminution in apparent brightness still exactly cancels this effect (since the two effects involve the same geometry), and the paradox remains unaltered.

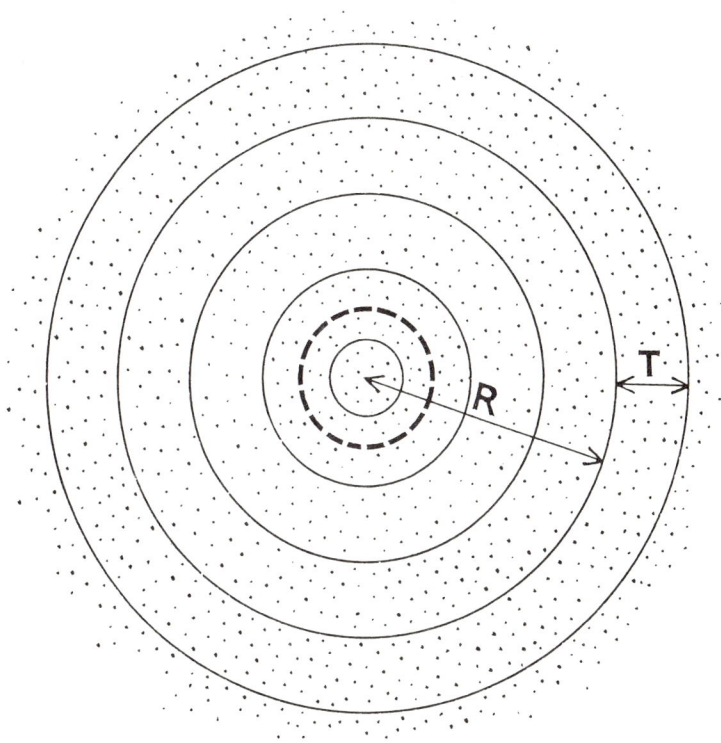

Figure 1.6. A schematic diagram to illustrate the paradox concerning the darkness of the night sky. We divide the universe around our earth up into shells of thickness T, each shell so distant as to contain many, many galaxies (represented by dots). As we go farther away, successive shells obviously contain more and more galaxies (dots); on the other hand as we go farther away the individual galaxies appear to us to be less and less bright. It turns out that these two effects — the increasing number and the decreasing individual apparent brightness — exactly cancel, so that the total light received at the earth is the same for all shells, however distant. Thus we could add shell to shell indefinitely, and conclude that the sky is infinitely bright, day or night.

The fallacy of the argument lies in the fact that the universe is expanding. Consequently all galaxies outside the dashed circle in the diagram are in effect speeding from us so fast that their light never reaches us. We can only ever receive light from the galaxies inside the dashed circle: the total of this light is insignificant, so that (in the absence of the sun) the sky is dark at night.

The fallacy of the paradox is that the universe *is* expanding. Thus as one goes to further and further galaxies the speed of recession gets greater and greater, until eventually the point is reached when the galaxies are receding at the speed of light itself and are consequently no longer observable. This happens at a distance of some 10,000,000,000 light-years from us, a distance which puts a limit on the size of the observable universe.

It is to be emphasized that this expansion is cosmic in scale. The expansion refers to the universe of galaxies, and not to the details of behaviour *inside* any one galaxy. Thus the galaxy M81, eight million light-years away, is indeed speeding from us at 80 miles per second; but Alpha Centauri and the other stars in our galaxy, much less Mars and Venus, have motions relative to the earth which are of purely local origin. Galaxies themselves are in a sense the smallest unit of intelligible study to the cosmologist.

(2) The Universe is homogeneous and isotropic

An assumption which is implicit in the above description, and which underlies all modern cosmologies, is that on a grand scale the universe presents the same aspect from every point. This assumption that the present universe is essentially the same at all places (i.e., that it is "homogeneous") and in all directions (i.e., that it is "isotropic") is sufficiently important to merit the title of "cosmological hypothesis", or, less tentatively, "cosmological principle". It says that our corner of the universe is in no way special. Clearly this idea of homogeneity and isotropy on a cosmic scale is philosophically a very compelling and satisfying one, and it is not surprising that one way or another it is woven into the very fabric of all modern cosmologies.

Over the past few years the cosmological principle has been elevated from its pristine status as an attractive hypothesis to that of an experimental fact. To describe how this metamorphosis has come about we must turn to the history of an experiment.

As we shall see in a moment, the modern relativistic "Big Bang" cosmologies envisage our universe as still expanding from an originally much denser and much hotter state. This primeval fireball of radiation and matter must have been at a temperature in excess of 10,000,000,000°C. As the fireball has expanded and cooled it is pertinent to ask what has happened to the original radiation. Theory suggests that as the universe has expanded, all the frequencies present in the radiation spectrum will have been red-shifted: this is to say the radiation will still be present, distributed as a uniform background throughout the entire universe, but cooled to a much lower temperature. The decrease in radiation temperature is in fact inversely proportional to the expanding radius of the universe.

The details of this prediction that the universe is filled with radiation from the original Big Bang, *if* there ever was a Big Bang, were first worked out by Alpher, Bethe and Gamow (α, β, γ) in 1948: they concluded that the present temperature of this all pervasive equilibrium radiation would be between 5 to 25°C above the absolute zero of temperature (at —273°C). More recently a group at Princeton University led by the brilliant experimentalist Robert Dicke independently did this calculation, with greater accuracy, to decide that the present radiation temperature would be less than 10°C above absolute zero. This group then proceeded to build an experimental device to search for the radiation, and check their prediction.

In 1965, while the Princeton group was busy constructing its instrument, another group from Bell Telephone Laboratories was encountering odd difficulties in an experiment aimed at eliminating background noise from communication satellites such as Telstar. These latter people found that even when all known corrections were taken meticulously into account they were left with an inexplicable residuum of noise, as if there were a universal background of radiation with a temperature 3°C above the absolute zero of temperature. When acquainted with the Princeton prediction, the Bell Laboratories people realized that, in the tradition of the three Princes of Serendip, they had in fact performed an experiment of much greater significance and interest than the experiment they meant to perform.

Subsequently the Princeton people and several other groups have made measurements. Each group has dealt with a different part of the wavelength spectrum of the radiation. The results are shown in *Figure 1.7* where it can be seen that all are in remarkable agreement with the conclusion that space is filled with equilibrium (or "blackbody") radiation corresponding to a radiation temperature of about 2·7 degrees Absolute.

An independent check on the existence of this background radiation can be provided by observations on the particles called "cosmic rays". If the Universe is indeed filled with 2·7° black-body radiation, then we should not observe any primary cosmic ray protons with energies in excess of 10^{20} electron Volts; such energetic protons will long ago have lost their energy in inter-actions with the pervasive background radiation. A group in the Sydney University School of Physics, under the leadership of Professor C. A. B. McCusker, is currently conducting an experi-

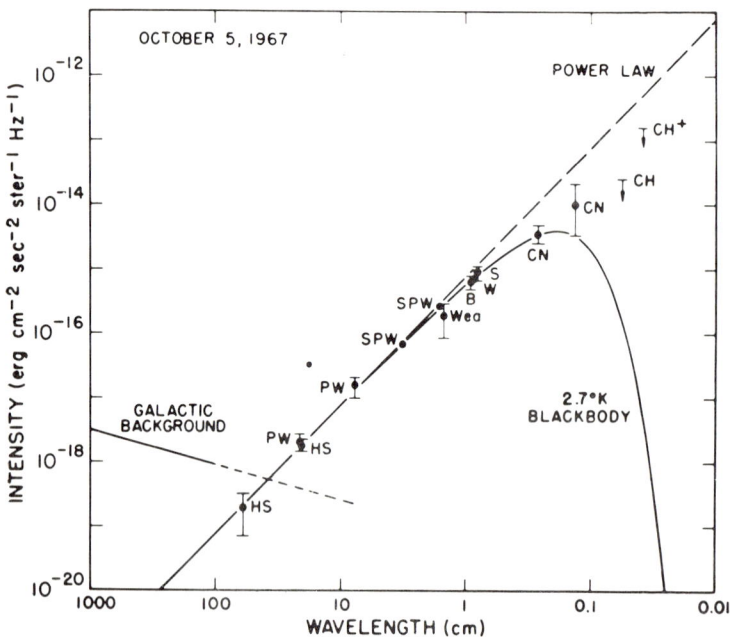

Figure 1.7.   Measurements of the cosmic background radiation by different experimental groups detecting radiation at various specific wavelengths. The solid curve labelled "2.7 blackbody" represents the intensity versus wave-length plot for equilibrium or blackbody radiation with a temperature of 2.7 degrees Absolute; the present radiation surviving from the primeval fireball (if there was a Big Bang) would indeed be expected to conform to some such blackbody shape of curve. Theory and experiment are seen to be in close harmony.

ment to determine whether the cosmic ray "energy spectrum" does in fact exhibit a cut-off at around 10^{20} electron Volts. The detection (or non-detection!) of such highly energetic cosmic ray particles at ground level requires detection apparatus spread over some 10 square miles: such apparatus has been set up in the "Pilliga Scrub", outside Narrabri.

This background radiation is of enormous interest for its own sake, and we shall have occasion to return to it later.

However, the background radiation also serves as a splendid means of verifying the cosmological hypothesis that the universe is homogeneous and isotropic. This can be done simply by looking in all directions in the sky, and verifying that the radiation temperature is exactly the same wherever our antennae look. Such experiments have been done by Partridge and Wilkinson from Princeton, by Conklin and Bracewell from Stanford, and by Penzias and Wilson who did the original lucky Bell Laboratory experiment. The results are that, to within the accuracy of the experiments, no departure from the uniformity postulated by the cosmological principle has been found. For example, comparing different pieces of the sky which are one degree wide (i.e., pieces subtending an angle of one degree at the eye), Conklin and Bracewell find no variations, although their experiment would detect piece-to-piece variations in radiation temperature as tiny as $0 \cdot 09$ per cent.

Thus current observations of the background radiation not only suggest that the universe originated in a Big Bang, but also serve to place the conventional cosmological assumption that the universe is homogeneous and isotropic on an experimental foundation.

Indeed this recent body of experiment fully deserves the position I have accorded it here as one of the two factual cornerstones of contemporary cosmology.

New Cosmologies

(1) Big Bang cosmologies

"Big Bang" is the colourful title popularly given to the class of evolving universes suggested by general relativity. In more drab language, scientists call these evolving general relativistic universes "Friedmann cosmologies" after one of the earlier mathematicians who investigated them.

The central theme of the theory of general relativity, derived by Albert Einstein in the first two decades of this century, is that the intrinsic geometry of space is intimately connected with the matter in it. That is, Einstein suggested that the space of the universe may not be described by the ordinary Euclidean geometry with which we are all familiar both from high school and from everyday experience, but rather that the gravitational fields associated with the presence of matter may "curve" space. In Euclidean geometry the shortest distance between two points is of course a straight line, but in more generalized curved spaces the shortest distance between two points will not necessarily be a straight line, but rather will be some curve — such generalized "shortest distance" curves are called geodesics. Of course all this only pertains on a large scale; on the local scale of the everyday world any differences from the ordinary Euclidean geometry to describe the space we live in are entirely negligible.

It goes without saying that we would not accept these tenets of general relativity were it not for the fact that some of their predictions have been verified by experiments. Thus the theory predicts that when the light from a distant star happens to pass very close to the sun on its way to us the path along which the light travels is bent (because the gravitational field of the sun curves the space in its close vicinity, turning the light's geodesic from a straight line to a curve). This effect has indeed been observed. Another test relies on the observed fact that the orientation of the major axis of the orbit of the planet Mercury rotates, or precesses, more rapidly than the old Newtonian gravitational theory predicts; Einstein's concept of the universe is mathematically able to account for this peculiar motion, which Mercury displays more than any other planet because it is so close to the sun.

These two tests are fairly conclusive, although their accuracy leaves a little to be desired.

At the present time several other tests of the theory are being planned. One involves a more accurate version of the bending of starlight test mentioned above; here people are bouncing radar signals from the surface of Mercury or Venus at times when these planets are just about to hide from us behind the sun. The radar reflection paths will be bent in the vicinity of the sun in the same way described above for the case of starlight. A second test will use a very accurate form of clock orbiting in a satellite, and will compare it with a similar earthbound clock. The difference in the two timekeeping rates will provide a test of the general theory of relativity. A third test will use a spinning gyroscope mounted in a satellite on a very accurate polar orbit: general relativity makes a specific prediction about the rate of precession of the axis of the gyroscope (this is a fiendishly difficult experiment to carry out to the required accuracy). You will probably see accounts of these experiments in the newspapers when they produce results.

Now if the mass of the sun is enough to curve space in its immediate neighbourhood, perhaps all the mass in the universe is enough to curve all the space in the universe. Maybe a geodesic in the universe is not a Euclidean straight line. Perhaps ordinary Euclidean geometry seems so fundamental to us only because our everyday straight lines and planes are so extremely small compared with the size of the universe.

To investigate a non-Euclidean universe we must use our imagination.

Let us imagine a one-dimensional creature which can only move and conceive of motion along a straight line, or alternatively along the circumference of a circle. If he lived in a straight line, then he could move backwards and forwards along this line (but never sideways); he could, if he wished, travel for an infinite distance along the line without retracing his steps. Conversely if he happened to live on the circumference of a circle he could again move only backwards and forwards along this curved line, but now to travel an infinite distance he would of necessity have to retrace his footsteps for an infinite number of revolutions. Not only can our creature not travel along a radius of the circle, he cannot even conceive a radius, for he knows only one dimension, along the circle.

Next let us imagine a two-dimensional creature which can dwell either on a Euclidean flat plane or alternatively on the surface of a sphere. Again if he lives in the plane he can embark on infinite journeys without recrossing his path, whereas if he lives on the sphere he is fated to retrace his steps repeatedly on any infinite journey. And again he cannot travel on, nor even conceive the existence of, the radius of the sphere.

Now the creature living in a straight line or on a flat plane is inhabiting a Euclidean universe of one and two dimensions respectively. A creature living on the circumference of a circle or on the surface of a sphere is inhabiting a positively curved non-Euclidean universe of one and two dimensions respectively. (Geodesics in these non-Euclidean universes are clearly not simple Euclidean straight lines.)

The question posed by the general relativistic cosmologies is do we, as three-dimensional creatures, inhabit a Euclidean universe one step up from the line-plane sequence? Or do we inhabit a non-Euclidean universe such as that one step up from the circle-sphere sequence? The Euclidean universe is trivial to imagine, for it is simply three-dimensional and infinite in extent. On the other hand a non-Euclidean universe of three dimensions is impossible to envisage explicitly. Just as the two-dimensional creature on the surface of his spherical universe cannot possibly conceive the existence of a third dimension (namely the radius of the sphere, a line perpendicular to the two directions the creature does know), so we cannot envisage a fourth dimension mutually perpendicular to the three directions we do know. However, it is an altogether straightforward matter to construct the mathematics of such three-dimensional non-Euclidean universes. That is to say although it is impossible to draw little pictures of non-Euclidean geometries such as the three-dimensional member of the circle-sphere hierarchy, nevertheless the study of such universes presents no problems.

It remains to enunciate some geometrical properties of non-Euclidan universes.

In an ordinary Euclidean geometry, consider the volume bounded by a sphere of radius R; as R increases the surface area of the sphere increases exactly proportionally to R^2, and the volume increases exactly as R^3. In a non-Euclidean universe these exact proportionalities no longer hold. *If areas increase more slowly than* R^2 *and volumes more slowly than* R^3, *the non-Euclidean geometry is said to be positively curved. If areas and volumes increase more rapidly than* R^2 *and* R^3, *the geometry is said to be negatively curved. Euclidean space then represents the special limiting case of zero curvature.*

Some insight into these geometrical properties may be obtained by returning to our creatures inhabiting one- and two-dimensional universes. First let us randomly sprinkle dots on the flat plane, and then draw several equally spaced concentric circles on the plane *(Figure 1.8a)*. The number of dots included within ever larger circles in this two-dimensional Euclidean space increases as the square on their radii.

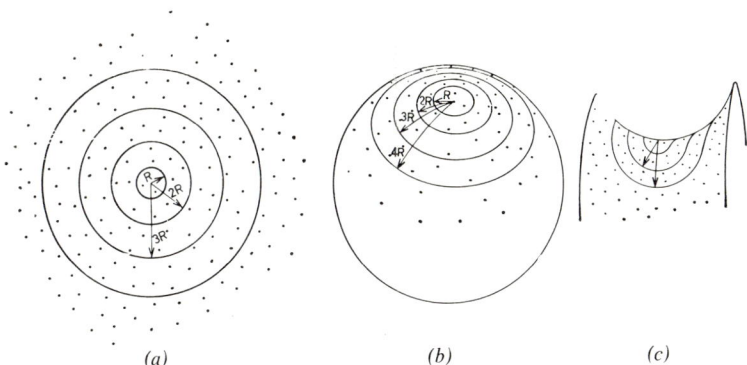

(a)   *(b)*   *(c)*

Figure 1.8.   (a) In Euclidean space, the number of randomly spread dots included in ever larger circles increases as the square on the radii.

(b) In a two-dimensional positively curved space, the number of randomly spread dots included in ever larger circles increases less rapidly than the square on the radii.

(c) In a two-dimensional negatively curved space, the number of randomly spread dots included in ever larger circles increases more rapidly than the square on the radii.

If we repeat this process of dot sprinkling and counting on the surface of a sphere, the creature who dwells there will find that their number increases less rapidly than the square of the distance from him, as can be seen from *Figure 1.8b*. To be explicit, whereas the plane dwelling creature finds the number of dots to increase as πR^2, the spherical-surface creature finds the number to increase as $2\pi\varkappa^{-2}$ $(1-\cos \varkappa R)$ where $\varkappa$ is the curvature of the spherical surface (namely $\varkappa$ equals the reciprocal of the radius of the sphere itself); as R increases this expression increases more slowly than πR^2. Notice that the Euclidean case (the result πR^2) is regained as the limiting case as $\varkappa$ tends to zero. That is to say the spherical-surface dwelling animal lives in a two-dimensional non-Euclidean geometry with positive curvature; the Euclidean geometry follows as the special case of zero curvature.

Alternatively a two-dimensional non-Euclidean space of negative curvature can be illustrated by the saddle-like surface of *Figure 1.8c*. Here the number of dots counted by our convenient creature increases more rapidly than the square on the distance from him.

The foregoing discussion of non-Euclidean spaces was motivated by the basic idea of general relativity, that the intrinsic geometry of space is inextricably connected with the distribution and density of matter in the universe. This relation is provided by *Einstein's field equations,* which have the character of being an appropriately generalized version of the sort of conservation equation or continuity equation which in other contexts describes the flow of fluids or of electric charges. The field equations of general relativity relate the intrinsic geometry of space to the density of matter and radiation in it.

The early work on Einstein's field equations was complicated and obscured by attempts to derive *static* universes from them; the equations admit in a natural way only expanding (and contracting) universes. This is fine for us, with Hubble's observations behind us, but early work on the field equations *preceded* the knowledge that the universe is in fact not static but expanding.

In 1922 the Russian mathematician Friedmann deduced two non-static, homogeneous models of the universe from Einstein's original equations. One class of mathematical solutions corresponds to a universe ever expanding from an initial singular state of hyperdense concentration: these universes all have either zero (Euclidean) or negative curvature. The second class of solutions corresponds to an oscillating universe, which again expands from a singular state, but which eventually stops expanding and contracts back into its initial state, repeating this oscillatory cycle endlessly: these universes all have positive curvature. Both classes of model universes are compatible with the observation that the universe is presently expanding.

In more precise and mathematical terms, we can say that Einstein's original field equations provide a set of relations between the density of matter in the universe at time t, $\varrho(t)$, and the radius of the universe at the same time, $R(t)$. Other quantities such as the rate of expansion (proportional to $dR(t)/dt$) and the rate at which the expansion is accelerating or decelerating (proportional to $d^2R(t)/dt^2$) follow once $R(t)$ is known. *Moreover, a complete and unique solution of the field equations can be written down once we know the value of any TWO of these quantities at some specific time.* (This is because the equations are second order differential equations.)

In particular, our cosmology is uniquely defined within the framework of general relativity if we know the values *now, at our present epoch*, of any two of the following quantities: the rate of expansion, the overall density of matter in the universe, the rate at which the expansion is accelerating or decelerating, or the radius of the universe.

The two quantities which are usually chosen for a specific characterization of a unique Big Bang cosmology are the "expansion parameter", H_o, and the "deceleration parameter", q_o. The subscript zero symbolizes the fact that these quantities are measured now, at our epoch in time.

As its name suggests, the expansion parameter measures the rate at which the universe is presently expanding. Indeed H_o is just the proportionality constant in Hubble's Law relating the velocity of recession, v, with the distance, d, of relatively nearby galaxies: $v = H_o\,d$. (H in fact stands for Hubble, and is often called the Hubble constant.) Thus from the observations of Hubble and later astronomers (notably Baade and Sandage) the expansion parameter can be determined. H_o clearly has the physical dimensions of an inverse time, and its currently measured value is more conveniently expressed as

$$\frac{1}{H_o} = 13{,}000{,}000{,}000 \ years \qquad (1.3)$$

For the mathematically inclined reader, we may mention that in general the expansion parameter H at any epoch is defined in terms of the radius at that time as

$$H \equiv \frac{1}{R}\frac{dR}{dt}$$

The deceleration parameter q_o is not so simple to describe in words. Basically, q_o is a dimensionless number which measures the current rate of change of the expansion parameter itself. The larger the value of q_o, the greater the rate at which the expansion of the universe is presently slowing down or decelerating. (In mathematical terms, q at any epoch is written as

$$q \equiv -\frac{1}{RH^2}\frac{d^2R}{dt^2} = -1 - \frac{1}{H^2}\frac{dH}{dt}.$$

This is just the mathematician's way of saying that q measures the deceleration of the expansion.) As we shall soon see, an experimental measurement of the current value of the deceleration parameter is a great deal more difficult than that of H_o.

Recalling that the main theme of general relativity, expressed in Einstein's field equations, is that the intrinsic geometry of space is intimately connected with the density of matter therein, we would expect to be able to specify the geometry and matter density of our universe in terms of the two chosen parameters q_o and H_o. This is easily done.

In particular, the average overall density of matter in the universe at our epoch is given by

$$\rho_o = \frac{3H_o^2 q_o}{4\pi G} \qquad (1.4)$$

where G is the gravitational constant. Using the observed value of H_o [Equation (1.3)] this reduces to

$$\rho_o = 2 \times 10^{-29} q_o \text{ gm/cm}^3 \qquad (1.5)$$

The intrinsic geometry of space is related simply to the deceleration parameter q_o, as follows: if q_o exceeds the value $\frac{1}{2}$ ($q_o > \frac{1}{2}$), then we live in a positively curved non-Euclidean space, corresponding to an oscillating universe. If q_o exactly equals $\frac{1}{2}$ ($q_o = \frac{1}{2}$) we live in a Euclidean space, corresponding to an (only just) ever-expanding universe. If q_o lies between zero and $\frac{1}{2}$ ($0 \leqslant q_o < \frac{1}{2}$) then we live in a negatively curved non-Euclidean space, corresponding to an ever-expanding universe.

These three kinds of geometry and universe are illustrated in *Figure 1.9*, where we plot the radius of the universe, $R(t)$, as a function of time t for various values of q_o. The Euclidean cosmology ($q_o = \frac{1}{2}$) is the critical special case which divides the ever-expanding, negatively curved, "open" universes from the oscillating, positively

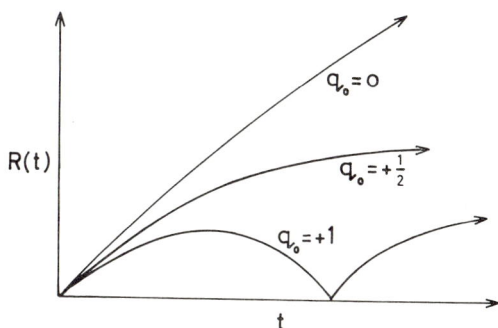

Figure 1.9. The radius of the universe, R(t), as obtained from Einstein's field equations for various values of the deceleration parameter q_o. $q_o = 0$ is a geometry with negative curvature, $q_o = +1$ a geometry with positive curvature, and $q_o = \frac{1}{2}$ is the special case of Euclidean geometry (that is zero curvature).

curved, "closed" universes. We can remark that the total time, T, for each complete cycle in the oscillating universes ($q_o > \frac{1}{2}$) is given by the simple formula

$$T = \frac{1}{H_o} \frac{2\pi q_o}{(2q_o - 1)^{3/2}}. \qquad (1.6)$$

Notice that as q_o tends to $\frac{1}{2}$, T tends to infinity, again illustrating that the Euclidean geometry ($q_o = \frac{1}{2}$) follows as a limiting case.

The cosmological roles played by the expansion and deceleration parameters, H_o and q_o, can be illustrated by a rough analogy. Suppose we fire a rocket towards the heavens from the surface of some unexplored planet of unknown mass. Three things can happen: the rocket can have insufficient velocity, and fall back on us; or it may *just* escape from the planet's gravitational pull; or it may escape with energy to spare. Not knowing the properties of this hypothetical planet, we do not know *a priori* what the critical "escape velocity" is. A measurement of the rocket's initial velocity (analogous to a cosmological measurement of the expansion parameter H_o) will not tell us its subsequent behaviour. We need also to measure the initial deceleration of the rocket due to the planet's gravitational field; the future history of our rocket can then be predicted. If its deceleration is sufficiently severe compared with its velocity (if $q_o > \frac{1}{2}$) it will reach an apogee and then return to us; if its deceleration is small enough compared with its velocity (if $0 \leqslant q_o < \frac{1}{2}$) it will not only escape but will have energy to spare; if the deceleration has a certain critical value in relation to its velocity (if $q_o = \frac{1}{2}$) the rocket will just barely escape. This analogy is superficial, but not too bad.

Finally, it is interesting to ask how far back in time was the apocalyptic explosion in which any one of these universes was born? How far back to the "creation" of our universe? At first glance one may think that this time is simply the inverse Hubble constant time, given by *Equation (1.3)*. In fact this is true only in the limiting case when q_o tends to zero; more generally the time since the Big Bang is somewhat less than this time in *Equation (1.3)*, with the two times being related by a specific mathematical formula involving only q_o. *However, in all cases the answer is that roughly ten billion years have elapsed since our universe began.* Of course in the oscillating universes this time refers to the beginning of the present cycle.

It is interesting to note that these cosmologies had been propounded *before* the key cosmological experiments described above had been performed. The theory has taken these observations in its stride, and indeed has even predicted them; it agrees with the fact that the universe is expanding, homogeneous and filled with a residuum of radiation from the primeval fireball.

In review, we see that the Big Bang theories of cosmology derived from general relativity state that at some time in the remote past the universe began as a dense, hot fireball of radiation and matter which proceeded to explode outwards, growing progressively cooler as it expanded. There are two variations on this basic theme. In one, the universe continues to expand ever outward from some singular point in time. In the other, the universe pulsates in cycles of expansion and contraction, expansion and contraction. Each cycle begins again in a fireball of radiation and elementary particles; as expansion takes place the universe cools, and chemical elements are built up from the elementary particles; eventually in a cold, tenuous universe the expansion slows down, halts; and then the universe contracts back to the fiery furnace which decomposes the elements back to radiation and elementary particles so that the next cycle can go on to rise afresh from the ashes. If this be the case, then we are at present in the comparatively early phase of an expansion; in either case roughly some 10,000,000,000 years have elapsed since the present expansion began.

All such cosmologies, whether singular explosions or ever-recurring oscillations, are conveniently lumped together under the heading of Big Bang because they all share the characteristic of implying an evolving universe, a universe which at the moment is expanding from a hotter, denser past into a colder, less dense future.

(2) The Steady State cosmology

Of recent times, the major rival of the Big Bang relativistic cosmologies described above has been the so-called Steady State theory, which envisages a universe which always has been, and always will be, just as it is now. There is no beginning, no end, no continuous evolution in the Steady State picture.

This idea was first introduced some 20 years ago by a group of young British physicists: Hermann Bondi (now Director-General of the European Space Research Organization), Thomas Gold (now Joint Director of the Cornell-Sydney University Astronomy Center) and Fred Hoyle (now Director of the Institute for Theoretical Astronomy at Cambridge).

This theory is undeniably philosophically very attractive; it goes beyond the cosmological principle met above to enunciate as it were a "perfect" cosmological principle, in which the universe is homogeneous and isotropic not only in space but also in time. It claimed many afficionados because it offered a scheme in which questions such as how and why the universe began were meaningless. It offered a steady universe with no beginning and no end. However, it should be borne in mind that this property is shared by the oscillatory versions of the Big Bang cosmologies, which can pulsate through endless cycles of expansion and contraction.

In order to reconcile such a steady state with the observed expansion, it is necessary to postulate that new matter is continually being created. If such matter were *not* continuously created, then clearly as the universe expanded the average density of matter in it would fall. To preserve a steady state new matter needs must come into being at just such a rate as to keep the average density a constant despite the expansion.

This state of affairs is illustrated in *Figure 1.10,* which shows the history of (a) a Big Bang universe and (b) a Steady State universe, both in one dimension. The vertical axis represents time, the horizontal axis the distance co-ordinate in this one-dimensional universe, and the lines show the paths or "world lines" followed in time by particles in these universes. In the Big Bang case (a) all paths begin at the beginning ($t = 0$), and they diverge as the universe expands. Thus if we look back from *our* time (the dashed line across the figure) we see the world lines closer together, that is we see a denser universe, whereas the future is a less dense universe. On the other hand, in the Steady State picture (b) the pattern *must*, by definition, look the same at all times. Our time (the dashed line again) is in no way special. Clearly the only way for this constancy of the pattern to be preserved along with the observed fact that the universe is expanding, that the world lines are diverging from each other, is for new world lines to continually originate at all times. Thus the Steady State theory automatically implies continuous creation of new matter.

Since the rate of expansion is known, the rate at which matter need be created is calculable. A colourful way of framing the result of the calculation is to say that one hydrogen atom per century is created in a volume of space equal to that of the Empire State Building. While this may appear a small number, and while it is certainly not detectable directly (!), the number is not small when added up throughout the whole observable universe.

If one accepts the Steady State picture, one can choose to regard the creation of matter as the significant thing, and construct the basic equations of one's cosmological theory with a corresponding term in them. Hoyle has done this by adding to Einstein's original field equations a new term, called by him the "C-field"

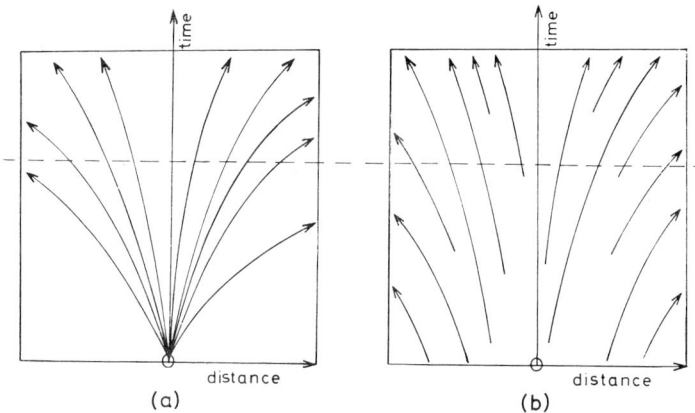

Figure 1.10. Schematic pictures of a one-dimensional universe, with particles' distance plotted against time, for

  (a) a Big Bang universe, and

  (b) a Steady State universe.

In case (a) all the particle trajectories originate at t = 0, and the pattern evolves as the universe expands. In case (b) the pattern is the same at all times, but since individual particle paths are diverging due to the expansion of the universe, new particles must come into ·being to preserve the pattern. An observer at any one time, along the dotted line, cannot distinguish between the two pictures.

(C for Creation). Upon solving these equations the expansion of the universe follows as a consequence of the C-field: space must readapt itself to accommodate the new matter, and this appears as an expansion. This mathematical elaboration of the Steady State theory differs from the original version of Bondi and Gold mainly in that it makes a specific prediction of the overall density of matter in the universe (about 2×10^{-29} grammes/cm^3).

This is as much as need be said about the Steady State theory. We have dismissed it comparatively rapidly, because by now there exists quite a bit of evidence against it. In subsequent chapters we will mention in passing various bits of circumstantial evidence, but the real body blow, the fact which would move almost any jury to convict, is the observed presence of the background blackbody radiation. This radiation simply has no place in the Steady State theory. Indeed the accumulation of evidence against the theory has moved Hoyle* to write "it now seems likely that the Steady State idea will have to be discarded, at any rate in the form in which it has become widely known".

(3)˙ Other modern cosmologies

Apart from the Big Bang and Steady State theories outlined above, several other cosmologies have been propounded during the past 30 years or so. Although these more exotic cosmologies are consonant with the observed homogeneity and expansion of our universe, none of them has ever enjoyed much popular favour with scientists, mainly for reasons outlined below.

The most significant of these assorted "other" cosmologies† is that due to Dirac, who, in 1937, predicated a universe wherein some of the fundamental constants of physics are changing with time.

---

* Hoyle's "recantation" referred to in the text is to be found in a quite readable article by him in the scientific journal *Nature*, Volume 208, page 111 (October 9, 1965).

† Most of these theories are motivated by a striking set of numerical coincidences. First, if we express the strength of the electromagnetic or coulomb force between electron and proton as a ratio to the strength of the gravitational force between electron and proton (both forces depend on distance as r^{-2}), we arrive at the vast number 10^{40}; that is electromagnetic forces are 10^{40} times as strong as gravitational ones. Next, the ratio of the Hubble constant time, *Equation (1.3)*, which roughly measures the age of the universe, to the characteristic time for nuclear forces (which happens to be given by e^2/mc^3, with m the electron mass) is again approximately the number 10^{40}; that is, the ratio of the largest time scale in the universe to the smallest is about 10^{40}. Finally, a crude estimate of the number of protons in the universe is that there are 10^{80} of them, and $10^{80} = 10^{40} \times 10^{40}$! This last number is arrived at by estimating the current mass of the universe as $(4\pi/3) R_o^3 \rho_o$ (where ρ_o is the density, *Eq. (1.4)*, and R_o is the present radius of the universe), and simply dividing by the mass of a proton.

These numerical coincidences are quite possibly no more than coincidences.

However, most of our "other" cosmologies are based on the idea that this remarkable series of approximate equalities reflects some deep relation between cosmic quantities and microsopic ones. Thus Dirac's $G \frown t^{-1}$ and Gamow's $e^2 \frown t$ are postulated to *preserve* the rough equality between the two different factors which presently have the value 10^{40}. As cosmic time ticks on, one of these (namely that involving the age of the universe) will change and the coincidence will be lost unless the fundamental constants involved also change with cosmic time.

Whether cosmology should be asked to take these coincidences seriously, and to endeavour to explain them, is a very debatable question.

In particular, in Dirac's cosmology the fundamental gravitational constant G (the proportionality constant in the basic formula, gravitational force $= Gm_1m_2/r^2$) is changing in inverse proportion to the cosmic time t elapsed since the universe began: $G \sim t^{-1}$. Derivative from this theory is the rather arbitrary cosmology of Jordon. More recently Teller (1948) and Gamow (1967) have sought to revive Dirac's basic ideas, with the alteration that it is the fundamental charge on the electron, e, rather than G which changes in time; in Gamow's cosmology $e^2 \sim t$ and in Teller's $e^2 \sim \log t$, where t is again the time elapsed since the beginning of the universe.

Yet another cosmology was put forward in 1935 by the mathematician Milne (no relation to Winnie the Pooh); here again certain of the fundamental constants of Nature end up being time dependent.

The fact upon which all these theories founder is that the fundamental constants of physics (such as G, e, c, and the fundamental constant of quantum mechanics h) show no evidence of having changed as cosmic time has rolled on.

Thus a change in the gravitational constant G over the past few billion years, that is over significant times on a cosmological scale, would for example have produced changes in the dynamics and scale of our solar system. Such changes would be discernible in the geological history of our earth, which is itself reckoned as being at least two billion years old. Geophysicists assure us that no such evidence for a changing G exists. Other astrophysical consequences, for example in the evolution of the sun (whose dynamical structure involves G intimately), can be imagined, yet such indications of a changing G have been sought in vain.

If G does not change, perhaps other physical constants such as e^2 change instead. This suggestion was made by Gamow last year in an attempt to revive Dirac's cosmology, and was promptly answered by a rash of letters in the leading scientific journal *Physical Review Letters*. Dyson and Peres pointed out that changes in e^2 would be reflected in changes in the decay processes in radioactive elements, whereas strong evidence suggested no such anomalies. More crushingly, Bahcall and Schmidt observed that familiar "doublet", "triplet" and more general "multiplet" spectral lines can be discerned in the spectra of far distant galaxies. Such very small separations between members of a spectral multiplet are known to be simply proportional to a dimensionless quantity called the "fine structure constant", namely $e^2/\hbar c$. Some of these galaxies are so far away from us that their colour spectra are red-shifted by as much as 20 per cent, which means that if the value of e^2 was proportional to cosmic time (Gamow's suggestion) then the value of the fine structure constant which we would observe now for these distant galaxies (the light from which has been travelling to us for long ages) would differ by 20 per cent from its present value at our epoch in time. Experimental observations of the multiplet spacing in these galaxies distant from us in space and time reveal the ratio of the fine structure constant then to that now to be $1 \cdot 001$ with a probable error of $0 \cdot 002$, a far cry from any 20 per cent difference.

In short, all theories which envisage the fundamental constants of physics changing with time have so far no sooner taken wing than they have been shot down, leaving the arena clear for the contest between the Steady State and the various Big Bang cosmologies.

Which cosmology?

Let us recall where we stand. Beginning with a digression on earlier cosmologies, we sketched the pieces of our current picture of the universe as it is today. Then the two main experimental facts of cosmology, that the universe is homogeneous and isotropic and that it is expanding, were presented. This eventually has left us with the choice between the Steady State cosmology and an infinite class of Big Bang or evolutionary general relativistic cosmologies which can be discriminated among by a determination of the present value of the expansion parameter q_o.

How shall this choice be made? What experimental facts can be marshalled to bear upon these cosmological questions?

To begin with, the theory of the structure and evolution of stars can be used to give results, albeit rather indirect ones, of interest to cosmology. This story will be told in Chapter 2.

Next, we could move one step up the ladder, to the study of the evolution of galaxies. However, it is not too much to say that the problem of understanding why there are different kinds of galaxy, and of how galaxies originate and evolve, is the biggest unsolved problem in present-day astronomy. The properties of individual stars fall largely within the classical study of astrophysics, whereas the phenomenon of galaxy formation touches directly on cosmology. Indeed, the study of galaxies forms a bridge between conventional astronomy and astrophysics on the one hand, and cosmology on the other. Thus it is particularly unfortunate that the subject is so ill-understood that any discussion of it here would be fruitless.

Finally, we can move on to the viewpoint where galaxies them-selves constitute the smallest intelligible unit of study. This we do in Chapter 3, where cosmological information is extracted by using very distant galaxies, and their like, as indicators and counters to find out how things were in past time.

The way of the cosmologist is beset with difficulties both conceptual and experimental. If we live in a universe evolving from a Big Bang, our task has been stated strikingly by one of the seekers, Lemaître: "The evolution of the world can be compared to a display of fireworks that has just ended: some few red wisps, ashes and smoke. Standing on a well-chilled cinder, we see the slow fading of the suns, and we try to recall the vanished brilliance of the origin of the worlds."

CHAPTER 2

The Evolution of Stars

As mentioned in the preceding chapter, the excuse for the present discussion of the life history of stars is that it sheds a certain amount of oblique light on cosmological questions. For another thing, it contains a good deal of concrete down-to-earth physics, which reassuringly illustrates the relation between the further reaches of astrophysics and everyday laboratory work.

The scheme in this chapter is as follows:

First, we shall give a detailed account of the chain of nuclear reactions whereby hydrogen is burned to helium to provide the energy source in stars such as the sun. (Professor Bracewell has made passing mention of this process earlier in the book.)

Next we shall consider the morphology of stars, the study and classification of the forms of the different stars we see around us. This empirical study culminates in the so-called "Hertzsprung-Russell" diagram.

Then these first two sections will be synthesized, to arrive at an *understanding* of the Hertzsprung-Russell diagram. Some observational tests to check this understanding will be described.

Finally, the cosmological information which can be extracted from this work on the evolution of stars will be presented. In this context the age of the oldest stars is clearly of interest. Less directly, pertinent knowledge about how the chemical elements are cooked from the starting ingredient of hydrogen can also be gleaned.

Hydrogen burning in stars

As Professor Bracewell has pointed out earlier in this book, in his first chapter on our sun, until quite recently one of the major enigmas of astronomy has been the explanation for the source of energy whereby our sun and other stars replenish the celestial flame which they continuously pour into space. As long ago as

Roman times, Lucretius wrote in *On the Nature of Things*: "We must believe that sun, moon [!] and stars emit light from fresh and ever fresh supplies rising up."

But until the recent discovery of the nuclear energy locked in atomic nuclei, only two sources of heat and fire were known*; chemical energy (as derived, for example, from coal, gas and oil) and gravitational energy (as utilized when falling torrents of water have their gravitational energy tapped to generate hydroelectricity).

Around the turn of the century physics had advanced to the point where it became possible to make calculations to prove (as some had long suspected) that neither chemical energy nor gravitational energy would enable our sun to shine at its present rate for more than the blink of an eye on an astrophysical time scale. Thus, if the energy sources were gravitational, the sun could not have been radiating at its present rate for more than 20 million years, whereas the emergence of life on earth dates back at least 500 million years. (This limit to the gravitationally-fed lifetime comes from assuming the sun to have contracted from infinity to its present dimensions, releasing gravitational energy thereby; such a calculation gives an extreme upper limit to the lifetime.) And if chemical energy were the supply, then the sun would be burned out in a mere few thousand years.

Following the discovery of radioactivity at the end of the 19th century, the existence of the atomic nucleus built up from protons and neutrons began to be recognized, and the science of nuclear physics was born. For the reasons given above, some people, notably the great astrophysicist Eddington, began to assert that

---

* On a more fundamental level, we can remark that there exist only *four* basic forces in nature, from which all other interactions are derived directly or indirectly. These four fundamental forces are the gravitational force, the electromagnetic or coulomb force between charged particles, and two sorts of nuclear forces, the strong nuclear force and the weak nuclear force, which have very short ranges and in effect only act inside nuclei. All of chemistry and biology derives from the coulomb forces, and thus so does all chemical energy. Similarly, gravitational energy derives from the gravitational forces. These two basic forces were understood long before the nuclear forces were realized to exist; even today nuclear forces are far from fully understood. For a much more detailed discussion of these matters, see the 1966 Nuclear Research Foundation Summer Science School book *Atoms to Andromeda*, pp. 251 et seq.

the sun's "reservoir can scarcely be other than the sub-atomic energy which, it is known, exists abundantly in all matter".*

By this time Aston had shown that, in principle, energy could be released when heavier nuclei were built up out of lighter nuclei, as for example when helium is built up from two protons and two neutrons. (This is because in general heavier nuclei tend to be more stable, at least until one gets as far as iron, with a nuclear mass of 56 times that of hydrogen, after which the trend is reversed, and yet heavier nuclei are progressively *less* stable. Today we refer to this process of building more stable nuclei out of lighter nuclei as nuclear fusion, and we are distressingly familiar with the energy release so obtained.)   Moreover, by 1920 Sir Ernest Rutherford had demonstrated experimentally that indeed atomic nuclei could be transmuted; Rutherford broke up the atomic nuclei of oxygen and of nitrogen by bombarding them with energetic protons.   Eddington† summarized the position by pointing out that "what is possible in the Cavendish Laboratory may not be too difficult in the sun".

Until the 1930's there remained one substantial objection to Eddington's idea that our sun is powered by nuclear fusion, and that in general stellar fires burn nuclear fuel.   In classical physics, the positive charges on nuclei provide a mutually repulsive force (like charges repel!) which is strong enough to keep them from colliding, and thus fusing, unless their kinetic energies are much higher than they will be in the centre of the sun, where the temperature is too "low".   That is, in classical physics, the sun's inner temperature of 15,000,000°C is too low to overcome the charge or coulomb barrier between nuclei.   This difficulty was resolved by the advent of the new "quantum" mechanics, which

---

* This is an extract from Eddington's address to the Physical Society in Cardiff, in 1920.

† This also comes from Eddington's Cardiff address in 1920. At the same time, well before these matters were fully understood, he remarked with chilling prescience, "If, indeed, the sub-atomic energy in the stars is being freely used to maintain their great furnaces, it seems to bring a little nearer to fulfilment our dream of controlling this latent power for the well-being of the human race—or for its suicide."

superseded classical mechanics for describing atomic systems, and which allows particles to "tunnel through" barriers that would be impossibly high in classical physics.*

Thus in the sequel Eddington has turned out to be right. But in 1925, when things were not looking so well for his ideas, Eddington turned to his critics with the suggestion that if they did not think stellar interiors were hot enough to burn nuclear fuel, then they should "go and find a hotter place".

Once given a knowledge of nuclear physics and of quantum mechanics, it remains to work out the details of the nuclear reactions which supply energy inside stars at temperatures not greater than 20,000,000°. This is not a trivial task.

By the late 1930's the answer had been provided, largely by last year's Nobel Prize winner, Hans Bethe, of Cornell University. There are two possible chains of nuclear reactions, both of which end up converting four hydrogen nuclei (protons, ^1H) into a helium nucleus (^4He), with a release of energy in the process. That is to say stellar energy comes from the burning of four hydrogen nuclei to form one helium nucleus. These two chains of reactions have come to be called the proton-proton chain and the carbon-nitrogen or CN cycle.

Of these two cycles, the proton-proton chain seems to be the more generally relevant, and in particular the one occurring in our sun. We shall now embark on a quite detailed exposition of the p-p chain, which will illustrate explicitly how everyday laboratory physics makes contact with theoretical astrophysics†. Then we shall return to touch upon the CN cycle, and finally we shall describe a critical experiment which is about to give the first direct test of the validity of our ideas about the interior of the sun.

The p-p chain begins with the nuclear reaction

$$^1H + {}^1H \rightarrow {}^2H + e^+ + \nu \qquad (2.1)$$

---

* In classical mechanics, if you catch a lion and put him in a cage with a high enough impenetrable barrier, he will stay put. But a quantum-mechanical lion has a finite probability of suddenly just popping up outside the cage. Fortunately this kind of effect only manifests itself on the sub-microscopic atomic scale. Thus it is only relevant to *very* tiny lions. But all this is another story.

† I am also influenced by my own recent interest in this particular field.

where here ^2H is the stable hydrogen isotope, deuterium, consisting of a proton and a neutron bound together, and e^+ is a positive electron or positron. The symbol ν represents a neutrino, a species of phantom-like particle with neither mass nor charge, but able to carry away energy at the speed of light. Next the deuterons can combine with further protons to give the helium isotope ^3He:

$$^2H + {}^1H \rightarrow {}^3He + \gamma \qquad (2.2)$$

where γ represents a gamma-ray, a quantum of electromagnetic energy. The most common termination of the p-p chain is now arrived at by the combination of two ^3He particles according to

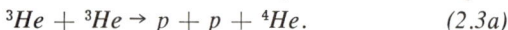

$$^3He + {}^3He \rightarrow p + p + {}^4He. \qquad (2.3a)$$

This series of reactions is illustrated schematically in *Figure 2.1*.

H- BURNING
THE FUSION OF ORDINARY HYDROGEN
IN MAIN SEQUENCE STARS (THE SUN)

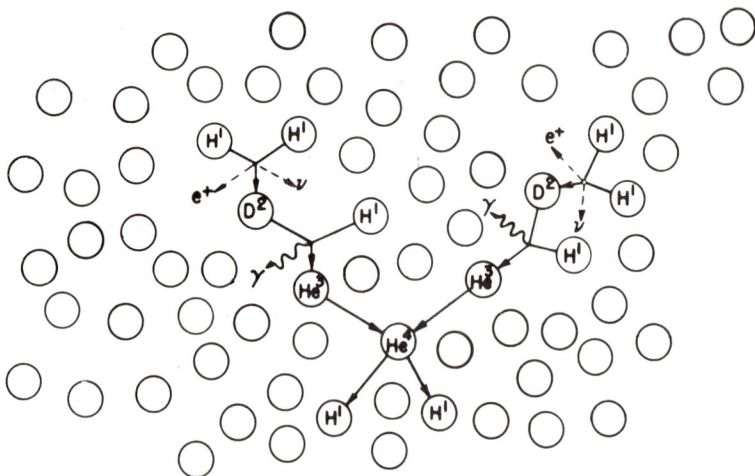

OVERALL RESULT: 4 HYDROGEN NUCLEI —► HELIUM NUCLEUS

ENERGY RELEASE = 100 MILLION KILOWATT— HOURS
PER POUND CONVERTED

Figure 2.1.

Note that if we consider only nucleons (neutrons and protons), the upshot of this series of reactions is that four protons have been transformed into one helium nucleus,

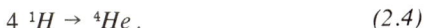

$$4\ {}^1H \rightarrow {}^4He. \qquad (2.4)$$

The resultant energy release stems from the fact that four separate protons have a total mass of $4 \cdot 0291$ mass units, whereas the very stable and tightly bound helium nucleus has a mass of $4 \cdot 0015$. The remaining $0 \cdot 0276$ mass units are converted into energy, according to Einstein's familiar relation $E = mc^2$.

An easy calculation shows that on this basis the conversion of only 10 per cent of the sun's hydrogen into helium is enough to keep it shining at its present rate for some ten billion years, which is about twice the known age of our solar system.

The completion of the *p-p* chain is complicated by the fact that, although *Equation (2.3a)* represents the most common ending, it is alternatively possible for the reaction

$$^3He\ +\ {}^4He \rightarrow {}^7Be\ +\ \gamma \qquad (2.3b)$$

to produce beryllium. The chain then proceeds to completion *either* by an electron capture,

$$^7Be\ +\ e^- \rightarrow {}^7Li\ +\ \nu\ +\ \gamma, \qquad (2.5)$$

followed by

$$^7Li\ +\ {}^1H \rightarrow 2\ {}^4He, \qquad (2.6)$$

or by the sequence of reactions

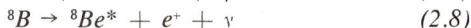

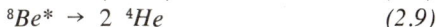

$$^7Be\ +\ {}^1H \rightarrow {}^8B\ +\ \gamma \qquad (2.7)$$

$$^8B \rightarrow {}^8Be^*\ +\ e^+\ +\ \nu \qquad (2.8)$$

$$^8Be^* \rightarrow 2\ {}^4He \qquad (2.9)$$

(The asterisk denotes that the ${}^8Be$ nucleus is not in its usual "ground" state, but rather it is in an unstable "excited" state.) In either case, the final result is as before simply that four protons are transmuted into one helium nucleus, according to *Equation (2.4)*. The reason for the setting down of these subsidiary methods of completion of the *p-p* chain will be clear later.

At this point the distinction between two different questions must be emphasized. On the one hand, to know the *total amount* of energy which can be released by hydrogen burning according to

Equation (2.4), one need only know the mass differences involved, as given above. On the other hand, the *rate* at which energy is released per second depends entirely on the details of the individual nuclear reactions *(2.1)-(2.9)*; to explain the *rate* of energy generation in the sun one must know the rate at which each of these reactions proceeds at the temperatures which prevail inside the sun.

We shall illustrate how this sort of detailed information as to the reaction rates is obtained by considering two examples, namely the reactions *(2.1)* and *(2.3a)*.

The reaction *(2.1)* which inaugurates the *p-p* chain is a particularly interesting one. Unlike the other major reactions, *Equations (2.2)* and *(2.3a)*, it is a so-called β-decay reaction, which proceeds under the aegis of the weak nuclear force. The other main reactions involve strong nuclear forces, which (reasonably enough) are stronger than the weak nuclear force.* Indeed the forces involved in the two-proton β-decay are so weak as to preclude any possibility of any laboratory measurement of the reaction rate at energies corresponding to those of relevance inside the sun. However, this circumstance is mitigated by the fact that obviously only two-nucleon forces are involved in *Equation (2.1)* (in distinction from all the other reactions, which involve at least three nucleons), so that an accurate theoretical calculation is feasible. Since this reaction is the weakest, and thus the slowest, of all the major reactions in the *p-p* cycle, its rate is of particular importance in setting the overall rate for the hydrogen burning process.

Bethe and Critchfield first estimated this reaction rate in 1938, but since the β-decay process was at that time ill-understood and the two-nucleon forces inaccurately determined, their estimate was too low by a factor of about 10. Later, in 1952, Edwin Salpeter of Cornell University made a much more accurate calculation of the rate. Since then there have been minor revisions in the details of the two-nucleon forces, and advances in other technical details pertaining to such calculations, so that last year John Bahcall and I further revised this reaction rate, in a calculation which should be accurate to within 3 per cent.

---

* For further discussion of the weak and strong nuclear forces see the footnote on page 173 and the reference therein.

Unlike the two proton β-decay reaction *(2.1)*, the other reaction rates in the *p-p* chain are *almost* amenable to direct laboratory measurement. Actually, technical difficulties still stand in the way of making laboratory observations of these reactions at energies

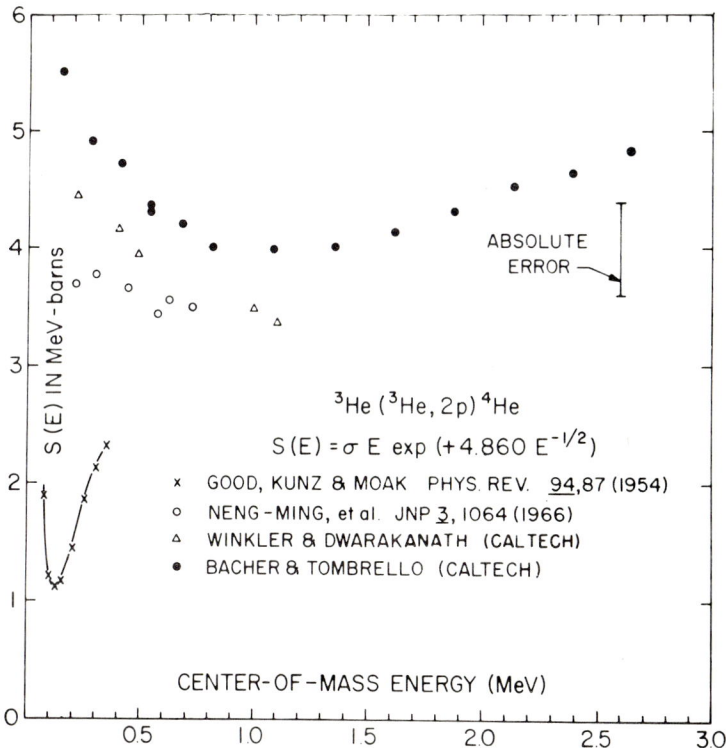

Figure 2.2. The results obtained by various laboratory groups (as shown) studying the reaction $^3He + {}^3He \rightarrow {}^4He + p + p$. The horizontal co-ordinate axis represents the relative energy of the 2 ^3He nuclei, and the vertical co-ordinate axis is S(E), a quantity which measures the rate at which the reaction proceeds.

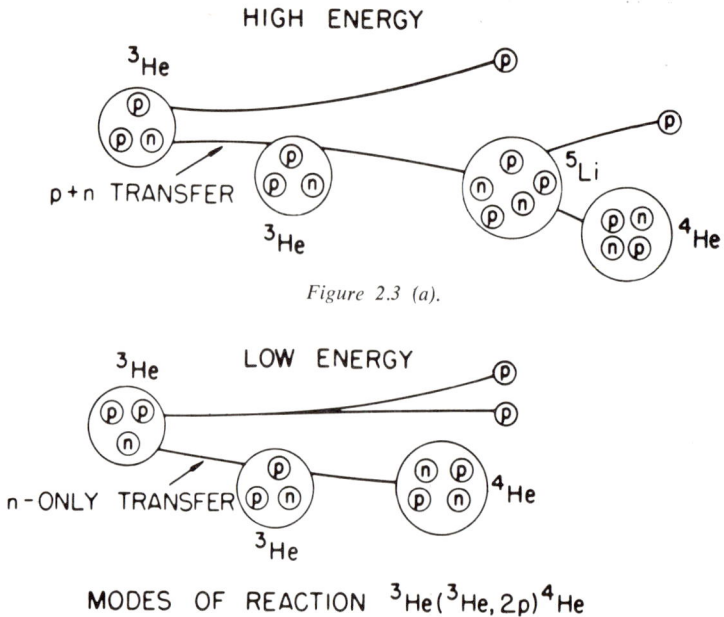

Figure 2.3 (a).

MODES OF REACTION ^{3}He(^{3}He, 2p)^{4}He

Figure 2.3 (b).

A schematic illustration of the mechanism whereby this nuclear reaction proceeds (a) at "high" temperatures, and (b) at "low" temperatures.

as *low* as those corresponding to the temperatures inside the sun,* but one can generally get fairly close, and then hope to derive the desired information by projecting downwards.

_____

* At first glance it seems downright odd to refer to the particle energies inside the sun as "low". However most terrestrial nuclear physics experiments are performed with aligned *beams* of particles, which have been accelerated to energies about the coulomb barrier referred to earlier. Once one goes to energies much below the coulomb barrier, the reactions begin to proceed so slowly as to make for great experimental difficulties. Yet the *thermal, or random,* energies of protons in the sun, even at a temperature of 15,000,000°, correspond (as mentioned earlier) to energies well below the coulomb barrier. The point is that although *thermal*, or random, energies corresponding to millions of degrees are difficult to achieve in the laboratory, it is a comparatively easy matter to achieve much larger energies in beams with a specific orientation in space, where no containment problems are encountered.

Such a procedure of projecting, or "extrapolating", from one's experimental data can only sensibly be done if one understands what is going on in the particular reaction, that is (to relapse into comfortable jargon) if one can identify the reaction *mechanism*.

An interesting example of this type of work is illustrated in *Figure 2.2*, which summarizes experiments on the reaction *(2.3a)* above. The reaction *(2.3a)* is of particular importance, not only because it is the dominant mode of completion of the *p-p* chain, but also because its competition with the alternative *(2.3b)* serves to determine the frequency with which the subsidiary reactions *(2.5)*-*(2.9)* will occur. It is thus extremely relevant to the neutrino experiment described below. In *Figure 2.2*, the horizontal axis is the relative energy of the colliding ^3He nuclei (measured in the conventional nuclear physics units of millions of electron volts, or MeV), and the vertical axis is the quantity $S(E)$ which determines the rate at which the reaction proceeds. The energy region of relevance to our sun and like stars is a mere $0 \cdot 01$ MeV.

Now at comparatively high energies (above 1 MeV) the reaction presents few experimental difficulties, and not only can its rate $S(E)$ be well determined, but also a detailed understanding of the reaction mechanism can be derived from experiments. At "high" energies the mechanism is in fact as illustrated in *Figure 2.3(a)* namely:

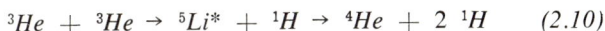

$$^3He + {}^3He \rightarrow {}^5Li^* + {}^1H \rightarrow {}^4He + 2\ {}^1H \qquad (2.10)$$

where $^5Li^*$ is a transient or "virtual" state. On this basis, one would extrapolate the curve obtained from the "high" energy experiments above, down to arrive at a guess of 1 or 2 for $S(E)$ at $0 \cdot 01$ MeV. And indeed the experiments of Good and co-workers in 1952 tend to support such an extrapolation.

However, this reaction is sufficiently important that the difficult task of pushing the experiments down to around $0 \cdot 1$ MeV has recently been undertaken, both by two different groups at the California Institute of Technology and by a group of Russians and Chinese. The earlier incorrect points of Good *et al.* bear eloquent testimony to the difficulties of this experiment. As can be seen from *Figure 2.2*, the experimental data points begin to swing upwards at these "low" energies, with the overall curve changing slope. Moreover, it can be shown experimentally that the mechanism *(2.10)* no longer prevails at "low" energies.

How then do we extrapolate down to $0 \cdot 01$ MeV with confidence? Last year Don Clayton of Rice University and I suggested that the mechanism illustrated in *Figure 2.3(b)* prevails at low enough energies. This mechanism, whereby a neutron "tunnels" directly from one ^3He nucleus to the other, to produce ^4He and leave behind two protons, has the advantage that the charged protons do not have to come so close together as they do in the other mechanism shown in *Figure 2.3(a)* and *Equation (2.10)*. Thus it is entirely plausible that this new mechanism should win out at low enough energies, below the coulomb barrier. Preliminary numerical calculations based on this neutron tunnelling model suggest that it does accord with the upward slope of the experimental points at low energies, and thus we can fairly confidently extrapolate down to $0 \cdot 01$ MeV to get a value of about 5 for $S(E)$ there.

The above detailed discussion of the two *p-p* chain reactions *(2.1)* and *(2.3a)* is aimed at making explicit the relation between ordinary, mundane laboratory physics and our understanding of astrophysical questions. I think it tends to be difficult to believe in the intimate connection between the two without such a detailed exposition of a few examples. Furthermore, the discussion has served to make clear the difficulties that beset work in this particular little realm of physics.

There is another method of burning hydrogen to helium, as summarized in *Equation (2.4)*, mentioned above as the *CN* cycle. In this cycle, one needs some ^{12}C to be present to begin with. These carbon nuclei then serve essentially as catalysts in a cycle whereby successive additions of protons, followed sometimes by β-decays which change a proton into a neutron, build from

^{12}C to ^{13}N to ^{13}C (β-decay) to ^{14}N to ^{15}O to ^{15}N (β-decay).

At this point the addition of a further proton can result in the breakup

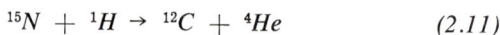

$$^{15}N + {}^1H \rightarrow {}^{12}C + {}^4He \qquad (2.11)$$

thus producing helium out of a total of four protons [*Equation (2.4)*] and regenerating the "catalytic" ^{12}C nucleus. The details of the *CN* cycle show that at stellar temperatures *above* about 20,000,000° it will tend to go faster than, and therefore dominate, the *p-p* cycle. However in cooler stars (and our sun's internal temperature is about 15,000,000°) it is the *p-p* cycle which will burn the nuclear fuel.

Other chains for the burning of heavier elements, at even higher temperatures, can be similarly worked out, and such cycles have their place in older stars with hotter interiors.

On such a foundation of theory and terrestrial experiment, astrophysicists thus construct a towering edifice of theory as to what goes on deep inside stars. In contrast, astronomy only provides information about the *surface* of stars. To deal with the transport of energy from the nuclear furnaces deep inside to the radiating stellar surface requires in itself yet another body of theoretical knowledge.

In brief, there is at present *no direct evidence* whatever bearing upon our complex web of ideas about the deep interior of stars such as the sun. Oddly enough, one can lift from Shakespeare a quite uncannily apt quotation:

> The Heaven's glorious sun,
> That will not be deep-searched with saucy looks;
> Small have continual plodders ever won
> Save base authority from others' books.

One must hope that the fact that this quotation comes from *Love's Labour's Lost* is in no way prophetic.

The position can alternatively be stated in somewhat less elegant prose as, "We got to get all this theory out of things." This quotation (here taken wildly out of context) comes from the infamous ex-Governor of Alabama, George Wallace.

Although it would seem quite impossible in any way to see directly to the sun's interior, human ingenuity has recently devised a method. The crux of this exciting experimental idea is that the neutrino particles, which occasionally appear in these nuclear reactions, since they have no electric charge nor rest mass, are able to travel at the speed of light through cold or hot, dense or tenuous matter with equal ease. Thus the neutrinos produced deep inside the sun escape easily into space. This boon has of course its accompanying bane, in that it is extremely difficult for the experimentalist to stop these neutrinos in order to detect them.

We now turn to the reactions *(2.1) - (2.9)* and note that neutrinos pop up at three places, namely in reactions *(2.1)*, *(2.5)* and *(2.8)*. Those from reactions *(2.1)* and *(2.5)* have energies too low to make their detection feasible as yet, but the neutrinos from the β-decay of ^8B, reaction *(2.8)*, not only are of comparatively high energy, but also have their energy rather exactly defined.

Furthermore this energy corresponds rather precisely to that of a neutrino which can induce the reaction whereby chlorine is transmuted into argon,

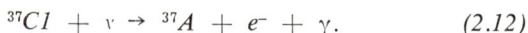

$$^{37}Cl + v \rightarrow {}^{37}A + e^- + \gamma. \qquad (2.12)$$

Thus an experiment has been under way for some three years to detect the solar neutrinos from the reaction *(2.8)* by catching them (according to the reaction *(2.12)*) in a vast 100,000 gallon tank which contains chlorine in the form of the common cleaning fluid tetrachloroethylene. This experiment is buried 4,500 feet underground in a disused mine in South Dakota to screen out cosmic rays. The detection of the argon atoms formed by the solar ^8B neutrinos is a *tour de force* of nuclear chemistry, since one only expects some three events per day in the whole tank! (At least if the experiment does not work, Bahcall and Davis can open a dry cleaning factory.)

This very significant experiment should present its first results any day now.

If the results agree with the predictions based on our current understanding of stellar interiors, the experiment will provide a fine capstone for this elaborate theoretical structure. Conversely, if the experiment disagrees with the prediction, it will be back to the drawing board for nuclear astrophysicists. In either case, the prospect is an exciting one, and the experiment has aroused much interest in the scientific community; many bets lie on its outcome.

To summarize this section, we have seen how at the beginning of this century people were forced to the conclusion that the sun burns nuclear fuel to derive its radiance, and how these general ideas have subsequently been borne out by an explicit scheme of reactions. This theoretical picture of the goings-on at the centre of the sun is about to be subjected to the critical test of a direct experiment.

A general description of stars

Turning from the nuclear fires deep inside stars, we shall now discuss some of the more directly observable stellar properties. More explicitly, we shall pay attention to the following characteristics of stars: luminosity or brightness; distance from us; surface temperature or "colour"; radius; mass; and chemical composition.

In this section we shall be concerned mainly with cataloguing empirical facts; the *understanding* which emerges therefrom will be deferred to the next section.

The most readily discernible characteristic of the stars visible in the night sky is that they differ in *apparent brightness*. This was recorded by the Alexandrian Greeks. Hipparchus, in the course of compiling the first star catalogue of some thousand stars, devised a scale (called *apparent magnitude*) for comparing the apparent brightnesses of stars. Since astronomers are in many ways conservative creatures, this scale of Hipparchus is still with us, albeit in somewhat strained form. Indeed such pervasive retention of historical forms of classification lends the science of astronomy a certain quality of exasperating charm. In any event, the scale today is one in which if two stars have a difference of five units in apparent magnitude then they differ by exactly a factor of 100 in their apparent brightness. A difference of *one* unit in apparent magnitude corresponds to a difference of a multiplicative factor of $(100)^{1/5} = 2 \cdot 512$ in apparent brightness; in general each *increase* of one unit up the scale of apparent magnitude corresponds to a *decrease*, of about $2\frac{1}{2}$ times, in the star's apparent brightness. To belabour this point, we further note that if an apparent magnitude difference of five amounts to a difference in apparent brightness of 100, then an apparent magnitude difference of ten amounts to a difference in apparent brightness of 100^2 or 10,000, and an apparent magnitude difference of fifteen to a brightness difference of 100^3 or 1,000,000, and so on. Such a scale, where magnitude changes additively while brightness changes multiplicatively, is called a logarithmic one, and we can summarize by saying:

Apparent magnitude $=$ Log (Apparent Brightness) $+$ constant *(2.13)*

Stars down to the "sixth magnitude" are visible on a clear night
(away from the polluted air of modern cities), and the largest
telescopes can see down to the "twenty-third magnitude".

However, as we well know, the apparent brightness of a star
will depend not only on its intrinsic characteristics, but also on
how far away from us the star happens to be. For a study of the
nature of stars we are clearly much more interested in their
intrinsic brightness, or luminosity, which we will henceforth denote
by L. The intrinsic luminosity is related to the apparent brightness
(denoted by l) by the familiar inverse square law:

$$l = \frac{L}{d^2} \qquad (2.14)$$

where d is the distance of the star from us. By a variety of
ingenious techniques (which we have not space to discuss here)
this distance d can be determined, to arrive at a determination
of the intrinsic luminosity of the various stars. Again this
luminosity is generally measured on a scale of *"absolute magnitude"*;
absolute magnitude and luminosity are logarithmically related in
just the same way as were apparent magnitude and apparent
brightness, *Equation (2.13)*.

The range in the stellar luminosities is quite remarkable; the
most luminous stars are some 10^{11} times more luminous than the
least. The sun sits somewhere in the middle of this range, with
the most luminous stars being about 100,000 times more luminous
than it, and the weakest some 1,000,000 times less luminous.

In short, the luminosity of a star measures how much energy
the star is actually pouring out into space; this quantity L is
conventionally expressed in terms of the star's "absolute
magnitude".

Another of the comparatively easily observed characteristics of
stars is their surface temperature. This can be deduced indirectly
from the colour spectrum of the star.

As has been mentioned earlier, if the light from a star is passed
through a spectrograph and broken up into its various constituent
wavelengths, the resulting "spectrum" has a great deal of information about the surface of the star coded into it. Such spectra
will contain both bright and dark lines ("spectral lines") as well as

a continuous background of radiation ("continuous spectrum"). The line spectrum comes from the details of the energy levels within the various atoms which are emitting and absorbing radiation, and the continuous spectrum comes from the equilibrium or blackbody radiation, which is radiated in essentially the same way as that radiated by the filament of an ordinary electric light or by the glowing element of an electric stove or heater. Moreover this continuous spectrum has a characteristic shape,* with a maximum intensity at a colour or wavelength which is universely proportional to the temperature of the radiating surface. (Indeed the explicit formula for this relationship is exactly $\lambda = 2 \cdot 9 \times 10^7 \, T^{-1}$, where λ is the wavelength measured in Angströms and T the temperature in degrees Absolute). That is to say the hotter the surface of the star, the shorter the wavelengths of the continuous radiation, and conversely the cooler the surface, the longer the wavelengths. Thus hotter stars tend to be blue and cooler stars red.

The line spectrum contains information not only about the surface temperature but also about the pressure and chemical composition of the surface, for all these factors enter into the determination of the precise character of the absorbed and emitted spectral lines.

When telescope and spectrograph are combined to obtain stellar spectra, the main types illustrated in *Figure 2.4* are obtained. Any one star can then have its spectrum readily classified as one of these nine characteristic types, which are denoted by the letters O B A F G K M N S. This extraordinary collection of letters labelling the spectral sequence has its origin in the dim history of the subject. Regardless of its historical origin, this sequence can be conveniently remembered as the first letter of each word in the sentence Oh Be A Fine Girl Kiss Me Now, Smack. Most students find no difficulty in remembering this.

Today these spectra are understood, and the spectral sequence is associated with surface temperatures as follows: the O stars are the hottest, with surface temperatures in the general vicinity of 50,000°C or more; then we run down through B stars (15,000°C),

---

* By way of illustration, see *Figure 1.7*.

Figure 2.4.   Actual photographs of the colour spectra of stars exemplifying the principal types in the spectral sequence, O B A F G K M N S. In each case, standard laboratory spectra are included above and below the stellar spectra, for comparison. (Mount Wilson and Palomar Observatories.)

A stars (10,000°C), F stars (6,600°C), G stars (5,500°C), K stars (4,400°C), and M stars (3,400°C); ending with the coolest N and S stars at about 3,000°C to 2,000°C. Again space does not permit a digression upon the details of the atomic physics whereby the various members of this temperature sequence give their characteristic spectral signatures.

Notice that the temperature range from the hottest to the coolest stars is not much more than a factor of 10; this is in contrast to the enormous range in stellar luminosities. We may also call attention to the fact that these surface temperatures are very much less than those encountered deep in stellar interiors, as mentioned in the previous section.

There is a very general law, first derived from thermodynamics by Stefan and Boltzmann in the last century, which enables one to relate a star's luminosity L to its surface temperature T_s and the stellar radius R:

$$L = (4\ \pi\ \sigma)\ R^2\ T_s{}^4 \qquad\qquad (2.15)$$

Here σ is just a universal constant, which is well determined and can in fact be expressed in terms of the fundamental constants of physics. This relation says that for any set of stars with the same luminosity, that is with the same energy output, then cooler, redder stars will have larger radii. Alternatively, for any group of stars with fixed surface temperature or "colour", then those with smaller energy output will have smaller radii.

An evaluation of the stellar radius R is readily achieved if one knows both the distance of the star from us and its angular size, that is, the angle which the distant star subtends at our eye. This latter measurement is difficult to carry out, since for example one of the nearer bright stars, β-crucis (in the Southern Cross), subtends at our eye the same angle as would an old Australian penny at a distance of 4,000 miles! Recently, however, the very ingenious Brown-Twiss intensity interferometer at Narrabri (see *Figure 2.5*) has made possible such measurements of the angular sizes, and thus the radii, of about fifty stars. This precision measurement of R, for stars of accurately known L, can be used in *Equation (2.15)* to calibrate the methods whereby T_s is culled from the spectral sequence. This calibration is valuable, since there are uncertainties in the spectral information about very hot (O) and very cool (N and S) stars, arising from the fact that they radiate largely in the ultra-violet and infra-red parts of the spectrum respectively, and these parts of the colour spectrum cannot get through the earth's atmosphere to our terrestrial telescopes.

The above observational facts about stars can be summarized by plotting a graph of their luminosity L (conventionally expressed as an absolute magnitude, that is on a logarithmic scale) against their surface temperature (conventionally expressed in terms of the spectral classification).

Such a colour-luminosity graph was first constructed by the astronomers Hertzsprung and Russell, and it will henceforth be referred to as the *HR* diagram. *Figure 2.6* is a schematic picture of the overall *HR* diagram; in addition *Figure 2.13* shows an actual *HR* diagram for a particular (and not typical) cluster of

Figure 2.5. The stellar intensity interferometer operated by Sydney University at Narrabri. The two movable reflecting mirrors collect light from a star and, by cleverly correlating the signals received by the separate mirrors, a measurement of the star's angular size can be effected.

stars, and *Figure 2.9* is again a schematic representation of the results for an assortment of groups of stars.

One immediately notices that only certain parts of the colour-luminosity or *HR* diagram are significantly populated.

The most conspicuous feature is a narrow band of stars, extending from the top left to the bottom right, called the *main sequence*. Some 80-90 per cent of all stars, our sun among them, are main sequence stars, with most of them about as luminous as the sun or fainter. The bottom left contains a region of the so-called "white dwarf" stars, which account for about 10 or 20 per cent of all stars. Another region is the more diffuse area to the top right, where dwell the "red giant" stars, which constitute only a few per cent of all stars. Other types of star, which are very much minority groups, are vaguely indicated in *Figure 2.6*; such are the pulsating "variable" stars and the spectacularly explosive "novae".

We can see from *Equation (2.15)* that regions of high luminosity and low temperature (that is, the top right) correspond to comparatively large stellar radii, whereas low luminosity and high temperature (that is, the bottom left) imply comparatively small radii. This explains the christening of "giants" and "dwarfs".

Before going on to expound our present understanding of the *HR* diagram and the life history of stars, there are two remaining

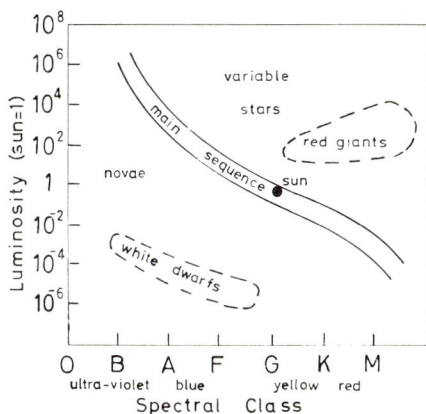

Figure 2.6. A schematic representation of the luminosity-colour or HR diagram for stars.

important stellar attributes to be mentioned, namely mass and chemical composition.

With the exception of our sun (whose mass can be inferred from the planetary orbits), the mass of a star is a difficult quantity to measure. However, there are in the heavens many pairs of stars which revolve around each other, and in a few favourable cases one can measure all the properties of the orbits of these "binary" stars. An analysis of these orbits then yields the mass of both stars. Although only a tiny fraction of all stars can have their mass measured in this way, we do have enough data points to see what masses are typical in the various parts of the *HR* diagram.

The chemical composition of a star's surface can be deduced from the stellar spectra, as discussed above. The conclusion is that, in general, stellar "atmospheres" consist mainly of hydrogen with a fair admixture of helium, and a sprinkling of heavier elements. For example, the atmospheres of our sun and like stars contain more than 80 per cent hydrogen, and less than 4 per cent of elements other than hydrogen and helium. (It is interesting that helium was identified by spectroscopy as a constituent of the sun before it had ever been found on earth.)

However, if we want to know the *internal* chemical composition of stars, we must of necessity forego direct observation and appeal to theoretical studies. As we shall see, it now seems well established that stars begin life consisting almost entirely of hydrogen, which they burn to form helium, so that the interior consists of these two elements in a proportion which depends on the star's age.

The history of the accumulation of the above body of empirical knowledge, partly summarized in the *HR* diagram, goes hand in hand with the development of human thought. Our current explanation and understanding of these facts will now be described.

The Life History of Stars

Stellar structure

The great majority of stars remain quite stable, neither expanding nor contracting as they radiate energy steadily into space. For this to happen, they must adjust themselves so that the rate at which energy is supplied from the burning of nuclear fuel is just equal to the rate at which energy is radiated away. Quite generally such a stable situation is referred to as an "equilibrium" one.

The bookkeeping in this equilibrium balance is fairly straight-forward. Stars are gaseous throughout, and so, in the familiar manner of all gases, they maintain equilibrium by balancing out the various pressure forces at every point. The two countervailing forces at work in stars are the inward *gravitational pressure*, which seeks to collapse the star altogether, and the outward *gas pressure*, due to the kinetic motions of the individual particles, which strives to expand the star. As we know from the ordinary gas laws, the gas temperature is proportional to its pressure, because the temperature is a measure of that random kinetic motion of the particles which indeed produces the gas pressure. This compromise between the inward gravitational pressure and the outward gas pressure is made at every point within the star. As we move inwards, the gravitational pressure increases as more and more matter overlies us, and consequently the magnitude of the gas pressure, and thus the temperature, also rises. This explains the fact, noticed earlier, that the maximum temperature is achieved at the centre of the star.

One further point remains. Given that the energy is produced by nuclear fusion reactions at the hottest place in the star, namely its centre, how is this energy transported to the stellar surface to be radiated away? This question is critical to the star's system of checks and balances, because it is this energy transport rate which will determine how fast the energy is lost from the star. The answer to the question is that the energy is carried from the centre to the surface mainly by radiation (in the same way as heat travels away from an electric heater, or from the sun's surface to us), although just at the surface itself, and generally also in a central inner core, the predominant transport process is convection (that is, large-scale motion of matter, as in the heat flow from the tropical oceans to Western Europe by the Gulf Stream). The physics of both modes of energy transport is fairly well understood.

We now have all the ingredients for an understanding of a star's anatomy and metabolism. We have seen in detail how energy is produced by nuclear reactions, at a rate which depends on the central temperature and density, and we have asserted that the rate at which this energy is transported to the surface, there to be lost into space, is also calculable. Throughout, the star is counterpoised between the compressive gravitational pressure and the expansive gas pressure. An explicit statement about the chemical composition of the star (and consequently about which particular nuclear reactions are taking place, and about the details of the energy transport) defines an explicit "stellar model".

Once a specific stellar model is adopted, then from a knowledge of the total mass of the star we can calculate all its other physical properties such as its size, luminosity, surface and internal temperature, and its pressure and density throughout.

Some of the details of these relations between a star's mass and its other properties merit discussion.

Consider first the total rate of energy production, as measured by the luminosity L. For more massive stars, the gravitational pressures will be larger; in turn this means larger balancing gas pressures and thus higher temperatures. Higher central stellar temperatures mean that the colliding nuclei will be more energetic, and will therefore find it easier to tunnel through the coulomb barriers which hinder fusion. Thus the upshot of a larger mass is an enhanced luminosity. Moreover, since a small increase in the energy of colliding nuclei makes it substantially easier for reactions to happen, a small increase in stellar mass M produces a comparatively large increase in luminosity. More explicitly,

$$L = aM^{4 \text{ or } 5} \qquad (2.16)$$

where a is some constant, and the exponent varies between four and five depending on the details of the stellar model. On the other hand the surface temperature T_s and the stellar radius R tend to be simply proportional to M. Previously we saw how observations showed that there are vast variations in L for stars, but comparatively small variations in T_s; this is explained by the above remarks particularly [*Equation (2.16)*] along with the comment that stellar masses do not vary over a very large range.

Returning to *Equation (2.16)*, we observe that it says that the rate of energy production per unit mass (L/M) is proportional to M^3 or M^4. But the total nuclear energy reservoir per unit mass is *fixed*, depending only on the physics of nuclei. Now if the nuclear capital per unit mass is fixed, and the rate at which this capital is spent is greater for more massive stars, it follows that massive stars must live shorter lives before going bankrupt. Moreover this effect, which after all goes as the third or fourth power of the total mass, is no small one. Thus massive stars are the heaven's prodigals — bigger, hotter, brighter, they burn out faster and go to an early grave like characters in a celestial cautionary tale.

The detailed discussion just given is explicit to the equilibrium of main sequence stars (which, remember, comprise some 80 to 90 per cent of all stars). The details differ for other equilibrium stars such as white dwarfs (which obey quantum-mechanical gas laws instead of ordinary classical ones), and red giants (which have non-uniform chemical composition). The whole discussion is largely irrelevant to the spectacular minority of non-equilibrium pulsating or exploding stars.

One final word need be said, before turning to explain the *HR* diagram. Even the equilibrium stars are, in the long run, undergoing slow changes as they burn their hydrogen and lose energy. What form does this change take? As energy is lost, the star contracts further under gravitational forces, and consequently gets hotter. This is actually rather odd; in a purely gravitational system (balanced by gravitational and gas pressures) a loss in energy means an increase in temperature! This fact is dignified with the title of the "Virial Theorem"; it is in arresting contrast with more everyday systems where a loss in energy means a decrease in temperature.

All this means that as stars age, losing energy as time goes on, their central temperatures and densities increase, thus enabling them to tap fresh sources of nuclear energy which are only available at exaggerated temperatures, until eventually all nuclear fuel is exhausted.

Stellar evolution

The embryonic star is born when, for motives not yet clear to us, a cloud of gas detaches itself from its surrounding area of gaseous nebulosity, and begins to contract. As the protostar contracts under gravity, it heats up (in accordance with the dictates of the Virial Theorem), until the star's nuclear fire is lit by the switching on of the *p-p* reaction. For more massive stars the *p-p* reaction is too slow to halt the gravitational contraction, which continues until the central temperature is high enough to ignite the faster *CN* cycle.

The infant star enters the *HR* diagram on the right-hand side, moving upwards as its temperature and luminosity increase, until eventually the nuclear reactions in the centre produce enough energy to balance the gravitational pressure. At this point the star arrives on the main sequence at a place which is determined by its mass. Heavier, large, bright, blue stars lie to the top left of the main sequence, and the more common lighter, less luminous, yellow or red stars such as our sun lie towards the right. This part of the star's "evolutionary" track is illustrated in *Figure 2.7*.

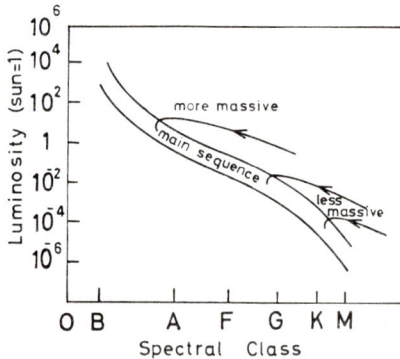

Figure 2.7. The evolutionary tracks of stars of various masses in the initial stages of their lives.

This initial contraction onto the main sequence occupies only a tiny fraction of a star's total life, most of which is spent *on* the main sequence. This is why most stars are observed to be on the main sequence, and essentially none presently evolving onto it. The absence of stars from any part of the *HR* diagram does not necessarily mean that stars do not traverse that part, but may only mean that they traverse it quickly, and therefore very few are observed there at any one time. Similarly, only a tiny fraction of the human population is to be found at an one time as newly born babies in hospitals, yet all of us have been there once.

Once established on the main sequence, a star settles down to burn hydrogen, obeying the Virial Theorem by growing a little hotter and more luminous, and thus moving a little up the main sequence, as it radiates energy into space. All stars start off composed very largely of hydrogen, but slight differences in chemical composition do occur even in very young stars; these slight variations are the explanation for the main sequence having a finite width, being a band rather than just a line.

The actual length of time spent on the main sequence will depend on the rate at which a star consumes its hydrogen fuel. As noted earlier, this rate increases markedly with the initial mass of the star. Thus our sun, an average sort of character, has been sitting happily on the main sequence for about 5 billion years, and is scheduled to go on radiating energy from the *p-p* reaction for another 5 billion years. Less massive stars will live even longer on the main sequence. Conversely more massive stars will burn their hydrogen fuel more rapidly, to the extent that the brightest blue giants will consume their sustenance in a fleeting few million years.

Why do stars quit the main sequence, and where do they go? This question can be answered, particularly for stars of moderate mass, although we do not have quite the same degree of confidence in our calculations here as we had for the main sequence phase.

The time has come for a star to move off the main sequence when it has converted some 10 to 15 per cent of its hydrogen to helium. As helium accumulates at the core of the star, it at first produces slight changes in the equations which govern the overall equilibrium, but eventually (at this critical value of 10 to 15 per cent) the strain of juggling things to adjust to the helium-rich core becomes too much for the star. The resulting instability manifests itself in a contracting (and therefore hotter) core and an expanding (and therefore cooler) outer region. As the star's outer region cools, the surface temperature falls, and the star becomes redder. However, the dramatic increase in radius more than offsets the decrease in T_s so that [*Equation (2.15)*] the luminosity increases, being fed from hydrogen burning in the shell surrounding the helium core, and from some helium burning in the core. Thus the star moves off the main sequence upwards and to the right. At some stage in this puffing-up process, the star may establish a temporary equilibrium as a "red giant". From the comparative scarcity of the red giant population on the *HR* diagram, we can confirm the calculations which suggest that this stage is a short one in a star's life.

It is estimated that our sun will evolve into a red giant in about 5 billion years, at which time the earth will suffer the doom predicted in the Koran, when it is grilled by the sun at a distance of "a spear and a half". But that is the least of our worries.

Evolution to a red giant does not solve the star's difficulties in trying to balance its energy production and loss budget. The helium core continues to grow, and to become hotter by gravitational contraction. Eventually, at some $100,000,000°C$, helium burning sets in in earnest. The resulting "helium flash" produces very rapid burning and very high temperatures which can ignite the burning of even heavier nuclei such as carbon and oxygen (themselves produced by helium burning) and subsequently even magnesium and sulphur. During the helium flash the time scale for significant changes shrinks to a few thousand years, and the star does all sorts of odd things. For one thing its luminosity falls slightly and its radius greatly, while its surface temperature rises and the evolutionary track is consequently across towards the left and somewhat downwards on the *HR* diagram. The star's unstable character is manifested by variations in its energy output, making it a pulsating or variable star. Moreover, in trying to solve its problems, the star commonly sloughs off some of its outer skin, steadily or spasmodically losing mass.

It is this mass-loss process which brings uncertainty to present-day calculations of the later stages of stellar evolution. The complexities of the internal instabilities can be handled with modern computers, but the mass-loss, which is ill-understood and probably depends on details such as the star's rotation and magnetic field, introduces ineradicable uncertainties into contemporary calculations.

Nonetheless it is probably correct that stars which lose little of their initial mass as they evolve end up in a cataclysmic explosion

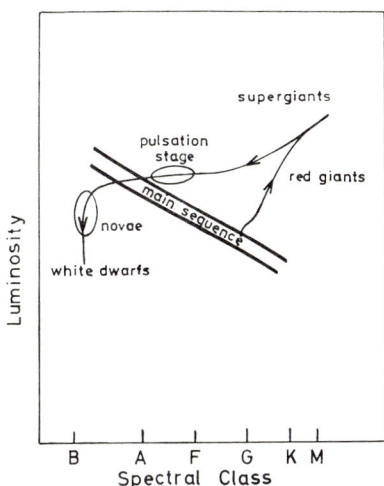

Figure 2.8. The evolutionary tracks of stars of moderate mass after they have left the main sequence.

as supernovae. Stars which have lost much of their mass relatively quietly and steadily can eventually end their days placidly as white dwarfs.* These objects have gravitationally contracted down to radii of about 1 per cent that of the sun (although their masses are approximately equal to the sun's), and have burned all their nuclear fuel, so that they are cooling down towards invisibility, their present radiance being from the fading embers of the extinguished fires.

This later evolution of stars is illustrated in *Figure 2.8*.

---

* Other endings to the stellar story can be, and have been, imagined. If a heavy enough star survived the burning of all its nuclear fuel, it is mathematically possible for it to collapse gravitationally in upon itself indefinitely, contracting to an unobservable, singular point. An alternative ending is a star so dense that it has throughout its entire volume the incredible density of the atomic nucleus; all the electrons would be bound to protons to give nothing but neutrons. Such a "neutron star" would contain the mass of our sun in a sphere with a radius of about one mile. Neither of these theoretical and mathematical entities has ever been observed, although as I write (April, 1968) there is much interest in a most strange radio object, which some have conjectured to be indeed a neutron star.

All in all, things add up to a fairly clear understanding of the HR diagram depicted in *Figure 2.6*. Stars are born in clouds of gas, come to stable maturity on the main sequence, and eventually puff-up in neurotic tantrums which end in either a spectacular explosion, or else in the senility of a white dwarf.

Observational confirmation

In the manner just described, it is possible to calculate the explicit evolutionary path, at least up to the red giant stage, along which a star of any particular initial mass will move as time goes by. Alternatively this can be turned around, and *we can calculate where on the* HR *diagram we could find a group of stars of different masses which all have the same age.*

To say this again in a different way, we notice that to each point in the history of a star of given mass there corresponds a unique point on the *HR* diagram. The evolutionary track of a particular star can be obtained by adding up all these points over different times; whereas the area occupied on the *HR* diagram, that is the colour-luminosity curve, for a group of stars all of the same age comes from adding up all the points for different masses at that particular time.

Now there are in the sky a large number of groups or clusters of stars, and it is plausible to assume that all the members of a particular cluster are more or less the same age. Thus each such cluster should give a set of points on the *HR* diagram which corresponds to one or other of these theoretical colour-luminosity curves. Such an agreement between theory and observation will give us not only confidence in the theory, but also the actual age of the cluster considered.

Such a series of colour-luminosity curves is shown in *Figure 2.9*, which follows the original work of Alan Sandage. The titles "Pleiades", "NGC 188" and so on refer to the names somewhat vagariously applied to certain specific clusters, and the curves have been calculated for the ages which have been found to fit the observed colour-luminosity curves for these clusters. The agreement between theory and observation has been found to be quite good.

Moreover, once an age has been so assigned to a cluster from the theory, we can check to make sure that the physical conditions in the cluster's environment are appropriate. Thus the very young groups, such as those in Perseus, should be associated with clouds of hot gas of the sort from which stars form; adolescent groups such as the Pleiades are also embedded in dust; and the older groups, such as the globular cluster M 13, seem to be quite free of surrounding nebulosity.

Cosmological Inferences

All the foregoing discussion has been aimed at showing how observational data on stars, as summarized in the *HR* diagram, can be understood on the basis of theoretical stellar models, in which the enormous energy supply derives from nuclear fusion. The ostensible reason for all this was to end up getting cosmological information from stars. So we now come, in what briefly follows, to the tail which wags the dog of Chapter 2.

(1) **The oldest stars**

One question which emerges from the preceding discussion and which is clearly of relevance to cosmology is, *how old are the oldest stars?*

To answer this question, we turn to the comparison between theoretical and observed colour-luminosity curves for the various

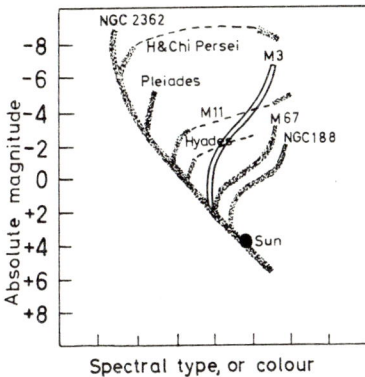

Figure 2.9. Colour-luminosity curves for several clusters of stars. The dashed lines represent parts of the HR diagram through which stars would be expected to evolve too quickly for us to be likely to observe any there.

Figure 2.10. The "lagoon" nebula in the constellation of Sagittarius. Such environments of gaseous nebulosity are where we would expect to find stars forming, and indeed many very luminous young stars are to be found here. (Mount Wilson and Palomar Observatories.)

Figure 2.11. The Pleiades and surrounding nebulosity. These are a group of young, hot, blue stars in the constellation of Taurus. (Mount Wilson and Palomar Observatories.)

Figure 2.12. The globular cluster M13 in the constellation of Hercules (the famous Hercules Cluster). Some 50,000 odd stars are present. (Mount Wilson and Palomar Observatories.)

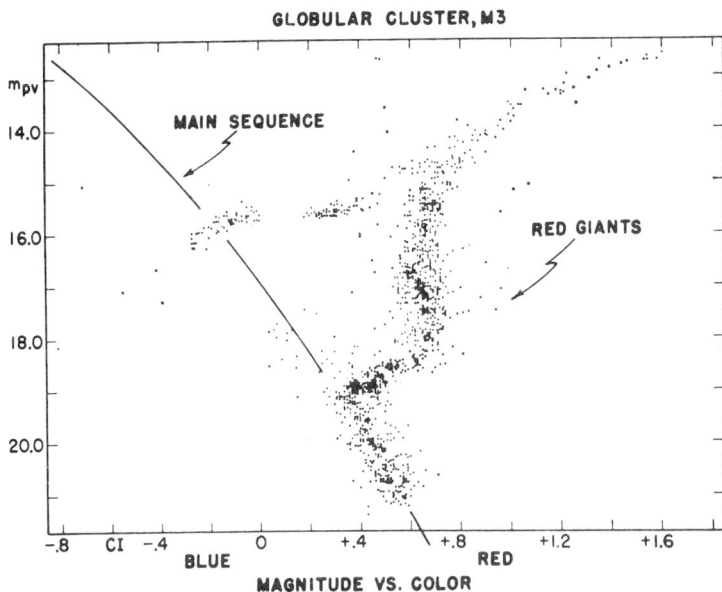

Figure 2.13. An HR diagram showing the observed colour-luminosity curve for the globular cluster M3, as measured by Arp, Baum and Sandage.

groups and clusters of stars, and look for the greatest age thus
deduced. The globular clusters such as M 13 (*Figure 2.12*) or
M 3 (whose observed colour-luminosity curve is shown in *Figure
2.13*) are fairly typical of the oldest clusters. In fact the greatest
age obtained by the method so far is for the group NGC 188
(whose colour-luminosity curve is included in *Figure 2.9*); the age
is 10,000,000,000 years. M 3 and several similar clusters are
almost this old.

Alternatively we could look at the ages which the theory of
stellar evolution assigns to the oldest individual white dwarfs.
But here there are difficulties, not only in the theory of the later
(post red giant) stages of evolution, but also in the experimental
determination of masses. (Notice that the more usual age dating
method, via colour-luminosity curves of clusters, cunningly side-
steps the need to determine masses.) Nevertheless we may mention
that rough estimates suggest the oldest white dwarfs are also about
10 billion years old.

What is the cosmological significance of this age?

Turning back to the discussion of Big Bang or general relativistic
cosmologies, we recall that they all require a universe which is
at most as old as (and more generally less old than) the inverse
Hubble Constant time, currently estimated at 13 billion years
[*Equation (1.3)*]. Now we note that the universe is at least as
old as the oldest stars in it (!) so that this most antique stellar
age sets a lower limit to the age of the universe. Had this lower
limit been much more than 10 billion years, so that it exceeded
the oldest possible age of the Big Bang universes, then all such
cosmologies would have been eliminated.

As a matter of fact, until Baade (and later Sandage) revised
its value in 1953, the inverse Hubble Constant time given by
Hubble himself put an upper limit to 2 to 3 billion years on the
age of a Big Bang Universe. The impossibility of reconciling
such a small age with facts such as the lifetimes of the oldest stars
provided stimulus for the development of the Steady State theory
in the early 1950's; however, the current value of H_o, *Equation
(1.3)*, presents no such problem. Notice that in the Steady State
theory, with no beginning and no end, the ages of the oldest stars
cannot possibly pose any problem.

So much for the absence of a negative inference. Is there any positive deduction we can draw from the 10 billion year old ages of the oldest stars? Can we use this time to choose among the variants of the Big Bang cosmologies, to decide whether the universe is ever expanding or oscillating (that is, to assign q_o)?

The answer is that we can use the observations to reject those oscillatory variants of the general relativistic cosmologies which have large values of the deceleration parameter q_o; values of q_o in excess of about 2 lead to universes which were born significantly less than the required 10 billion years ago. Beyond that we cannot go, partly because the oldest stellar age of 10 billion years is not all that accurate, and partly because there is after all nothing to stop the universe being *older* than the oldest stars in our galaxy.

Nonetheless, at present the tentative viewpoint is that the first stars probably began to form shortly after our galaxy was born, and that in turn the birth of all galaxies followed shortly after the primeval Big Bang, so that in fact we can infer from stellar ages that the universe is roughly around 10 billion years old. But all this is quite tentative.

(2) **Hydrogen to helium ratio**

One of the tasks of any cosmology is to account for the presently observed relative abundances of the chemical elements in the universe. In this respect the Big Bang theories enjoy a considerable advantage over the Steady State theory, in that the fantastically hot primeval fireball, and its close aftermath, provide an environment in which a great deal of element "cooking", or synthesis from simpler material, can take place. On the other hand, in the Steady State theory the chemical elements must be continuously built up from hydrogen (since, by definition, the universe is maintained in its Steady State by the continuous creation of hydrogen), and galaxies such as our own must of necessity serve as the workshops for this synthesis. As we shall now see, this poses difficulties for the Steady State hypothesis.

If the chemical elements are to be synthesized from hydrogen in environments not substantially different from those which presently exist, then, in particular, helium must be produced from hydrogen by thermonuclear processes inside stars. From the theory developed earlier in this chapter, we can estimate that the helium to hydrogen ratio to be expected from current stellar activity is roughly 0·01. This then is the expected helium to hydrogen ratio in the Steady State theory.

In contrast, general relativistic cosmologies cook most of their helium in the hot, dense world immediately after the Big Bang, which leads to an expected helium to hydrogen ratio of about 0·14 for them.

Experimental determinations of the helium to hydrogen ratio in stars and gaseous clouds within our galaxy yield ratios from 0·08 to 0·18. The ratio seems to be just as high in old stars as it is in young ones. In particular, two quite different methods of measuring the ratio for our sun lead to concordant values of 0·09.

These experimental results are roughly in harmony with the figure of 0·14 for the Big Bang theories, although the smaller observed ratios (such as the sun's) are a bit bothersome. On the other hand the observed values are too high to be reconciled with the Steady State figure of 0·01 derived from stellar evolution studies.

Staunch advocates of the Steady State viewpoint argue that perhaps stellar activity was much greater in the earlier stages of the evolution of our galaxy, and consequently that the ratio of 0·01 arrived at on the basis of *current* activity would be misleadingly small. However, the failure to observe any significant class of objects with a low (0·01) helium content tends to invalidate this evasion.

It is necessary to add a cautionary postscript. The figures of 0·14 and 0·01 presented above, as well as the experimentally observed ratios, all depend on complicated theoretical calculations. The present state of the art is so beset with uncertainties that none of these calculations is above reproach.

*   *   *   *

I would like to end the Chapter by harking back to the solar neutrino experiment referred to earlier.

This experiment, results from which are imminent, constitutes the first *direct* check on what goes on inside a star.   If it should give a result in disagreement with the theoretical predictions, then the foundation will have been wrenched from beneath the towering, elaborate edifice of theory described in this Chapter.

CHAPTER 3

Looking Backward in Time

The simplest way to answer cosmological questions as to the nature of our universe would be to travel back in time to inspect the universe as it was in the past. If the past universe turned out to be the same as ours, we would accept the Steady State theory; if it proved to be denser we would choose one of the Big Bang cosmologies, and the specific choice would be dictated by the rate at which the density increased as we travelled back, that is, by the rate at which the current expansion is decelerating. If the density increased very quickly as we went back in time we would infer an oscillating Big Bang universe, whereas a slower increase would imply an ever-expanding variant.

Recalling from Chapter 1 how the present rate of change of the cosmological expansion is measured by the deceleration parameter q_o, we see that our imaginary Time Traveller can summarize his discoveries in this one number, and we shall be content if, on his return, he lives long enough to gasp out the number. His value of q_o will specify our actual universe from among the various alternative cosmologies, as described in Chapter 1.

Lacking a Time Machine, can we look back into the past? In fact we can.

The ability to see backward in time stems from the fact that light, and all other electromagnetic radiation, travels at a finite speed. We do not observe a galaxy as it is now, but as it was at the moment when the light reaching us now started on its journey. In the case of very distant galaxies, the light began travelling several billion years ago, and so we have direct evidence bearing on the state of things several billion years ago.

It remains, however, to extract cosmological information from these light signals from earlier epochs.

Let us suppose for the moment that space is bestrewn with *Standard Hypothetical Objects* (which we shall promptly christen SHOs), all of which are identical, and whose properties do not change with time. Now as we consider SHOs at greater and ever-greater distances, various effects will be observed. At first their *apparent* brightness will decrease inversely as the square on their distance from us (*Equation (2.14)*), and the red-shift of their spectra will increase proportionately to this distance (Hubble's Law). However, at greater distances departures from these simple relations will become manifest, due partly to the strange non-Euclidean geometry of the spaces in which the general relativistic Big Bang cosmologies are embedded, and partly to the fact that the light from far distant SHOs will have come from the past, where the world is somewhat different. Of course, the apparent brightness will continue to decrease, and the red-shift to increase, as the SHOs are more distant from us, but the effects will not be the simple proportionalities applicable to close distances.

A specific relation between the apparent brightnesses and red-shifts of any set of intrinsically identical objects, such as our SHOs, can be calculated from any one particular cosmology. The different cosmological theories yield different relations, as is illustrated in *Figure 3.1*.

In *Figure 3.1* the curve labelled D displays the Steady State theory's relation between apparent magnitude (logarithmically related to apparent brightness as in *Equation (2.13)*), and the logarithm of the red-shift, for a set of intrinsically identical objects at different distances. The curve labelled A is the similar relation for a typical oscillating general relativistic universe, with $q_o = + 1$; B is that for a Euclidean Big Bang cosmology, $q_o = \frac{1}{2}$; and C is that for one of the ever-expanding Big Bang universes, with $q_o = 0$. It will immediately be noticed that the differences among the various theories do not become appreciable until one goes to large enough red-shifts (that is, large enough distances from us in space, and consequently in time).

Also, bearing in mind that our SHOs are intrinsically distributed uniformly densely throughout space, we could count the *apparent* number of SHOs at increasing distance from us, or alternatively with increasing red-shift. Such a number versus red-shift relation would give another set of curves such as those in *Figure 3.1*, with the cosmologies manifesting characteristic differences at large enough red-shifts. Conversely, we could similarly plot number versus apparent magnitude for our convenient SHOs.

Thus if our SHOs existed, we could merely sit quietly at the eyepiece of a telescope and yet gather all the information available to the Time Traveller.

All this is very fine, except that Nature has not endowed us with such a set of Standard Hypothetical Objects. Our SHO creatures, with the crucial property of being *intrinsically* identical, exist only in our imagination.

Our next task is to search the cosmos for sets of objects which are approximately intrinsically identical. Alternatively one can hope to use statistics to average out intrinsic differences, the point here being that although the height of any one person in Sydney cannot be predicted with confidence, the average height of any random sample of (say) a thousand people can be so predicted.

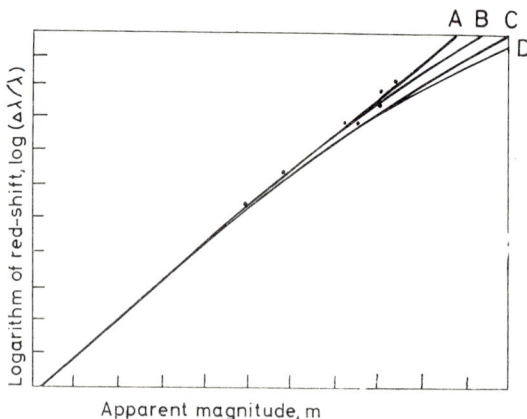

Figure 3.1. Apparent magnitude versus red-shift curves for various cosmologies. The curve labelled A is for a typical oscillating general relativistic or Big Bang universe ($q_o = + 1$), B is for a Euclidean universe ($q_o = + \frac{1}{2}$), C is for an ever expanding Big Bang universe ($q_o = 0$), and D is for the Steady State theory ($q_o = - 1$). The points shown are derived from Sandage's observations on the brightest members of clusters of galaxies.

In what follows we shall see how first galaxies, then radio objects, and most recently "quasars", have been used as approximate SHOs. Each such attempt has its faults: galaxies because the farthest discernible are still too close; radio objects because they are not fully understood nor accurately measured; and quasars because their very nature is most controversial.

Before going on to discuss these assorted approximate SHOs one by one, we return to make some more quantitative remarks about the relations between apparent magnitude (m), red-shift $(z = \Delta\lambda/\lambda)$, and the cosmological parameters H_o and q_o. These remarks amplify the more qualitative discussion above.

First we recall that the general relativistic cosmologies are discriminated by the choice of q_o: $q_o > \frac{1}{2}$ gives an oscillating universe with positively curved geometry; $\frac{1}{2} > q_o \geqslant 0$ an ever-expanding universe with negatively curved geometry; and $q_o = \frac{1}{2}$ is the special (expanding) Euclidean case (see *Figure 1.9*). In addition, the Steady State theory corresponds to $q_o = -1$, a fact which follows from the definition of q_o (on page 162 following *Equation (1.3)*), along with the remark that $dH/dt = 0$ in the Steady State theory, by definition.

Next it may be seen that for not too far distant objects, the apparent magnitude (m) versus red-shift (z) relation takes the form

$$m = 5 \log z + (1 \cdot 086) (1 - q_o) z + A \qquad (3.1)$$

Here A is a constant which depends on H_o, on the intrinsic luminosity of the object, and on some other physical constants. Terms of order z^2 have been neglected, so that *(3.1)* is only an approximation, and it stops being valid when z becomes as large as, or larger than, unity; under these circumstances a more complicated equation can be derived. Notice that for very small red-shifts, when the term containing $(1 - q_o) z$ can also be neglected, we get the very simple relationship, $m = 5 \log z + A$, which follows from the close distance laws that apparent brightness decreases as distance squared, and red-shift increases as distance.

Thus departures from the close-distance laws give a direct measure of q_o, via *Equation (3.1)*, and thence a unique cosmology can in principle be specified.

Galaxies as Cosmological Indicators

A point which was glossed over in Chapter 1 is that galaxies tend to occur not singly but rather in clusters, which contain typically a score or so member galaxies. It is these clusters of galaxies, rather than galaxies themselves, which are distributed in a homogeneous and isotropic manner throughout space. *Figure 1.3* illustrates one such cluster of galaxies near us. Moreover, galaxies exhibit a wide variety of types, differing in shape and brightness. Indeed so many quite different types of galaxies exist as to make it impossible to use any randomly chosen set of galaxies as even roughly approximate SHOs (whose crucial property, remember, is that they are intrinsically identical).

However, it so happens that the *brightest members of clusters of galaxies* tend to be very uniform creatures.

The typical group of galaxies comprises twenty or so members dominated by a bright elliptical galaxy, which in the galactic classification scheme bears the title EO (E for elliptical, O for brightest among the class of E's). As far as we can tell from those which are near enough for us to be able to make a detailed study, all these EO galaxies are very nearly identical in their intrinsic luminosity and other characteristics.

Consequently one may hope to use brightest members of clusters of galaxies as cosmological indicators, or approximate SHOs, to resolve the nature of our universe in the way described above. Such a programme has been under way for well over a decade now, led mainly by Alan Sandage of the Mt. Wilson and Palomar Observatories. In this project, the observational data comes from man's farthest seeing eye, the great 200-inch telescope at Palomar in the mountains of lower Southern California. A drawing showing the various components of this giant instrument is shown in *Figure 3.2*.

The *apparent magnitudes* of the large elliptical galaxies which are the brightest members of clusters are obtained more or less straightforwardly (although there are a few tricky technical details which we shall not dwell upon here).

The *red-shift* in the spectrum of light arriving from these galaxies can be found directly by comparison with standard laboratory

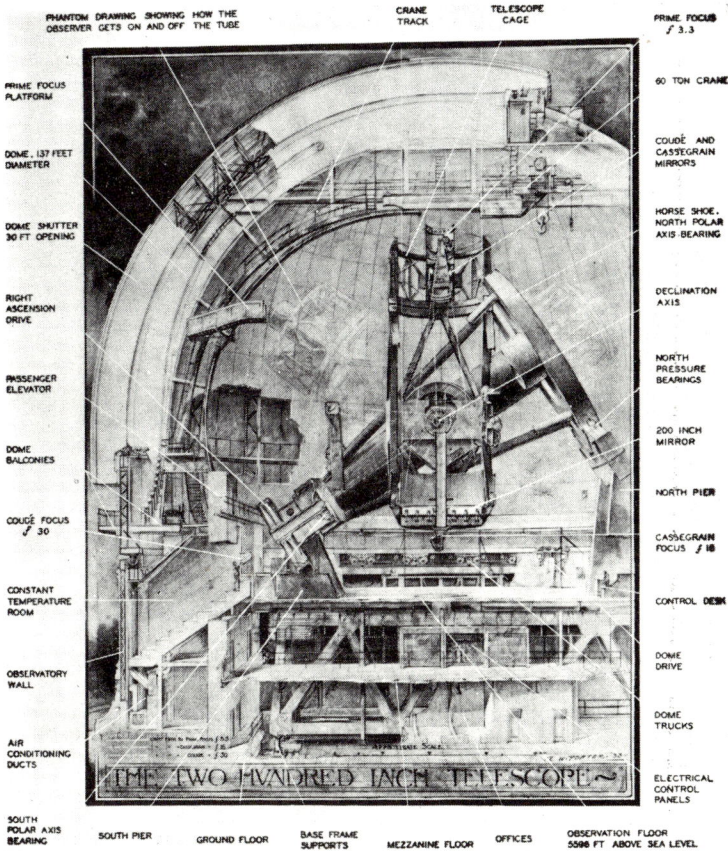

Figure 3.2. A detailed drawing of the world's largest telescope, the great "200-inch" at Palomar. (Mount Wilson and Palomar Observatories.)

spectra, once some of the characteristic spectral lines have been identified.

All this is illustrated quite explicitly in *Figure 3.3* which shows the apparent brightnesses and red-shifted spectra of a series of these brightest members of clusters of galaxies at ever-greater distances from us.

RELATION BETWEEN RED-SHIFT AND DISTANCE FOR EXTRAGALACTIC NEBULAE

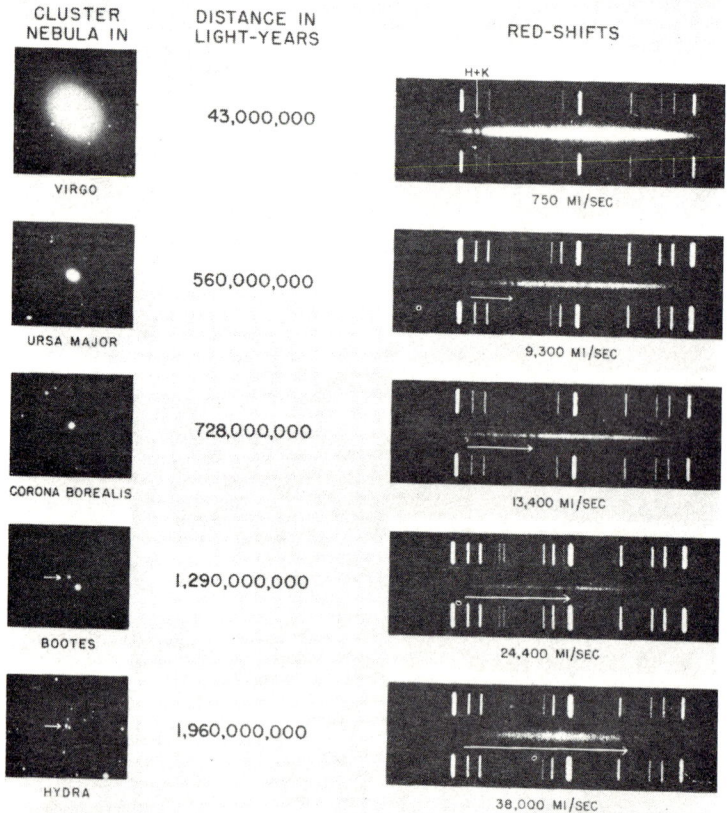

CLUSTER NEBULA IN	DISTANCE IN LIGHT-YEARS	RED-SHIFTS

VIRGO · 43,000,000 · H+K · 750 MI/SEC

URSA MAJOR · 560,000,000 · 9,300 MI/SEC

CORONA BOREALIS · 728,000,000 · 13,400 MI/SEC

BOOTES · 1,290,000,000 · 24,400 MI/SEC

HYDRA · 1,960,000,000 · 38,000 MI/SEC

Red-shifts are expressed as velocities, c dλ/λ.
Arrows indicate shift for calcium lines H and K.
One light-year equals about 6 trillion miles,
or 6×10^{12} miles

Figure 3.3. The actual spectrograms of five galaxies showing the extent of the shift of the H and K lines of calcium to longer wavelengths (against the background of a laboratory comparison spectrum). The apparent size of the galaxies decreases with increasing distance, and they appear ever fainter. (Mount Wilson and Palomar Observatories.)

Once these observations have been made, the results are plotted on the apparent magnitude versus red-shift graph (*Figure 3.1* and *Equation (3.1)*), and then one looks to see which cosmology gives the curve that fits these data points best. A set of such observational measurements for brightest members of clusters of galaxies is shown in *Figure 3.1*.

The upshot of all this work is (1967) the conclusion that $q_0 = +1$. However, one must immediately add that this number is not to be taken as gospel, for several reasons.

For one thing, the farthest galaxies which it is possible to observe are at distances corresponding to a red-shift in their spectra of about 50 per cent, which is to say $z = \Delta\lambda/\lambda = \frac{1}{2}$. Now while this is admittedly a vast distance from us, it is unfortunately not far enough for the characteristic differences in the various cosmologies to show up significantly in the apparent magnitude versus red-shift relation. As we may see from *Equation (3.1)*, the difference between $q_0 = +1$ and $q_0 = -1$, which spans an enormous range of different cosmologies, corresponds to a difference of only one unit in apparent magnitude at $z = \frac{1}{2}$. This is not a large difference. Thus to assign a value to q_0 with a high degree of confidence, we need a large number of data points, so that small experimental errors and fluctuations may average out. It has been estimated that to fix q_0, and thence our cosmology, with 95 per cent confidence that we are right, would require something like 100 times as many data points as currently exist.

Another worry is that, if we live in a Big Bang universe, it is probable that these EO galaxies were somewhat different beasts in the remote past. Such evolutionary effects would mean that the brightest members of clusters of galaxies at large red-shifts (and in this sense $z = \frac{1}{2}$ is a large red-shift) are *intrinsically* different from nearby ones, which invalidates the assumptions from which *Equation (3.1)* and the curves in *Figure 3.1* were derived.

Furthermore, early this year John Bahcall of the California Institute of Technology and I pointed out that the theoretical magnitude versus red-shift relations should include a correction for the scattering of light by electrons in inter-galactic space. This space is unimaginably tenuous, but nonetheless not wholly empty, and over the vast distances travelled by light reaching us from distant sources this scattering effect can be significant. For the red-shifts involved in Sandage's work, this effect amounts to a 10 to 15 per cent correction to the value of q_o. At larger red-shifts ($z > 1$), it is a more pronounced effect, and moreover it tends to bring the magnitude versus red-shift curves for the different cosmologies closer together, counteracting their tendency to diverge at larger red-shifts. This effect is thus a most inconvenient one for Nature to have provided.

Despite these various gloomy cautionary notes, it remains true that the work of Sandage and his collaborators on the brightest members of clusters of galaxies probably does lead to the conclusion that q_o is roughly somewhere between $+ \frac{1}{2}$ and $+ 2$, and consequently that we live in an oscillating Big Bang or general relativistic universe. Taking $q_o = + 1$ literally, the explosion took place some 10 billion years ago, and the entire cycle of expansion and contraction back to the primeval fireball will take 80 billion years (a number which follows from substituting $q_o = + 1$ in *Equation (1.6)*).

Radio Objects as Cosmological Indicators

Our narrative so far has paid almost exclusive attention to experimental evidence about the universe which has been gathered by observing the radiation emitted in the visible part of the electromagnetic spectrum, that is, by observing visible light. In fact the earth's atmosphere is impenetrable to most kinds of electromagnetic radiation; the atmosphere has only two "windows", one of which allows that part of the spectrum corresponding to visible light to reach the surface of the earth, and the other of which allows certain radio waves to reach us. Observations of the radiation arriving from outer space through the second of these "windows" began in the 1930's, and have generated the rapidly growing science of radio astronomy.

Probing the past density of the universe by counting extragalactic radio objects is in some ways a more clear-cut task than observing optically visible galaxies. In particular, the use of such radio sources as approximate SHOs makes it easier to accumulate a large enough number of objects for a statistical reduction of the data to be meaningful.

Since it is in general not possible to determine the red-shift of a radio object (because no analogue of the optical wavelength spectrum is available), the number-of-objects versus apparent magnitude graph must perforce be used. This provides a less sensitive discrimination between different cosmologies than does the apparent magnitude versus red-shift graph of *Figure 3.1*. Nevertheless the results of Martin Ryle and his colleagues at Cambridge, first published in 1961, seem to indicate that the universe was more dense in the past than it is now; indeed Ryle's results seem difficult to explain on the basis of *any* of the cosmological theories so far described.

It may well be that accurate cosmological information from radio source number versus magnitude diagrams will have to wait until the newer generation of radio telescopes begins churning out such information. Preliminary results from the very accurate one mile by one mile Mills Cross radio telescope (*Figure 3.4*) already

Figure 3.4. A view of part of the one mile by one mile Mills Cross radio telescope at Captain's Flat outside Canberra. This very accurate instrument, operated by the Physics Department of the University of Sydney, has only recently come into full operation.

conflict with Ryle's, and are consistent, at this early stage, with virtually all Big Bang cosmologies. The compilation of larger and statistically more significant samples of such accurate counts seems to be needed before any more is said on this topic.

There remains the further objection that our present under-standing of radio sources is simply not adequate to allow us to make a confident grouping of sources into various classes, such that all members of a given class be deemed intrinsically similar. This is a cogent objection, and a dramatic illustration of this lack of certainty as to the nature of radio sources is provided by the re-cently discovered "quasi-stellar" objects or quasars, to which we shall now turn.

Quasars as Cosmological Indicators

History

Until a few years ago, the objects now called quasars were thought to be quite ordinary faint stars within our own galaxy. Closer examination has, however, shown that these objects have enormous red-shifts; they are receding from us with the largest speeds of recession so far measured. Indeed some of these quasars have red-shifts so large as to put them well out towards the edge of the universe.

One of the difficulties which besets radio astronomy is that it is hard to make confident optical identifications of radio sources; to extract detailed information, such as the red-shift of a radio source, one needs to make correlations between radio and optical objects. The difficulty stems from the fact that it is hard to pin-point radio sources to the necessary accuracy.

A cunning method of achieving the desired precision in the location of radio objects was devised by a British physicist, Cyril Hazard, who as it were put the moon between his radio telescope and the source. This method of "lunar occultation", or observing the source just before and after it appears to us to hide behind the moon, gives a very accurate "fix" on the source's position in the sky, because the moon is a fair way off yet its position is very accurately known.

One of the first radio objects so located was 3C273, the 273rd source listed in the third Cambridge catalogue of radio sources.

Hazard, who at this time was on the staff of the Physics Department at Sydney University, did this work on the CSIRO radio telescope at Parkes.

Armed with this information, Maarten Schmidt, a young Dutch astronomer at the California Institute of Technology, pointed the 200-inch telescope at this precise point in the sky and found the optically visible object corresponding to 3C273. This seemed to be just a faint nearby star, which was most surprising; such stars are not radio sources. Persevering, Schmidt made a careful examination of the lines in the colour spectrum of this star-like object. At first the spectral lines were unrecognizable, but then Schmidt realized this was because the spectrum had suffered an unprecedentedly large red-shift. (See the diagram on page 148, *Figure 1.5*.) The first quasar had been recognized.

Since that time, in 1963, some forty or so other like objects have been identified. Also Sandage has found a class of objects which have similar features with regard to apparent brightness and red-shift, without the accompanying radio emission. It is convenient here to lump all these objects together as quasars.

If these red-shifts are taken to be cosmological in origin, if they are assumed to be due to the expansion of the universe (as are the red-shifts of ordinary galaxies), then the quasars are very distant from us. Yet their light and radio output is such that at first glance they appear to be local objects within or near our own galaxy. For this to be so despite the vast distance away that they really are, quasars must be shining *each with the radiance of roughly 100 ordinary galaxies!*

If quasars are so interpreted as super-luminous objects on a cosmological scale, then clearly the possibility exists that they may serve as tools for probing the past. In this role they have excited much attention and speculation since the first one was recognized for the pathological creature it is.

However, as to the question of the nature of quasars and whether they are at cosmological distances, there is considerable controversy.

The quasar controversy

The controversy as to the exact nature of quasars is as follows:

If, as above, we assume that quasars' red-shifts arise from their participation in the expansion of the universe, then it follows that they are far away, that they are (like galaxies) "cosmological objects", and furthermore that they pour out energy at a rate so great as to be very difficult to explain.

Conversely, if we assume that quasars are local objects, of quite ordinary luminosity, then their red-shifts must derive from an exceptionally violent explosion in the nucleus of our own galaxy or a neighbouring galaxy, in the past. On this premise, quasars are local phenomena, and they are speeding away with the kinetic energy acquired from some local galactic explosion. This latter explanation has been put forward by Terrel and elaborated by Hoyle and others. These people point out that, before quasars were discovered, independent evidence already suggested that such an explosion had taken place in the centre of our galaxy early in its history.

In the words of George Gamow, is a quasar "an elephant a mile away, or a mouse two feet away?"

Favouring the local, or "mouse", viewpoint is the fact that the energy output seems hard to explain otherwise. A more serious point is that observations suggest that the optical and radio emission from some quasars is varying significantly over times of less than a year. The implication is that such quasars, or at least their more luminous parts, are less than one light-year across in size. This is quite remarkable, and difficult (though not impossible) to explain.

Favouring the cosmological, or "elephant", viewpoint is the fact (emphasized by Schmidt) that all quasars seem to be speeding radially away from us. Were they local, some sideways motion should be seen. But if quasars are the debris of a local explosion that took place sufficiently long ago, then Schmidt's argument is not valid. In addition, if we assume a local explosion, then from the observed fragments we can reconstruct an estimate of the energy of the explosion, which turns out to be implausibly vast. Indeed the magnitude of the energy put into such an assumed local explosion is at least as hard to explain as is the magnitude of the energy output on the alternative cosmological assumption.

A method of resolving the mouse-or-elephant dilemma is to examine carefully the detailed character (the widths and relative intensities) of the spectral lines received from quasars. This is very much a topic of current research. The results as yet are indefinite, with the balance of evidence probably favouring the cosmological viewpoint.

We should also mention briefly yet a third hypothesis which has been advanced to explain quasars. This third alternative postulates that the large observed red-shifts are gravitational in origin, arising from the warping of the geometry of space (according to the tenets of general relativity set out in Chapter 1) by, for example, aggregations of neutron stars. This explanation is highly speculative, and moreover has many difficulties in coming to grips with the observations. In particular, this theory must be strained to produce red-shifts in excess of 200 per cent (that is $z > 2$), yet the largest red-shift so far observed is as big as 223 per cent ($z = 2 \cdot 23$).

In short, the balance of argument and observation tends to favour the cosmological, or "elephant", viewpoint, but the verdict is by no means decisive.*

Cosmological inferences from quasars

Suppose we assume that indeed quasars are at cosmological distances, deriving their red-shifts from their participation in the expansion of the universe, what then?

The biggest red-shifts observed for quasars are over 200 per cent; that is, for example, the fundamental Lyman-α line in the hydrogen spectrum is shifted from 1,216 Ångströms, in the infra-red, way up to around 3,000 or 4,000 Ångströms, in the visible spectrum. As mentioned when discussing the difficulties in extracting cosmological information from galaxies, it is much easier to distinguish among the various cosmologies at such large red-shifts (z greater than 1 or even 2).

However, for this feature to be useful to us, it is necessary that quasars be intrinsically identical objects, at least to a good approximation. Until recently, it was not clear to what degree this was so. Unfortunately, a clever and careful analysis by Maarten Schmidt has shown that (assuming them to be at cosmological distances) quasars exhibit a most pronounced evolutionary effect, there

---

* A certain amount of rather unseemly, though admittedly entertaining, prejudice has entered this controversy. One may, with some exaggeration, find in Yeat's *The Second Coming* an almost pertinent quotation:
  "The best lack all conviction, while the worst
  Are full of passionate intensity."

being many more of them at early epochs in time than there are at later epochs. Quasars seem to be a species of adolescent exuberance in the evolution of the universe. This conclusion emerges in a way which is independent of the details of the cosmology one assumes.

Thus if one assumes quasars' red-shifts to be cosmological in origin, they are nonetheless useless as approximate SHOs to deduce details as to which Big Bang general relativistic universe we live in. On the other hand, since the Steady State theory rests on the premise that the universe was the same in the past as it is now, the more frequent occurrence of quasars in the past is definite evidence against it.

Conclusion

We see that no firm conclusion emerges from this Chapter.

The reader may by this time feel that the author is a quibbling pedant, unable to come to an unequivocal decision on any single issue. More time has been spent describing the uncertainties and ambiguities in the observations than in describing the observations themselves. This is how things are in cosmology. It would be misleading to pretend otherwise. The experiments are difficult and ambiguous because one cannot control the environment and isolate individual phenomena as one can in terrestrial laboratory experiments.

Nevertheless, the indications are that we inhabit an oscillating Big Bang universe, some 10 billion years from the beginning of the present cycle, which cycle takes some 100 billion years for completion.

The other evidence, from stellar evolution and from the background radiation from the primeval fireball, is similarly vague, while the evidence from galactic evolution is so tentative that it has not been discussed. Even so, such indications as there are tend to support the above conclusion.

In contrast, we have seen that several solid pieces of evidence add up to the demise of the Steady State theory.

For those who are displeased by the absence of a truly decisive verdict, one can remark that long ago the prophet said (Ecclesiastes, VII, 13): "Consider the work of God: for who can make that straight, which he hath made crooked."

Introduction
To Space Flight

PART I

(Five Chapters)

by

E. F. M. REES

Dr. E. F. M. Rees,
Deputy-Director, Technical, NASA George C. Marshall Space Flight Center,
Huntsville, Alabama, and Director, Apollo Special Task-Team at NR, NASA
Manned Spacecraft Center, Downey, California.

Introduction To Space Flight

Today the history of rocketry and its application to space flight **is** common knowledge, having been widely publicized during the last 10 years all over the world. The names of the early pioneers such as Tsiolkovsky, Goddard and Oberth, whose visions and fundamental works have been pace-making in this field are mentioned in every textbook. The story of the first orbiting objects, the Sputnik and the Explorer satellites, which started the era of space flight is still fresh in the memories of everybody. I, therefore, do not think I should elaborate on history and historical events too long. However, before entering the various subjects of this series of lectures it might be worthwhile to touch very briefly on some of the basic principles of space flight.

What is space flight actually? In general terms, space flight may be called the unmanned or manned travel out of the atmosphere into an orbit about the Earth or away from the Earth into space. The word "flight" is not quite correct in that it is in common use for the movement of insects, birds, aeroplanes, etc., up from the surface of the earth utilizing the lift on their wings and bodies created by aerodynamic forces.

In contrast, a rocket or a space vehicle is moving on a ballistic trajectory. It does not need the support of air, or more accurately expressed, the aerodynamic forces for lift, and, therefore, it does not "fly". The atmosphere with its drag, winds and weather is rather a disagreeable obstacle which constrains the designer, for instance, to limited aerodynamic shapes for the space vehicle, and to special control schemes — to mention only a few. It costs also large energies, which means sizeable amounts of propellants for the engines to overcome the resistance of the air and thus to gain the velocities necessary to insert a satellite or spacecraft into an Earth orbit.

What keeps a mass or a satellite in an orbit without the necessity of adding energy? If we throw a rock it will follow a trajectory due to the initial conditions given when it leaves the hand of the pitcher; namely, velocity and angular vector of this velocity. According to Newton's law of inertia, this trajectory would be a straight line if the event would not have occurred in the gravitational field of the Earth which makes it fall back on its surface. The higher the initial velocity, the longer the range will be, of course. If we increase the velocity further and further, we must of necessity arrive at a condition where the rock does not fall back on the Earth any more but circles it. It is kept in balance between the gravitational pull of the Earth and the inertial forces, in this case commonly called the centrifugal forces. At this point of balance of the forces the rock or the object is entirely weightless. The minimum velocity for this condition, the circular velocity, is approximately 25,700 ft./sec. for our planet.

What laws guide orbital motion? An increase of the circular velocity pushes the object farther away from the Earth and the orbit becomes more and more elliptic. Finally at a velocity of approximately 36,300 ft./sec. the gravity of the Earth is overcome and cannot hold the object captive any longer. It escapes the earth — therefore this velocity is called escape velocity — and circles in an orbit about the Sun. Johannes Kepler, the noted German astronomer of the late sixteenth and early seventeenth century described in his first law the orbit of the planets as being an ellipse with the Sun at one of the two foci.

In his second law of areas, he proposed that the radius vector of the planet, i.e., a line drawn from the Sun to the planet, in sweeping over the area of the ellipse as the planet orbited, covers equal areas in equal time intervals. To effect this, of course, the planet must increase in velocity as, on its elliptical path, it approaches the Sun.

Finally, in the third law, Kepler's harmonic law, he states that the ratio of the squares of the orbital periods P_1 and P_2 of any two planets is the same as the ratio of the cubes of the mean distances a_1 and a_2 of the planets from the Sun; thus:

$$\frac{P_1^2}{P_2^2} = \frac{a_1^3}{a_2^3}$$

Subsequently, Newton postulated that the third law was not quite accurate and that instead, the proper relationship should be shown as $\dfrac{(M + m_1)\ P_1{}^2}{(M + m_2)\ P_2{}^2} = \dfrac{a_1{}^3}{a_2{}^3}$ where M is the mass of the Sun and m_1 and m_2 represents the mass of the planets. Since, however, the masses of the planets are relatively quite small compared to the Sun's mass, the correction is generally small.

It is, incidentally, of interest to observe that it was from his studies of the Keplerian relationship that Newton derived his universal law of gravitation.

Kepler's laws, with Newton's modifications, are valid everywhere for comets, asteroids, natural and artificial satellites, double stars[1] and all other known celestial bodies. They apply to the motion of an artificial satellite of Earth although slight perturbation of the satellite orbits occur due to Earth's atmosphere[2] and to the fact that Earth is not exactly spherical and homogeneous.

How is the energy generated and converted to move an object in space and to reach the previously mentioned extraordinary velocity ranges? Here we talk, of course, about rocket propulsion. The rocket is the only practical means, known so far, to propel an object *into space* and to change its position and accelerate it *in space*, because it does not need air either for aerodynamic lift nor its oxygen for combustion. Let us review briefly some of the basic principles.

To begin, the fundamental law applying to rocket propulsion is Newton's Third Law which sets forth the inseparability of action and reaction forces. In reality, the Newtonian mechanics approach requires some modification since its law of equal momentum change applies to a flow of rigid particles whereas, in a rocket jet actually a compressible fluid; namely, the combustion gases, are

---

1 A double star is a system which appears as a single point of light to the eye but consists actually of two stars rotating about a common centre of gravity.

2 Although satellites are generally inserted into orbits above the earth atmosphere, there are still air or gas molecules in low orbits (90-100 miles) which brake the velocity and thus make the satellites spiral back into the denser atmosphere where they burn and break up. Low orbit satellites have, therefore, a shorter lifetime, sometimes of only a few days.

involved. Hence, there are areas of varying pressure during the exit of these high velocity gases. Forces resulting from this condition were not included among those of the particle jet assumed earlier in stating the relationship of Newton's Third Law. Thus, total rocket thrust is the sum of gas momentum thrust and, also, that thrust resulting from the excess pressure of the exhaust across the nozzle area over ambient pressure. This summation is shown by the relationship:

$$F = mv_e + (P_e - P_a) A_e$$

Where F is pounds thrust, m = gas flow in pounds mass per second; v_e = gas exit velocity in ft. per second, P_e = exit pressure across the nozzle area, A_e = exit area across the nozzle and P_a = ambient pressure at any given time of flight or point of the trajectory. From this equation, it is readily apparent that rocket thrust increases with altitude since P_a decreases with altitude and becomes practically zero above the atmosphere.

Figure 1 shows this relation of thrust increase versus flight time, for instance, for the case of the Saturn V first stage.

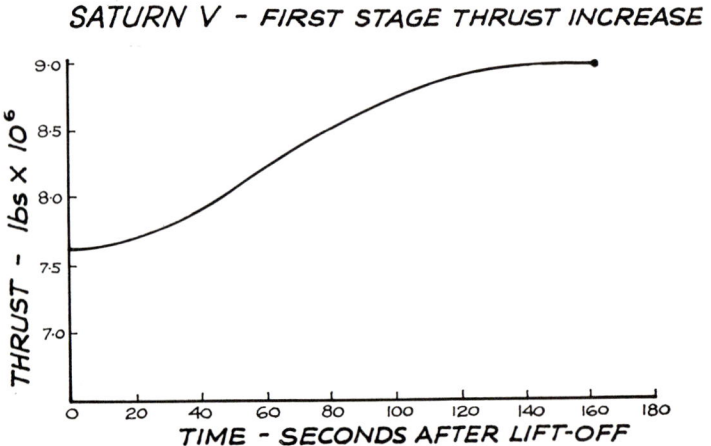

Figure 1.

At the same time, of course, the vehicle lightens as propellants are consumed with the first quantity of the equation remaining substantially constant. Thus, the thrust-to-vehicle-mass ratio at lift-off increases markedly. An additional favouring effect on the systems results from the decrease in loss from aerodynamic drag with altitude. The combined effects of these factors account for the increase in acceleration during the trajectory of the vehicle and thus for the inherent capability of rocket propelled vehicles to obtain the aforementioned very large velocities necessary for space flight.

The velocity at the conclusion of powered flight, also called the "vacuum burnout velocity", v_{vbo} is given by the expression:

$$v_{vbo} = v_e \, 1n \, \frac{m_o}{m} - gt$$

Where v_e is the gas exit velocity, m_o = initial mass, m = mass at burnout, $\frac{m_o}{m}$ = mass ratio, g = gravitational acceleration, t = flight time to burnout. This is the familiar form of the rocket equation in vacuum. If it is applied to the first stage of a rocket vehicle which lifts off from the surface of the earth, the aerodynamic drag as a time or altitude function has to be deducted.

This equation also states the case for the multi-stage rocket for space flight. Whereas a shorter range rocket-propelled guided missile can be built as a single stage vehicle, this is practically and economically not feasible for space flight requiring the high velocities at burnout. The necessary mass ratio for such a single stage to reach these velocities would confront the engineer with an impossible task as to the large size of this vehicle and other factors. Therefore, a space vehicle is composed of several smaller stages which are stacked on top of each other and whose burnout velocities or, in rocket engineer's terms, cut-off velocities are added to each other in sequence. Burned out stages are dropped as ballast in order to increase the mass ratio of the remaining stages.

Before closing this brief narrative on propulsion, it seems appropriate to discuss specific impulse, or Isp, a term that is used extensively in the language of propulsion.

Specific impulse is defined as the thrust produced by a rocket engine when a unit mass of propellant is burned per unit time. Stated another way, it is the ratio of thrust to propellant mass flow, as given by the expression $\text{Isp} = \dfrac{F}{w}$ where F equals pounds thrust and w indicates pounds of propellant per sec. The dimension of the specific impulse is seconds according to above formula. This may be somewhat confusing in that it does not completely reflect the definition.

The *specific impulse* is a measure of the efficiency of the propulsion system rather than a measure of its capability to produce a desired thrust. For example, an ion engine produces an extremely small propulsive force in the order of fractions of pounds, yet it has a high specific impulse of 1000 seconds or more, indicating a very low propellant consumption per time unit and long duration of the small force.

In tabular form, the Isp obtained in presently available systems of various propellant combinations or energy sources is generally as follows:

Solid propellants	approx. 250-300 sec.
Liquid propellants	approx. 260-430 sec.
Nuclear fissure	approx. 600-900 sec.
Electric propulsion (ion)	1000 plus seconds

The first three in the listing comprise thermal propulsion modes. In addition to these and the electric or ion propulsion, there are the so far theoretical concepts of the proton drive mode and the photon rocket which have an extraordinarily high Isp.

When it is possible to tap the energy of the solar wind through solar sails, there will then be a source of energy with an infinite Isp since no propellant will be involved.

How is a space vehicle directed to the orbit assigned for a particular mission or into a trajectory to the moon or planets? The problem of directing a rocket or space vehicle to accomplish a given mission is customarily discussed in terms of three separate functions: navigation, guidance and control. The boundaries between these three areas are to some extent arbitrary and conventional.

Navigation is the determination of vehicle state for initial conditions as well as during flight. This includes vehicle velocity and position in a reference co-ordinate system.

Guidance is the selection of a manoeuvres sequence to get from the present state to a desired state. The guidance function uses the navigation information to generate vehicle attitude or attitude rate commands.

Control is the execution of the manoeuvres called for by guidance.

The conditions for navigation are closely correlated to the guidance system to be selected. General viewpoints are that the navigational tasks should be solved preferably on earth or at non-vehicle-based stations. Onboard navigational methods should be considered only if onboard navigation assures benefits such as additional reliability and accuracy. Furthermore, a sound balance must be found between navigational accuracy and accuracy of the guidance system. Up-dating of navigational inputs into the guidance system must be considered as early as feasible to avoid accruing large error contribution.

Overall system performance requires that the guidance methods permit minimum propellant consumption for guidance manoeuvres and avoid excessive structural loads caused by such manoeuvres Thus, during high aerodynamic pressure regions of flight,[3] very low guidance constraint or none at all may be applied. This leads to comparatively large deviations from an unperturbed flight path. The guidance system must, therefore, be capable of accepting such large deviations and of determining a new optimum flight path rather than attempting to constrain a space vehicle to a standard trajectory.

The guidance system must also correct for numerous flight perturbations such as atmospheric perturbations from wind, unsymmetrical air flow because of vehicle dissymmetry, flight path deviations caused by nonstandard vehicle and engine characteristics and

---

3 During powered flights of a first stage through the atmosphere, the aerodynamic pressure increases with the square of the velocity. It reaches a maximum usually between 70 and 80 seconds of flight with almost all long range guided missiles and space first stages. Then, although the velocity continues to increase, the pressure drops rather rapidly to practically zero because of the diminishing density of the atmosphere.

performance, gravitational anomalies, navigation and control in-accuracies, and emergency situations.

Two basically different methods are used to provide navigation and guidance; one is based on inertial principles, and the other is based on information derived from electromagnetic measurements. Between the two extremes mainly represented by the inertial guidance system and the radio guidance system, various hybrid systems and modifications are possible and have been in practical use.

The pure inertial guidance system is a self-contained system performing navigation, guidance and control functions on board the vehicle. At the present state of the art, the most accurate result is obtained by acceleration sensors that are space-direction fixed on a gyro-stabilized platform, also called an inertial measuring unit. To provide mission flexibility and improve accuracy, navigation and guidance computations are best performed by a digital computer. The guidance scheme is programmed into the computer in the form of guidance equations; required aiming and end conditions are stored in the memory of the computer.

The most important advantage of the inertial guidance system is its independence from any reference outside the vehicle, such as ground stations on Earth. This independence provides inertial guidance with a high degree of flexibility and makes it immune to interference (jamming).

In a radio guidance system, the functions of navigation and guidance are performed in one or several stations. The vehicle carries a radio tracking/command beacon and the control system. Navigation (position and velocity measurement) is accomplished by radio tracking of the vehicle from one or more ground stations. The ground-based computer performs navigation and guidance computations and generates the guidance command, which is transmitted to the vehicle. The command is executed by the vehicle control system, which also provides attitude stabilization. The control system, sometimes called autopilot, uses three body-fixed gyros instead of a stabilized platform.

A main advantage of the radio guidance system is that the complex navigation and guidance computation equipment is located on the ground; therefore, the equipment can be designed for any technically possible accuracy without restrictions existing for onboard equipment such as, for instance, weight and volume limitations. Thus, for short duration missions, the accuracy of radio and inertial guidance methods is approximately equivalent; however, for flights extending over a long time, the accuracy of radio guidance is superior.

For lunar and planetary missions, the accuracy of the radio guidance system decreases with distance from the ground stations on earth and is therefore not sufficient for the terminal phases of the mission (achieving a certain orbit and landing). Therefore, a special terminal guidance mode is applied in such missions. Distance and velocity of the vehicle with respect to the moon or planet must be measured by passive radio (radar) methods using signals reflected from the surface of the celestial body, as long as no operational ground tracking and computation base can be established.

Space flight missions may be divided into a launch vehicle mission and a spacecraft (payload) mission. The launch vehicle mission generally ends with injection of the spacecraft into Earth orbit or into an escape trajectory. From the mission, aim conditions for the launch vehicle can be specified as a velocity vector (magnitude and direction) at a given altitude at the moment of final engine cut-off. These aim conditions may be functions of time, since the launch point on Earth and, for certain missions, also the target are moving through space.

The goal and the rules of the mission determine the flight phases and manoeuvres to be performed by the spacecraft. To accomplish its mission, the spacecraft may be required to perform one or several of the following manoeuvres: Earth orbital flight, rendezvous, lunar or planetary fly-by, orbit, landing, and take-off, and Earth re-entry and landing. These manoeuvres represent a large variety of navigation, guidance, and control problems. The solutions to these problems depend strongly on whether or not the spacecraft is manned.

Usually a large number of constraints apply to the trajectory of a space vehicle. These constraints result from mission rules, environmental conditions such as the atmosphere and location of launch point, operational requirements such as safety restrictions of launch azimuth and tracking requirements, and hardware limitations such as structural load limits and available propulsion means. The dominating and most severe requirement applying to the choice of trajectories is optimum propellant utilization.

Combining various constraints generates what is called in space flight a "launch window", i.e., a limited time period for launch to meet the mission goal for a given launch vehicle. Such a launch window may exist for each phase of a mission, for launch from Earth or from a celestial body, and for launch from orbit. In addition, there are geometric windows (e.g., restrictions in launch azimuth or in the re-entry angle).

Other basic features typical for space flight

After having touched on some of the main basics of space flight, other additional fundamentals should be of interest as part of this introduction.

Earth orbits.   There is, of course, an infinite variety of orbits possible.   They extend from circles[4] about the earth to very flat far out reaching ellipses.[5]

Based on the geographical location of the launching site, on the flight mission and on the location of the tracking stations spread around the earth, the azimuth[6] of the trajectory at the launch site is determined which results in a certain inclination of the orbit to the equator.   Special orbits are polar, equatorial and synchronous orbits.

Polar orbits are advantageous for certain missions, e.g. Earth observations, because, at least theoretically, the whole surface of the Earth can be covered in half a day.   The satellite is orbiting in a plane through the poles while the Earth is rotating underneath.

Equatorial orbits are achieved if the launching site is located on the Earth's equator.   If a satellite is put in a circular orbit high enough so that its orbital angular velocity is the same as the earth

---

4 The circle is a special case of the ellipse with both foci in the centre.

5 The point of an elliptical orbit closest to earth is the "perigee", the most distant point is the "apogee".

6 From a viewpoint of energy and payload weight, the optimum azimuth is "due East", i.e., in the direction of the rotation of the Earth.

angular velocity, this synchronous satellite[7] stands then motionless above the same point of the Earth. This scheme has good application for communication and weather satellite systems.

For manned space flight about the earth, generally low altitude orbits of 100-150 nautical miles are selected because long-time exposure of astronauts to the Earth's radiation belt[8] must be avoided.

Instrumentation and tracking systems. These systems are of utmost importance for space flight. Each element of a space vehicle is equipped with instrumentation systems which are functionally divided into three categories, namely:

a. Measuring systems
b. Telemetry systems
c. R. F. and Tracking systems

The purpose of a measuring system is to yield information about the instantaneous condition of the space vehicle, its status of movement about its centre of gravity, and the functioning of its subsystems and components. It also is designed to give data on scientific phenomena to be explored such as radiation pressures, temperatures, meteoroid impacts, etc. The results of these measurements are generally processed from sensors over transducers into electrical signals which in turn are transmitted by a telemetry system to receiving stations on the ground. In order to give an idea of the magnitude of these systems, it is interesting to note that during an unmanned development flight of the Apollo-Saturn V space vehicle, over 4000 separate measurements were transmitted and received, of which over approximately 98% were of excellent quality.

The R. F. and tracking system serves the purpose of voice communication with astronauts in case of manned flight, television, command transmission from ground to on-board and for determination of velocity and position of the space vehicle or satellite. For

---

7 A synchronous satellite, also called 24-hour satellite, has an altitude of approximately 25,000 nautical miles.

8 The earth's radiation belt is a region starting at an altitude of approximately 400-500 nautical miles and consists of cosmic particles trapped by the Earth's magnetic field.

constant contact with the vehicle, the United States has a world-wide network of tracking stations in operation.[9]

Environmental conditions. These conditions or criteria with which the designer of a space vehicle has to cope are very difficult to materialize especially since all flight hardware has to be of extremely light weight. In order to be brief, let me describe in a few catch words some of these conditions:

Launch Vehicle: Very deep temperatures from cryogenic propellants (liquid oxygen and hydrogen).

Very high temperatures from combustion gases.

High pressures in propulsion systems.

Extremely complex vibration patterns caused by excitation from rocket engines and from aerodynamic forces.

High dynamic pressures during flight through atmosphere.

Deep vacuum after departure from atmosphere.

Spacecraft: High dynamic pressures during flight through atmosphere.

Radiation hazard, especially solar flares, and earth radiation belt.

Deep vacuum in space.

Fire hazard.

Low temperatures in space.

Meteoroid hazard.

High accelerations and decelerations.

Extremely high temperatures upon re-entry into the earth's atmosphere.

Impact shock upon landing.

Unmanned and manned space flight. After it had been established that it was feasible to use man as navigator, pilot, observer and explorer, and for many other activities in space, it became conventional to categorize space flight into an unmanned and a manned area. Whereas the rocket or, better expressed, the launch vehicle is to a greater extent independent of the kind of payload it has to

9 One of these stations is Carnarvon in Western Australia.

carry, the design for a craft containing an unmanned payload is vastly different from that with men on board.

The main design features of unmanned payloads[10] are in the fields of sensors for scientific exploration, telemetry systems for transmission of data to earth, television cameras, tracking equipment, environmental conditioning systems for instrumentation, solar cells for power supply, reaction systems for trajectory control,[11] etc. All onboard systems have to be automatic. Their functions are initiated or stopped by commands from the ground, for instance, the correction of trajectories or the operation of TV cameras. It is important that scientific instrumentation or other devices are only in operation as long as they are really needed in order not to wear them out or deplete power and propellant supply unnecessarily.

In contrast, the design of manned spacecraft has to emphasize safety features for astronauts, human factors, medical aspects, etc., in the first place.

Subsequent lectures will deal with this particular subject in greater detail. Therefore, I don't think I should elaborate more on it here.

---

10 It is customary to call payloads for unmanned space flight "payload" or "satellite" and for manned space flight "spacecraft".

11 In space any movement of an object out of its prevailing inertial pace or about its centre of gravity can only be accomplished by reaction motors based on the rocket principle.

CHAPTER 1

Achievements in Unmanned Spacecraft

Introduction

Reviews of achievements in space flight are frequently being done by way of tabulating the number of satellites and their pay-load weights launched into space by a nation over the years of reference. In this chapter it is attempted to show the various missions and fields of application of satellites and some of the results. It should, however, be noted that this represents an over-view only rather than a source for information of results. It also should be emphasized that this chapter, by no means claims to give a complete story of all the achievements of the US in this area.

A. Satellite Meteorology

Space exploration has added a new dimension to meteorology and the daily weather report. Man now has the means of observing the Earth on a global scale. Atmospheric scientists, recognizing the potential of such observations, began immediate development of measuring devices that could be used to observe the Earth's weather from orbit, a highly advantageous observation point. Since early 1960, when the first weather satellite was placed into Earth orbit, considerable progress has been made in exploiting this new tool. Results of satellite meteorology have been applied primarily to weather observation and prediction, and, to a lesser extent, to studies of the Earth's atmospheric structure.

Observational data obtained from sixteen developmental and operational satellites have supplemented ground-based observations. (In themselves, the ground observations cover only 20% of the Earth's surface.) The satellite data have permitted such a marked increase in weather prediction capabilities over the past few years, that some scientists foresee the development of reliable, long-range weather predictions of two weeks or more.

Daily cloud cover pictures from the weather satellites have enhanced the ability to determine the location and intensity of major weather disturbances. In temperate zones, fronts and other weather features may be located and followed more precisely, a considerable aid to local and regional forecasters.

In the immediate future, the high-resolution, infrared radiometer, which measures the intensity of infrared radiation from cloud tops, will double the number of available pictures by providing night-time cloud observations. This greater coverage of the Earth will significantly improve short-range forecasting. A synchronous satellite will, for the first time, permit continuous observation of the complete life cycle of weather disturbances. This has recently been demonstrated over the Pacific Ocean area. This capability will be particularly useful in studying phenomena such as thunderstorms, hurricanes and cumulus cloud complexes, and will permit atmospheric motions as revealed by cloud systems to be closely monitored. Such a technical capability will constitute a major step in advancing the understanding of atmospheric processes.

B. Communications

Television broadcasts are relayed today by communications satellites to many millions of people in the world. In 1946, scientists reflected radio signals from the Moon. Then, in the early 1960s, radio signals were reflected from the passive, metallized balloon satellites, Echo I and Echo II. Using these satellites, scientists demonstrated that an orbiting sphere and improved ground techniques could be used to measure reflected radio signals from space. These experiments also provided the opportunity for the United States and the Union of Soviet Socialist Republics to take the initial steps toward a programme of scientific co-operation and data exchange. The initial activity of this co-operation was the optical observations of the Echo II satellite by stations in the USSR when the satellite was not in view of any of the US tracking stations.

The active communications satellite was developed through several different satellite programmes. An advanced phase of this development was successfully concluded in October, 1964, when a satellite was positioned in a synchronous orbit stationed over the equator in the Pacific Ocean just north-east of Australia. This phase of the communications satellite programme demonstrated the feasibility of television and high-quality voice transmission by satellite. The 1964 Olympic Games from Japan were transmitted both to the United States and to Europe. The European transmission was the first time two satellites were used in series: first, the programmes from Japan were transmitted to the United States using the satellite hovering over the Pacific, north-east of Australia; they were then retransmitted to Europe using another satellite. Although the events were not "live" transmission, it showed that satellites were a most effective means of international television transmission.

Satellite voice transmission, a new segment of the communications industry, is earning money from space. An operational satellite can transmit a trans-Atlantic telephone call at 11c a minute, a fraction of the cost of a cable. A satellite system, employing equatorial synchronous satellites, provides radio telephone relay links that simultaneously cover from all North America, Latin America, Europe and Africa.

Communication satellites may also be used to relay global television transmissions by using a large number of telephone channels. In the future, there will be full-time television relay satellites. If a 50-kilowatt station were placed into synchronous orbit, television transmissions could be directed into home television antennas on the ground. The signals could be directly received by dish antennas pointing at one of the satellites, with later rebroadcasting by conventional methods.

If a country wishes to participate in a world-wide television and telephone service, it can buy a ground station for about three to five million dollars. From that station, it can pick up from the satellite whatever was desired for local rebroadcast. By using the 50-kilowatt satellite transmitters mentioned earlier, TV broadcasts would go directly from orbit into the home antenna. Such a system would provide a broad variety of programmes from educational and commercial television. This technique is technologically within reach today. David Sarnoff, Chairman of Radio Corporation of America, once said that, with such a television system, illiteracy could be eradicated from the face of the globe within ten years.

C. Navigation and Air Traffic Control

A considerable number of scientific and technical improvements have been made in the field of navigation and air traffic control, and further studies are underway. Today, the US Navy uses simple satellites to track ship positions. For a ship to calculate its position, a procedure similar to that used by a surveyor is followed. To locate himself with respect to two mountains, the surveyor makes a triangulation, based on the distance between the mountains. That distance provides the baseline of a triangle. He then measures the subtended angle and determines where he is. In the same way, for ship navigation, the satellite knows its orbit and position with respect to time. So a ship-based computer makes a triangulation with a single, but moving mountain. The ship sends no signals to the satellite. Instead, the satellite produces a tone, and the doppler effect of the tone is measured. After the satellite has passed, a combination receiver-computer on the ship prints the longitude and latitude of the ship. For about $12,000, this equipment can be purchased for a seagoing yacht.

Traffic control is a more difficult area in which to apply satellite technology. Today, in the North Atlantic, for example, air traffic is massively increasing. During the same time of day or night, every trans-Atlantic airline wants to travel at the optimum route and altitude west bound or east bound, as determined by wind and weather conditions. For customer convenience, every line wants to fly during a few rush hours, so a lot of traffic is condensed into a very narrow airlane and a very narrow time frame. The airliners are spaced at a minimum separation distance of 120 nautical miles, at the same altitude. This is considered reasonably safe, but it appears to be necessary to reduce this distance to 90 nautical miles. This lower distance has raised a serious question in the minds of the airline captains on its safety from collision. If any one of the planes is not in its exact relative position, it will endanger the plane ahead or behind. This congestion of airline traffic has created a pressing demand for positive transoceanic air traffic control.

Air traffic experts would also like this traffic control to be integrated with better weather advisory service, preferably in the form of a weather map being cast directly onto the pilot's weather

Figure 1-1. Gemini IV: Nile Delta.

radar so that he can see his aeroplane's position, the location of other planes and the location of the bad weather on an integrated display. This appears to be a very promising method and if it works, it could even become attractive for aviation over the continent.

D. Earth Resources and Geodesy

The use of satellite observation for earth resources surveys is highly promising. High-resolution cameras can identify crop conditions, water potential from mountain snow and probably mineral locations. Pictures taken by the astronauts from the orbiting Gemini spacecraft have already been used to identify probable earth resources. Many examples are available but one relating to oil prospecting in Australia may be of particular interest. A photograph taken during the Gemini V flight, showed a part of the Amadeus Basin in Central Australia which has a gas-producing well. By using the photograph, seven partly or completely con-

Figure 1-2. Gemini IV: India, Ceylon, and the Himalayan Mountains.

cealed oil prospecting sites were identified. These had not been found on previous geophysical surveys. The photographs also showed surface features similar to those identified with other petroleum-producing areas.

Some of the pictures taken during the Gemini IV flight are extremely interesting and clearly demonstrate the amount of detail visible from space. Three of the many pictures are provided in the following sequence during a quick trip around the earth.

The sequence begins at the Nile Delta (*Figure 1-1*), the cradle of western civilization. The green area of the Nile Delta contains approximately 25 million inhabitants, none of whom is visible. From pictures such as these, and more detailed pictures that can be taken with high-resolution cameras, sociologists, public health agencies and urban planners can learn much about the management and control of densely populated areas on the earth.

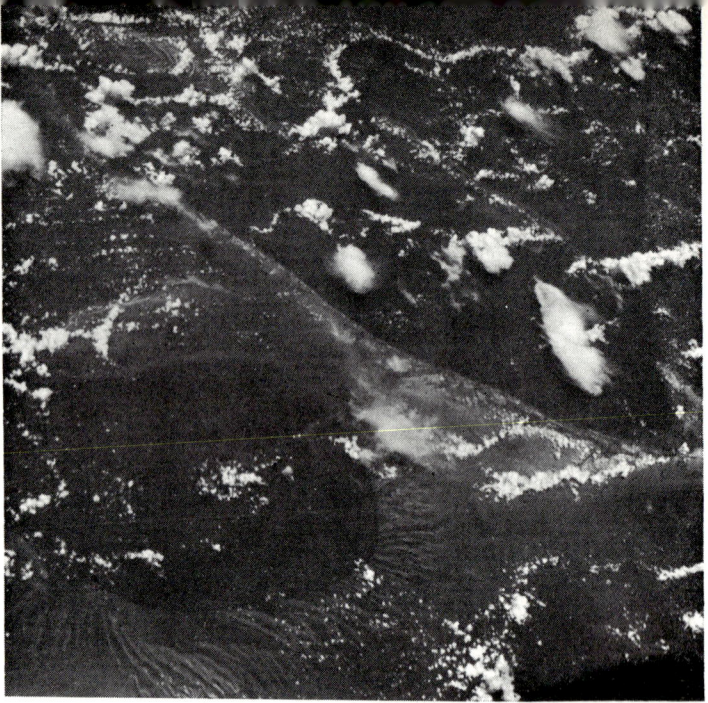

Figure 1-3. Gemini IV: "Tongue of Ocean" in the Atlantic.

Moving on across the Indian Ocean (*Figure 1-2*) shows a view of the complete Indian subcontinent from the Island of Ceylon up to the Himalayan Mountains. Of particular interest in this photograph is the clear area around the continent. There are clouds both over land and at the sea. Previous pictures from the Tiros weather satellites have not been detailed enough to show this clear area, and meteorologists have been quite interested in this phenomenon.

The last photograph (*Figure 1-3*) shows an area of the Atlantic Ocean called "Tongue of the Ocean" in the Bahamas. Most of what appears to be land is actually ocean bottom. The only land above water is the dark strip on the right edge of the picture. The rest is the bottom of the ocean which can be seen quite well from space. The ridges around the deep blue area are erosion in the ocean bottom, and oceanographers are quite interested in such pictures, particularly studies of such areas before and after hurricanes.

In summary, these pictures are extremely interesting and valuable to the geologists and geographers. They illustrate clearly what has already been accomplished with photography and how it can be applied in the future to identify and husband resources on the earth. With imagination and the application of high-resolution photographic techniques, the future of earth resources management holds real promise to mankind. (Chapter 5 contains a further discussion of this matter.)

E. **Technology Utilization**

The space programme is tremendous in size and complexity, and offers challenges both from a technical and management standpoint. The overall systems engineering and management techniques developed during the first decade of the space programme offer great possibilities for defining and resolving complex problems in areas other than the aerospace programme. These are problems involving not only technical aspects, but also social, economic and political facets that must be considered in their total resolution. The "systems" approach is now being applied to certain selected local, state and federal problems. Air and water pollution and city planning are a few of the areas where the systems approach has been used.

What is being done with other knowledge gained as a result of space research and development? So much of this information is available that it must be summarized into readily available form for general public use. The information ranges from concepts for improved screwdrivers to esoteric mathematical techniques, from welding practices to the synthesis of highly halogenated monomers and polymers, from computer programmes to improved drafting techniques.

A description of some of the developments is in order. One of these, a sight switch (*Figure 1-4*), permits a person to operate a switch merely by movement of the eyeballs. The switch is essentially a pair of eyeglasses carrying a small infrared generator and detector. It works by detecting the infrared energy which is reflected by the white part of the eyeball or absorbed by the pupil. By properly designing auxiliary circuits, it is possible to dial a telephone or to operate a motorized wheelchair merely by moving

*Figure 1-4.   The Sight Switch being used by Dr. Wernher von Braun,
Director of the NASA George C. Marshall Space Flight Center at
Huntsville, Alabama.*

the eyes. This device opens up many new possibilities for those
who are paralysed or bedridden and unable to move.

   Many other medical applications from the space programme
may be mentioned.   For example, miniaturized pressure sensors

are being developed that are small enough to insert into the heart through a hypodermic needle to measure cardiac activity directly; computer and photographic techniques are being developed that can clarify and increase the contrast on medical X-rays for improved diagnosis; the astronauts' pressure suit helmet is being adapted to allow analysis of the patient's oxygen consumption; this should be much more comfortable for the patient than the nose clip presently used in this diagnostic procedure. The "inflight" blood pressure, temperature and electrocardiogram monitoring systems utilized in manned spacecraft are being used in some hospitals for simultaneous monitoring of several intensive-care patients by a single nurse or doctor.

These are only a few examples of many engineering innovations that have been or could be applied to the general health and medical field.

During rocket launch and flight, it is very desirable to use on-board TV cameras to view such things as stage separation, fuel sloshing and dynamic operations of other sub-systems. These cameras should weigh as little as possible, to minimize effects on the weight of payload carried to orbit. As an example, a battery-powered TV has been developed which weighs only one and one-half pounds and is not much larger than a pack of cigarettes (*Figure 1-5*).

Examples like these illustrate the application of devices already developed. There is, however, another product of research and development which cannot be measured directly, although its influence can easily be even greater than direct applications. That product is basic research information which has not yet been directly applied. It may well be that the generation of new information about our world and the environment of space will be the greatest contribution which the space programme can make to mankind.

F. **Bioscience**

The environments of space offer unique research sites for studying living organisms and their relations with their physical environment. By studying Earth organisms in orbit, weightless and isolated from cyclic geophysical phenomena, we can identify the effects of these factors in life on Earth. If, in theory, the

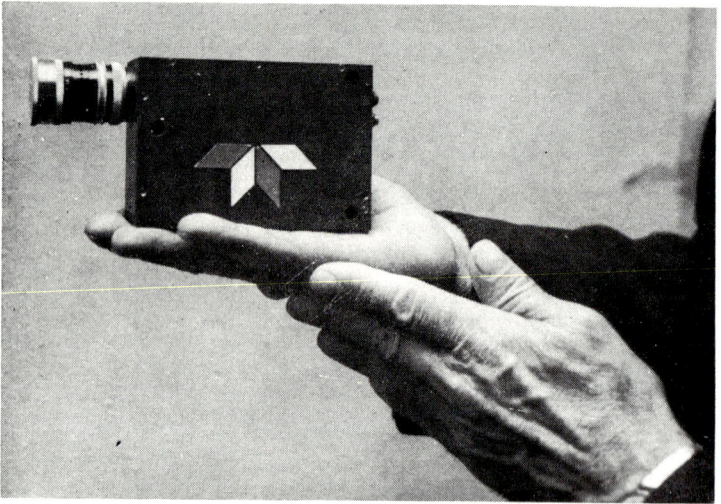

Figure 1-5. Lightweight television camera.

existence of extraterrestrial organisms of different character from those on earth is accepted, then the study of these organisms might well lead to a better understanding of Earth's ecology.

Space bioscience has three primary goals: (1) to detect extra-terrestrial life and assess its nature, origin, and level of develop-ment; (2) to study the effects of space environments on living Earth organisms; and (3) to develop fundamental theories and models of the origin and development of life, here and elsewhere in the universe.

In detecting extraterrestrial life, the research cannot be con-fined to looking for obvious signs of life. It is also important to discover paleobiological signs, such as "chemical fossils", i.e., organic compounds that originated from living systems in the distant past. Two powerful tools, gas chromatography and mass spectro-scopy, can be used to identify these signs. These tools have been tested on sedimentary rock samples known by their fossil content to contain substances of biological origin. Older rock specimens have also been examined, using these techniques, and signs of life have been identified in rock at least 3·1 billion years old.

The research not only looks backward in time but looks "forward" from the beginning of life on earth to gain knowledge on the evolution of life. Recently the National Aeronautics and Space Administration (NASA) apparently achieved the *de novo* synthesis of a molecule which is a member of a class of substance known as porphyrins. Special types of porphyrins are found in the chlorophyll molecule, the key to photosynthesis in green plants, and in the hemoglobin molecule, responsible for oxygen transport in higher animals. The formation of this molecule represents a major step in understanding the evolution of life from the non-living state.

Bioscience also examines effects of the space environment on terrestrial life. NASA's efforts in this area are best exemplified by the Biosatellite (*Figure 1-6*) which was launched on September 7, 1967, and recovered two days later. The satellite carried thirteen experiments, ranging from single-celled creatures, such as bacteria and amoebae, to green plants and higher evolved insects.

These experiments were intended: first, to examine the effect of weightlessness upon Earth organisms; and, second, to examine the effect upon these organisms produced by the interaction of weightlessness and a standard source of radiation.

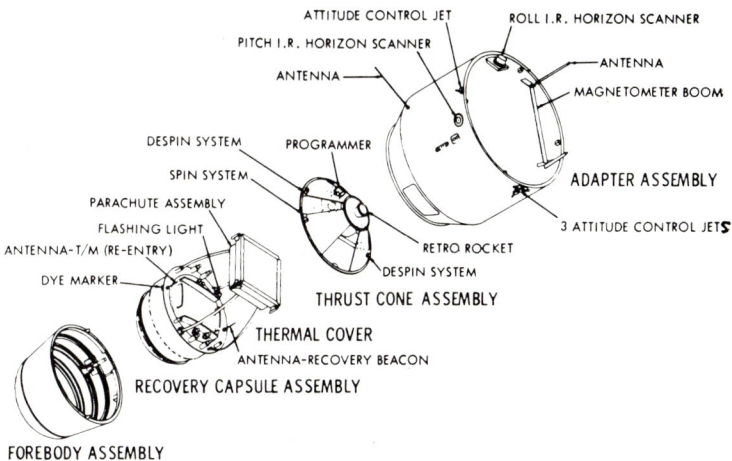

Figure 1-6. Biosatellite II spacecraft.

Much experimental work remains to be performed in support of the flight, such as cross-breeding of organisms to show the genetic changes, further analysis of returned samples, additional control studies, and final statistical evaluation of results. However, significant preliminary findings have been made.

There was only one case, *Neurospora*, a common mould, in which no response to the space environment was found. No explanation for this result has been formulated.

Weightless pepper plants showed an expected drooping of the leaves and an increase in sucrose and amino acids, paralleled by a decrease in starch content. One unexpected result was a peculiar twisting behaviour of the plants in zero gravity. This behaviour, seen in timelapse photography, was not observed in any of the ground controls and thus far remains unexplained.

Wheat seedlings showed the anticipated tendency for roots to angle upward. Starch granules in actively growing root tips of plants exposed to weightlessness showed random distribution in the cells, in marked contrast to certain Earth control plants where granules clustered at the bottom of cells. Some botanists believe that these granules may be a significant part of the gravity-sensing system in plants.

It was noted that the amoebae may have fed more actively in weightless conditions than on the ground, but the rate of cell division was not affected in flight.

In examining the combined effect of weightlessness and radiation, researchers sought both mutations and somatic effects such as change in morphogenesis and physiology.

The vinegar gnat (*Figure 1-7*), a standard laboratory insect for genetic studies, showed highly significant increases in mutation rates. One investigator found that test larvae showed abnormal chromosomes of a type which he had never seen before.

Experiments with a parasitic wasp gave results contrasting sharply with the additive effects of weightlessness and radiation seen in other species. While radiation did produce genetic and somatic damage, weightlessness apparently tended to counteract the radiation effects. In order to explain these unusual results, the experimentors postulate that weightlessness produced changes in cellular metabolism. Biochemical and biophysical breakdown

EFFECTS OF RADIATION & WEIGHTLESSNESS ON DROSOPHILA
Vinegar Gnat

HALF THORAX

VESTIGIAL WING

MISSING WING

"1:10,000 OCCURRENCE NORMALLY"

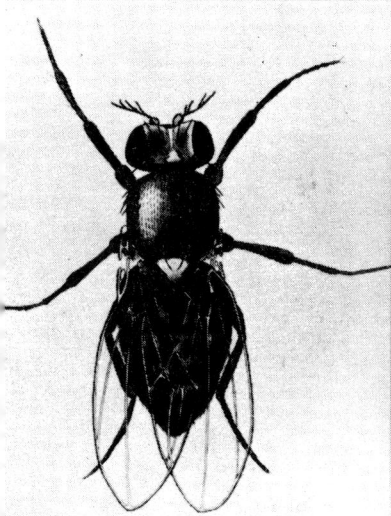

ADULT FROM NORMAL EGG (GROUND)

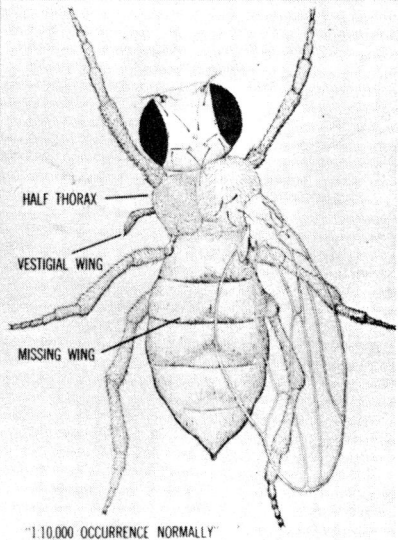

ADULT FROM EGG LAID IN BIOSATELLITE II
RADIATION AND WEIGHTLESSNESS

50% MORE ABNORMALITY IN BIOSATELLITE EGGS

Figure 1-7. The Vinegar Gnat, Biosatellite II project.

reactions due to radiation are assumed to have been slowed. Radiation repair mechanisms in the insect could then have overcome the original damage.

Thus, depending on the species used, development stage treated, or the biological effect sought (genetic, cellular, organ, or total organism damage), response to specific space environment factors will vary from none at all to those detrimental to the organism.

The results pose more questions than they have answered. These species and others must be flown for longer periods of time under new experimental conditions before researchers can make detailed comparisons or develop theories regarding the mechanisms involved.

G. Stellar Astronomy

Our atmosphere protects us in many ways, but, in doing so, it hides from us much of the universe. From the surface of the Earth, we can study the universe only through two windows

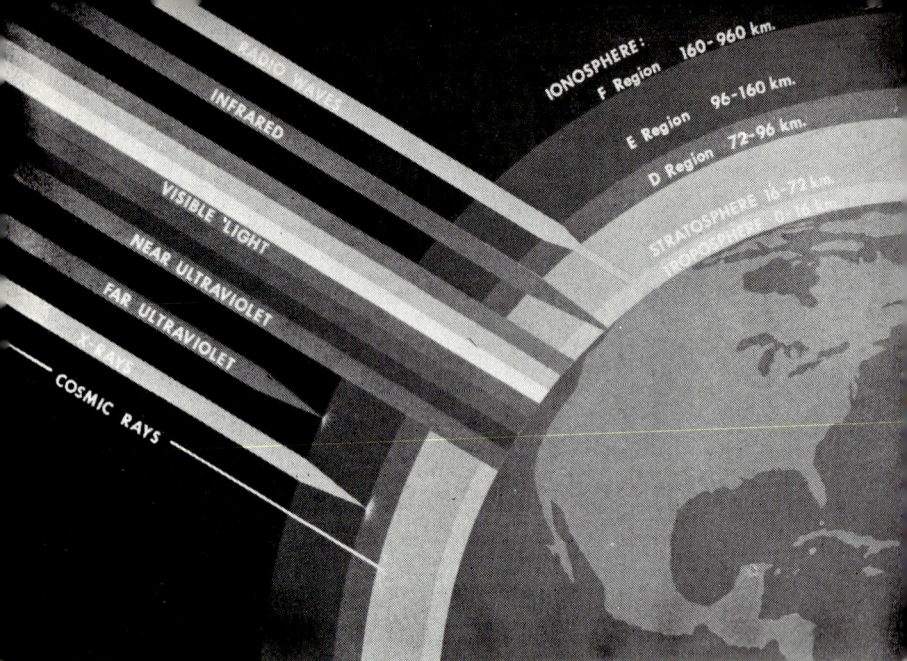

Figure 1-8. The Earth's atmospheric shield to electromagnetic spectrum.

in the atmosphere — the radio and the visual window. *(Figure 1-8.)* Even the visual window will not allow high-resolution observations because of the turbulences in the atmosphere. To study the universe with increased effectiveness, we must get above this protective shield. Man has begun to do this with sounding rockets and manned spacecraft. Here are some of the results.

In gamma-ray astronomy, rocket experiments indicate that celestial gamma-rays exist though no discrete sources have yet been detected. Gamma-rays of energy near 1 MeV were detected by an experiment in 1962. It has been determined that the background flux of cosmic gamma-rays is much lower than originally estimated.

X-ray studies in space were first carried out, using detectors to scan the sun. In 1962, the first X-rays from sources outside the solar system were detected. These X-rays were found to come from the direction of Scorpio in the general direction of the galactic centre. The total X-ray power from this source was observed to be 2×10^{-7} ergs per square centimetre per second which is equivalent to the visible light output of a fifth magnitude star, one just visible to the naked eye. Recently the source has been identified with a blue, variable, thirteenth magnitude star, a very, very weak image. Because more energy is emitted in the form of X-rays than in visible light, it is theorized that this is an old star which has lost its outer surface.

About 20 other X-ray sources have been discovered. Most of these are strung along the Milky Way, more or less concentrated within 29° on both sides of the direction toward galactic centre. However, before positive identification can be made, these stars must be studied with space-based instruments having much better pointing and resolving capabilities.

In ultraviolet astronomy, manned satellites have played the largest role. In July, 1966, the Gemini X astronauts Young and Collins spent 40 minutes taking the first ultraviolet spectral photographs from their spacecraft. They used a Maurer 70-mm camera with an ultraviolet transmitting lens and a refraction grating to obtain the spectra. All the exposures were about 20 seconds long. In the initial analysis, the spectra of 54 stars were identified. About half of these were bright enough to permit quantitative analysis of the ultraviolet energy curves.

On Gemini XI, astronauts Conrad and Gordon took photographs of ultraviolet stellar spectra with an instrument similar to that used in Gemini X. Objective-prism and objective-grating spectrograms were obtained in six star fields. Six exposures, ranging from 20 seconds to two minutes, were taken of each field. As an example of the results obtained, the grating spectra showed absorption lines in Canopus and Sinus and provided ultraviolet energy distribution data for approximately 50 stars.

Infrared studies, although plagued by technological difficulties, have just begun to provide data on stars and to confirm terrestrial observations on the existence of water vapour in stellar spectra. Extension of these techniques will permit the study of the energy distribution for all parts of the spectrum and thus improve the knowledge of stellar structure.

Some radio observation experiments have been conducted in space to study the cosmic background noise below 10 Mhz frequency. Radio astronomy experiments have not been the prime objective of past satellite missions, but radio astronomy Explorer satellites will soon be launched. They are designed specifically for medium resolution measurements of the cosmic background and observations of low-frequency radio bursts from the Sun and Jupiter.

This quick survey of astronomy emphasizes the need to be beyond Earth's atmosphere to observe, not only the stars, but even the sun and planets of our solar system, across the entire electromagnetic spectrum.

H. Planetary Environments

It has long been a major scientific objective to increase our knowledge of the other planets in our solar system, particularly those nearest the Earth. We have now developed the capability of sending probes to these planets, and the achievement of this scientific goal has become correspondingly more important. A detailed knowledge of the planetary atmospheres is a prime initial objective.

Before development of the planetary probe, planetary atmospheric information was based on optical, radiometric, and polarimetric telescopic observations. These observations provided only crude estimates of the desired environmental parameters. For example, estimates of the atmospheric pressure at the surface of Mars derived from polarimetric observations are now believed to be a factor of 6 to 10 too large.

When it became evident that spacecraft could, and would, be sent to other planets, it also became apparent that the success of these missions depended heavily on the accuracy and reliability of the onboard instruments. That accuracy and reliability depended in turn upon the knowledge of the various parameters to be observed — which is the ultimate goal of the programme.

Research efforts were directed toward improving measurement techniques and increasing the knowledge of the planetary environmental parameters. In 1963, from a series of spectrographic observations, it was determined that the atmospheric pressure at the Mars surface was 25 ± 15 millibars (a value which compares to $1013 \cdot 2$ millibars for Earth), that the atmosphere was composed primarily of carbon dioxide, and that there were 14 microns or less of precipitable water. Subsequent Earth-based observations have established an even lower surface atmospheric pressure of about 4 millibars for a 100% CO_2 atmosphere.

This lower pressure was confirmed in 1965 by an experiment carried on board Mariner IV. The spacecraft trajectory was such that radio signals from Earth were refracted as the vehicle, when viewed from Earth, passed behind the planet (*Figure 1-9*) An experiment to be flown in 1969 should provide much more refined information, since it will probe the atmosphere of Mars with a dual frequency radio signal: one frequency will be refracted by the neutral atmosphere and the other by the ionosphere.

The Mariner IV also returned 21 photographs of the Mars surface, revealing that it is similar to the Moon's surface with many craters of varying size. The resolution of these photographs, however, is insufficient to settle the question of whether or not there is life on Mars.

Prior to the space age, theoretical calculations had placed the temperature of Venus in the 300°K range. The occultation of Regulus by Venus had provided information on the temperature of the cloud tops (235°K) and the pressure at 55 kilometres above these tops. There were, and still are, no observations of the actual solid surface because of the thick continuous cloud cover. However, radar observations indicate that the surface temperature is approximately 700 $\pm$ 100°K. This, combined with previous

PARAMETER	PRE 1958 DATA	EARTH BASED DATA	MARINER IV DATA
SURFACE TEMP	230-270°K	211 $\pm$ 20°K	180 $\pm$ 20°K
SURFACE PRESSURE	85-100 mb	5-20 mb	4.1 - 5.7 mb
COMPOSITION	98% N_2, 2% CO_2	~100% CO_2	~100% CO_2
MAGNETIC FIELD	1/7 OF EARTH	—	1 1000 OF EARTH
IONOSPHERE	F REGION ON EARTH	—	F OR E REGION ON EARTH
RADIATION BELTS	MODERATE	—	VERY WEAK

FOR 50° S LATITUDE AT 1300 LOCAL MARTIAN TIME IN LATE WINTER
E REGION 60-100 KM ABOVE EARTH'S SURFACE
F REGION 100-600 KM ABOVE EARTH'S SURFACE

Figure 1-9. Mars data.

BI

PARAMETER	PRE-1958 DATA	EARTH BASED DATA	MARINER II DATA	MARINER V DATA
SURFACE TEMP	500-680 K	700+20 K	600 K	550 K
SURFACE PRESSURE	10-100 ATMOSPHERES	4-30 ATMOSPHERES		15 ATMOSPHERES
COMPOSITION	ONLY CO_2 IDENTIFIED	4 TO 96% CO_2 (?)% N_2		100% CO_2
MAGNETIC FIELD	SEVERAL TIMES STRONGER THAN EARTH		< 1/20 OF EARTH	< 1/300 OF EARTH
IONOSPHERE	SIMILAR TO EARTH		NONE INDICATED	WEAK
RADIATION BELTS	SIMILAR TO EARTH		NONE	NONE

Figure 1-10. Venus data.

estimates of surface pressures of 30,000 to 100,000 millibars and little or no oxygen and water vapour, presented a picture of an environment markedly unsuitable for the development of most of life forms prevalent on Earth. Radar observations from the Mariner II spacecraft confirmed the existence of these extremely high temperatures at the solid surface of the planet. In October, 1967, scientists of the USSR ejected a probe into the atmosphere of Venus from a fly-by satellite. Simultaneously, American scientists remotely probed the atmosphere from another fly-by satellite. The USSR probe, which descended into the atmosphere on a parachute system, confirmed that the surface temperatures were high, that the atmosphere was almost entirely CO_2, and that the surface pressure was approximately 27,000 millibars. The American satellite reconfirmed the high surface temperatures and also provided information concerning the charged particle and magnetic field environments. The surface pressure inferred from this data is 15,000 millibars (*Figure 1-10*), quite lower than the Russian value.

I. Lunar Technology

For many years, man has speculated about the surface characteristics of the Moon and the origin of its surface features. He has conducted a multitude of observations, using such diverse tools as radar, radiothermal measuring techniques, telescope, and experiments in the laboratory. To describe the creation of the lunar terrain, he has examined various terrestrial craters (man-made and meteor impact) and studied major types of geological phenomena. To this array of techniques and tools has now been added reconnaissance by probe. Let us now review some of the results of the Orbiter and Surveyor programmes and how these data have benefited lunar technology.

Even before the "hard" landings of Russian spacecraft, the "soft" landings of United States' Surveyor spacecraft on the Moon's surface, and the various photographic missions of the United States' Ranger probe, a large amount of topographic and geological data had been obtained from Earth-based analysis using telescopic methods. Although the maximum photographic resolution with the telescope was only 1 kilometre and the visual resolution was only 0·5 kilometre, a broad stratigraphic and structural framework evolved. This type of work led to tentative identification of large impact features, structural trends, and major sites of lunar volcanic activity. Laboratory cratering studies, and field studies of impact, volcanic, and man-made craters (produced by both nuclear and high-performance explosives) helped to predict the small-scale surface roughness and probable material characteristics for the upper layer of surface which could not be observed from the Earth. Although these data provided confidence that a space-craft could land on the Moon, there was still no definitive know-ledge of either the physical and chemical composition or origin of the lunar surface. To accomplish these objectives, the Orbiter and Surveyor programmes were initiated. Both programmes pro-duced great masses of data, still undergoing extensive analysis.

These programmes are characterized by the excellent quality of their photographs and other related information which enables man to examine large amounts of lunar data with various scales of resolutions. The Orbiter photographic environmental and tracking data and the "ground truth" data from the Surveyor spacecraft, have allowed man not only to locate suitable landing sites for the Apollo programme but to understand the lunar gravity field and the "figure of the moon".

In addition, these studies have given man greater confidence that the moon is sufficiently hospitable — from an engineering stand-point — for manned landings and exploration. Likewise, they have

Figure 1-11.) North Wall of the Crater Copernicus.   Lunar Orbiter photograph).

Figure 1-12. Oriental Basin (Lunar Orbiter photograph).

shown that'the Moon is a geologically diverse and interesting body, shaped by both impact and volcanism. Perhaps the most spectacular data obtained from the Orbiter programme were: (1) the photograph of the Earth from the vicinity of the Moon, (2) the almost

Figure 1-13. Ejecta material and Lunar ridges (Surveyor VII photograph).

complete photography of the far side of the Moon, (3) photography of the selected Apollo landing sites, (4) the oblique photographs of the crater Copernicus and many other geologically interesting science sites (*Figure 1-11*), (5) the photography of the polar regions, (6) the photography of the very large Orientale Basin (700 km in diameter) located on the edge of the Moon (*Figure 1-12*), (7) the photography of the Surveyor I spacecraft resting on the lunar surface and (8) the many structural features as seen in the Marius Hills area which suggests that convection and differentiation have occurred during the origin of the Moon.

Some of the interesting results from the Surveyor programme were (1) photography of the lunar terrain, both near and far afield (*Figure 1-13*); (2) the close, high-resolution (0·5 mm) photography of the lunar soil; (3) determination of soil properties, including bearing strength and analysis of the chemical composition; (4) manoeuvre of the spacecraft after the initial landing and operation of the spacecraft and its associated science equipment in the lunar environment; and (5) development of higher confidence for future landing attempts on other planetary bodies.

The ultimate benefits of the Orbiter and Surveyor programmes to lunar technology will not be realized until all the available data can be studied and digested.

The potential of space technology, which has already produced many tangible benefits, has been briefly discussed here. So far, it can be concluded that the space technology effort has been a strong motivator and catalyst for education; it has at the same time increased the rate of invention and the definition of basic knowledge, and has, in truth, created an exciting environment for the years to come.

CHAPTER 2

The Apollo Hardware

A. Purpose of The Apollo Programme

In 1961 President Kennedy launched the United States' manned lunar landing programme with his announcement that "this nation should commit itself to achieving the goal, before this decade is out, of landing a man on the Moon and returning him safely to Earth". With these words, Project Apollo[1] became a reality; its direct mission — a manned lunar landing and return.

The initial endeavour during this landing will be confined to a stay time on the Moon of about 18 to 20 hours and to exploration of the immediate vicinity of the landing site, conducting selected measurements, providing photographic coverage, collecting samples of lunar rock and soil, etc. Later lunar landings are planned for longer stay times on the surface and extended exploration of larger areas.

In addition to these objectives, the Apollo programme provides an expanded space flight capability for manned and unmanned operation in Earth orbit[2] and for penetration into deep space as to vehicle performance and hardware. It furthermore enhances training, engineering and scientific knowledge and capacity in space flight, provides an increased industrial base with facilities and the launching sites to support the development of this vast new frontier.

B. Mercury, Gemini and Early Saturns

The manned space flight programme of the United States had its inception with the Mercury project. Involving the flight of a relatively small single couch capsule weighing approximately 3500

---

1 The basic flight hardware of the Apollo programme; namely, the *Apollo space vehicle* (*Figure 2-1*), comprises the Saturn launch vehicle (*Figure 2-2*) and the Apollo spacecraft (*Figure 2-8*).

2 The uprated Saturn I (Saturn IB) launch vehicle can carry up to 40,000 lbs. of payload into an Earth orbit, the Saturn V up to 280,000 lbs.

pounds, the project passed an important milestone with the sub-orbital manned flight of 15 minutes with Commander Alan Shepard aboard. Its prime objective, however, was achieved with the successful accomplishment of a three-orbit flight by Colonel John H. Glenn,[3] followed by three additional orbital flights of longer duration. Upon its completion, Project Mercury had contributed a wealth of most important and valuable data to space technology and, more specifically, to the follow-on Gemini and Apollo programmes.

In addition to verifying that man could exist, observe and navigate in a space environment, it proved space suit and couch design, guidance and control systems, communication and tracking techniques and the Earth-landing system, to mention only a few.

As the next step leading to the Apollo programme, the Gemini programme was established. The two-man Gemini spacecraft, weighing approximately 7000 pounds, completed ten successful flights in demonstrating the feasibility of spacecraft rendezvous, docking, orbit change manoeuvres, astronaut extra-vehicular activities and many others.

On the longer duration missions, data were gathered on biomedical and behavioural characteristics of crewmen under weightless conditions.

In the field of launch vehicles, important contributions were offered by the early Saturn I programme. This multi-engine, two-stage launch vehicle, with approximately 25,000 pounds payload capability for Earth orbit, fulfilled important objectives. It proved, for instance, the feasibility and practicality of "clustering" engines and the possibility of using liquid hydrogen and liquid oxygen as high specific impulse propellants. Furthermore, it served as a test bed for navigation guidance and control mode and pertaining components, for launching schemes of very large rockets, for engine-out capability,[4] etc.

Of the ten flights in this programme, the last three orbited a micrometeoroid detection satellite for gathering, over a period of

---

3 The late Yuri Gagarin, Soviet Cosmonaut, was first to orbit the Earth.

4 If, in a multi-engine stage, one engine malfunctions during flight, it can be cut off without disturbing the flight goal as it is known, for instance, on four-engine aeroplanes. The Saturn launch vehicle stages are designed to have this "engine-out" capability.

time, statistics on meteoroid density in Earth orbit, information essential to the Apollo designers. After these ten flights, the Saturn I was replaced by the uprated Saturn I.

C. Some Design and Development Considerations

When the manned lunar landing goal was announced in May, 1961, there was no launch vehicle available that could meet the requirement to manoeuvre a 100,000 pound payload into a lunar orbit. The largest launch vehicle planned up to that point was the multi-engine Saturn I which, as discussed earlier, was then still under development and it was too under-powered to accommodate the needs of this new task. A much more powerful launch vehicle, therefore, had to be developed. It is called the Saturn V.

The selection of the Saturn V configuration was contingent, among other requirements, upon one important consideration which should be emphasized. In order to meet the stern demands of cost and established time schedule, maximum application of existing technologies was mandatory. The programme had to be so planned that extensive use would be made of available systems and components wherever feasible.

In the development of smaller rockets, for instance guided missiles, it is customary to test launch a sizeable number of prototype vehicles in order to prove the reliability of all systems until the weapon is considered "operational". This concept of programme planning is impossible for the Saturn launch vehicles and, in particular the Saturn V, because the costs per vehicle and launching are entirely prohibitive. Therefore, only very few vehicles for unmanned test flights could be programmed. After the first two or three test flights have demonstrated satisfactory flight performance, the Saturn V vehicle will be considered "operational" or "man-rated".

This drastic reduction in flight test effort in the Saturn V programme was made feasible by the greatest possible use of components and techniques proven in the Saturn I programme as mentioned before. As a result, the Saturn V third stage is generally patterned after the uprated Saturn I second stage with the same engineering team developing both stages. The Saturn V instrument unit is an outgrowth of the one used on Saturn I and IB.

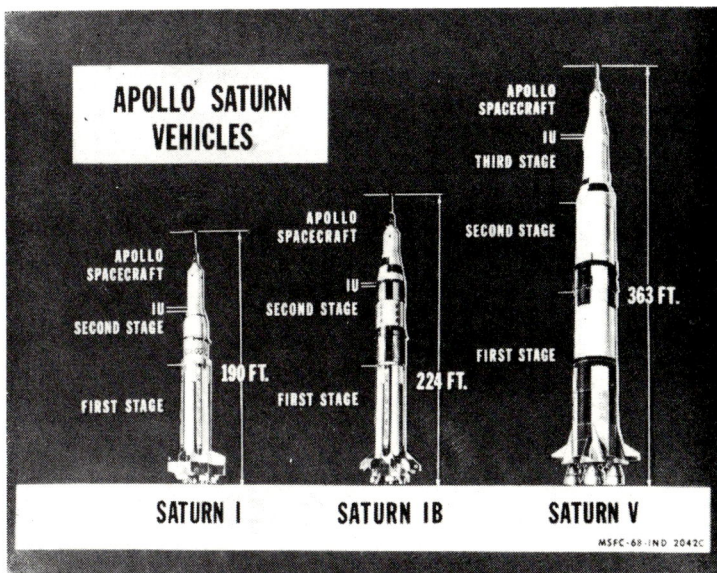

APOLLO SATURN VEHICLES

APOLLO SPACECRAFT
IU
THIRD STAGE

SECOND STAGE

APOLLO SPACECRAFT

FIRST STAGE

363 FT.

APOLLO SPACECRAFT
IU
SECOND STAGE

190 FT.

FIRST STAGE

SECOND STAGE
IU

224 FT.

FIRST STAGE

SATURN I   SATURN IB   SATURN V

MSFC-68-IND 2042C

Figure 2-1.

Figure 2-1 depicts this evolutionary development concept of Saturn V, from Saturn I through Saturn IB (also called the uprated Saturn I). Briefly explained:

The first stage of the uprated Saturn I was developed from the Saturn I and employs a cluster of eight of the same type engines.

The second stages of the Saturn I and uprated Saturn I are different in diameter, have different engines but use the same propellants. The IU, or instrument unit, of the uprated Saturn I is larger in diameter than the Saturn I but uses the same guidance and control principle and similar components.

With the Saturn V, new first and second stages are introduced whereas the third stage is the same as the second stage of the uprated Saturn I. The instrument unit (IU) is the same for both vehicles. The Saturn V will be described in more detail later in this chapter.

The spacecraft on the Saturn I is merely a mock-up. The uprated Saturn I and the Saturn V carry "live spacecraft". Here again only very few are launched as "unmanned" prototypes. The reason is not only cost but mainly the fact that an unmanned spacecraft has to carry much automatic equipment to replace the functions of an astronaut and, therefore, is not a good prototype to resemble a "manned spacecraft".

Although, for reasons stated before, only a few prototypes can be flight tested in the Apollo programme, the reliability of the flight hardware must be as high as possible. Therefore, as a design principle redundancy and failure mode and effect analysis has to be applied wherever feasible.[5] Furthermore, very extensive test programmes for qualifying systems, sub-systems, components and parts have to be conducted on the ground to an unprecedented degree.

In spite of all these concepts and evolutionary approaches of utilizing proven techniques and components, many new systems had to be designed and formidable problems overcome. A few of these complex design and development problems in the launch vehicle area may be of interest:

a. The heat output of the five clustered engines in the first and second stages of the Saturn V is enormous.[6] Particular care, therefore, has to be taken to shield the many sensitive components of the stage against the high-temperature combustion gases and flame radiation. This requirement is complicated by a compelling need to keep the heat and flame shields light in weight.

b. All Saturn V upper stages use a propellant combination of liquid hydrogen and liquid oxygen which provides a very high specific impulse. Although in the early days of rocketry scientists and engineers had pointed out the advantage of liquid hydrogen as a rocket propellant, the application is quite difficult. As a result, liquid hydrogen had never been used in big engines and big rocket stages until the Saturn I programme. One of the main technical difficulties and problems to be overcome is the very low temperature of liquid hydrogen ($-423°$F). Furthermore, hydrogen is very dangerous to handle because it has a tendency to explode viciously if in critical mixture with air. Moreover, as hydrogen is the lightest

---

5 These approaches to the reliability problem will be described in a later chapter.

6 One 1,500,000 pound thrust engine of the first stage represents what is probably the greatest controlled useful energy density ever contained in one package except for the nuclear weapon which, of course, is not comparable in terms of usefulness or control.

element and has a very low density, the propellant tanks have to be quite bulky and some of the advantages of the hydrogen are lost again. In recent years the technology has been mastered quite well and many static tests, as well as flights, have been performed with hydrogen/oxygen engines and stages.

c. The Saturn V mission requires engine re-ignition of the third stage after the vehicle has coasted in orbit for several hours. The unknown behaviour of liquid hydrogen in large quantities under *weightless* conditions created problems of proper settling of the liquids on the outlets in the propellant tanks. Test data obtained during the second flight of the Saturn IB vehicle verified the engineers' design assumptions.

d. The use of liquid hydrogen required the development of highly efficient cryogenic insulation materials. Since so little was known in this field, it was decided to proceed simultaneously with the development of interior as well as exterior insulations for the upper stages. Both the Saturn V third stage and the Saturn IB second stage use interior insulation whereas exterior insulation was selected by the designers of the Saturn V second stage. Successive development tests have proven the suitability of either method and both are still in use today on the respective upper stages.

The Apollo spacecraft confronts the design engineer with particularly complex but most interesting problems. As in the launch vehicle area, but more so in the spacecraft, "systems engineering"[7] has to be especially emphasized. Based on experience with Mercury and Gemini, and on the mission profile (see Chapter 3), a modular concept is applied on the Apollo spacecraft in that it is broken down into three primary systems, namely:

> Command Module (CM)
> Service Module (SM)
> Lunar Module (LM)

7 "Systems engineering" is a technical discipline where the designer considers the functional interplay or interface between the various subsystems of a design, for instance, between ground equipment and flight equipment or launch vehicle and spacecraft or, within a spacecraft, between astronauts' space suits and environmental conditioning systems or navigation, guidance and control system, etc. In a wider sense, "systems engineering" comprises the whole scheme or method of going to the Moon considered from the viewpoint of the engineer who has to translate it into designs and operational procedures.

The intricate functional matching of these systems points out the prominence of systems engineering in the spacecraft.

Some few examples of the many problems in this area may be illustrative:

a. According to the mission profile which will be described in detail in the next chapter, a transposition manoeuvre has to take place after the second burn of the third stage of the launch vehicle. This manoeuvre consists, among other operations, of a coupling of the combination command-service module (CSM) with the lunar module (LM) so that the astronauts can move through a tunnel back and forth between the two modules. The docking, coupling and latching mechanism, which has to connect the modules in an absolutely vacuum-tight fashion, has to be easily separable and has to function entirely automatically. The lightweight design of such a device is extremely complicated to devise, manufacture and test.

b. In case of a malfunction in the launch vehicle after lift-off and during the boost phase of the first stage, an abort system is available which would provide a means for removing the command module from the space vehicle. This would be accomplished by a rocket motor in a similar scheme as it was designed in the Mercury programme. The sensing system for danger (called the Emergency Detection System), the initiation of abort and its execution provided quite some headache to the design engineer.

c. Another difficult engineering problem constitutes the "Environmental Conditioning System" in the CM and the LM and its combined functioning in connection with the temperature, pressure and humidity-controlled space suits of the astronauts.

d. In order to prevent fire hazards in the CM and LM, all nonmetallic materials have been investigated as to their flammability and their capability to propagate fire. All nylon type and rubber-based materials, for instance, are unsuitable for use in manned spacecraft as are most of the polyurethane-type plastics. In order to decrease fire hazards further, Apollo has a 60% oxygen and 40% nitrogen atmosphere on the ground. While the vehicle is gaining altitude, the nitrogen is gradually replaced by oxygen until a pure oxygen atmosphere is reached at cabin pressure of 5 psi.

e. One of the most complex aspects of the whole lunar venture is the landing on the Moon because this cannot well be simulated, demonstrated and practised on Earth. Although some of the unknown existing a few years ago such as bearing strength and condition of the soil, surface structure, micrometeoroid density, temperature conditions on the surface, etc., have in the meantime been favourably clarified by the Surveyor landings and photographs from the Lunar Orbiter, the first landing will be a novel adventure as to navigation, landing spot, illumination on the lunar surface, lunar gravity, etc.

In general as stated before, technologies for chemical rocket engines, gyro-stabilized platforms, electronic computers for guidance and control, lightweight structural designs, and materials were proven and successfully used during earlier launch vehicle development. As typical examples for experience from earlier programmes can be mentioned the use of titanium for the main spacecraft structures, as well as continuing use of ablation heat shields, jet control systems for attitude control and stabilization, parachutes for recovery, design of astronaut couches, communication systems and others. Operation of the fuel cell[8] as a new power source was flight tested for the first time during Gemini flights. Its use is mandatory for the more demanding Apollo lunar land.

D. The Saturn V Launch Vehicle

Saturn V (*Figure 2-2*) is the largest member of the Saturn family. It is a three-stage vehicle capable of sending a 50-ton payload to the Moon or boosting into low Earth-orbit as much as 140 tons. The whole space vehicle configuration, including launch vehicle and spacecraft, stands 363 feet high and, when fully fuelled, weighs over six million pounds.

A more detailed description of the vehicle stages and the instrument unit follows:

Saturn V First Stage (S-IC) (Figure 2-3)

The Saturn V first stage is powered by five engines, providing a total thrust, at launch, of 7·5 million pounds. When completely

8 A fuel cell is basically a device for generating electrical current (dc) utilizing the reverse process of the common electrolysis. In the electrolysis, the chemical compound, water, is separated into hydrogen and oxygen by using electrical current through electrodes. In a fuel cell, hydrogen and oxygen in the presence of an electrolyte are combined, generating electrical current on electrodes and a by-product, water. This water can be used for drinking purposes by the astronauts.

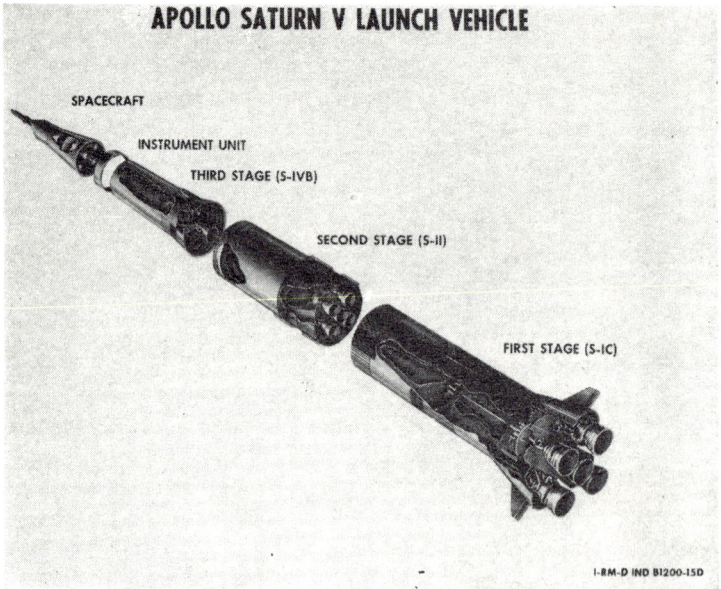

Figure 2-2.

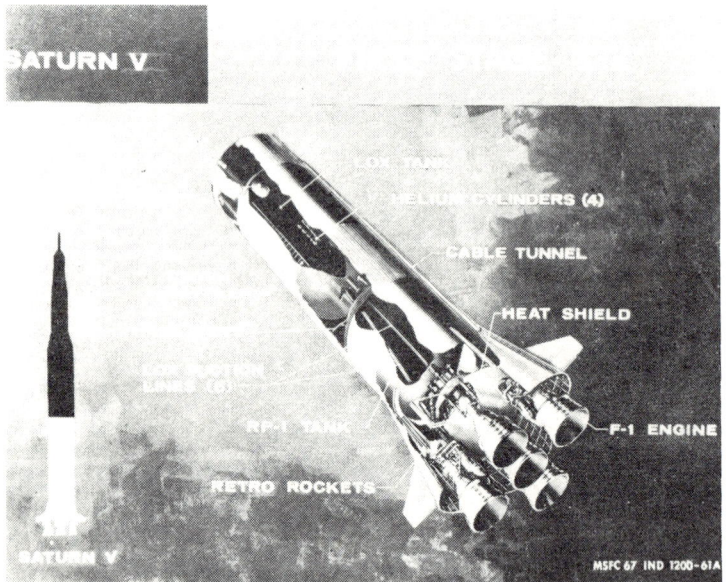

Figure 2-3.

assembled, this stage measures 138 feet in height and 33 feet in diameter. Without propellants, it weighs 304,000 pounds, and at ignition it weighs 4,800,000 pounds. The stage is composed of five major structural assemblies out of high strength aluminium alloy: the thrust structure, the fuel tank, an intertank structure, liquid oxygen (LOX) tank, and a forward skirt.

This stage boosts the space vehicle up to an altitude of about 40 miles and carries it 55 miles down range. The engines burn for about 155 seconds and at cut-off the vehicle speed is about 8900 feet per second (6100 miles per hour).

Saturn V Second Stage (S-II) (Figure 2-4)

The second stage of the Saturn V is the most powerful hydrogen-fuelled rocket stage yet built. It measures 33 feet in diameter and is 81 feet long. Completely assembled, the stage weighs about 88,000 pounds. It carries 790,000 pounds of LOX and about 150,000 pounds of hydrogen for a total stage weight at ignition of about 1,028,000 pounds.

For the lunar mission, the second stage takes over the launch vehicle propulsion from the first stage at an altitude of about 40 miles and boosts the third stage and Apollo spacecraft to nearly 115 miles and 935 miles down range. At the operating altitude, its five rocket engines produce a total thrust of more than 1,130,000 pounds, and increase the vehicle speed from an initial 8900 feet per second (6,100 miles per hour) to 22,500 feet per second (15,400 miles per hour). The engines burn for about 375 seconds.

The second stage structure, again of high strength aluminium alloy, is of quite different design from the first stage. The inter-stage is the mating structure between the first and second stage and houses the five engines which are attached to a thrust structure. The propellant tankage consists of an LO tank on top of which is mounted an insulated hydrogen tank.

An interesting feature of this stage is the common bulkhead which acts, simultaneously, as the upper portion of the LOX tank and the lower portion of the hydrogen tank. The common bulkhead consists of two thin aluminium bulkheads, separated by phenolic honeycomb insulation and bonded together by very precise and sophisticated operations. Use of a common bulkhead

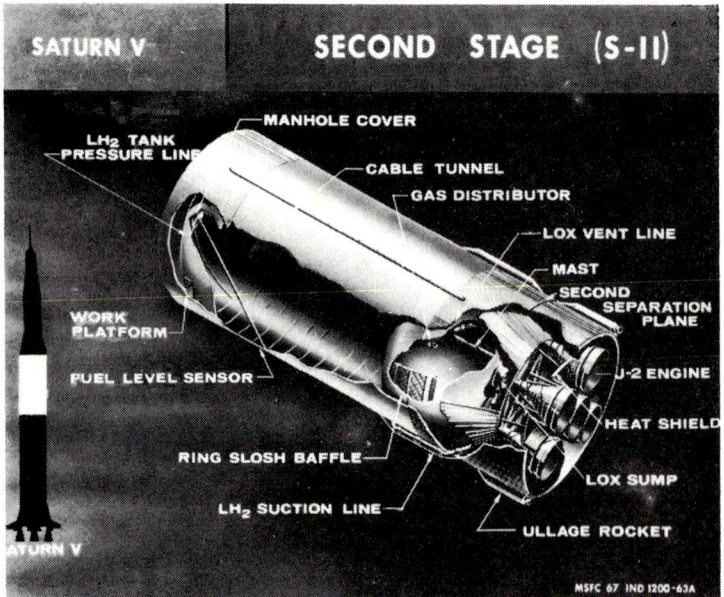

SATURN V — SECOND STAGE (S-II)

Labels: MANHOLE COVER, LH2 TANK PRESSURE LINE, CABLE TUNNEL, GAS DISTRIBUTOR, LOX VENT LINE, MAST, SECOND SEPARATION PLANE, WORK PLATFORM, J-2 ENGINE, FUEL LEVEL SENSOR, HEAT SHIELD, RING SLOSH BAFFLE, LOX SUMP, LH2 SUCTION LINE, ULLAGE ROCKET, SATURN V, MSFC 67 IND 1200-63A

Figure 2-4.

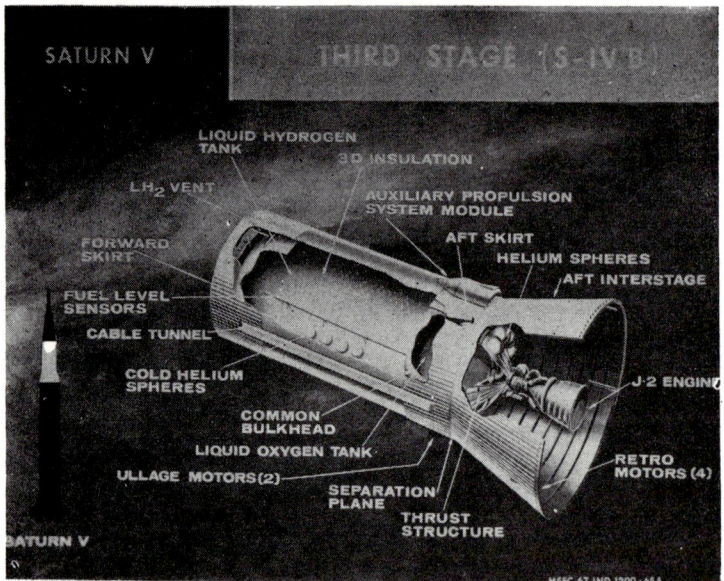

SATURN V — THIRD STAGE (S-IVB)

Labels: LIQUID HYDROGEN TANK, 3D INSULATION, LH2 VENT, AUXILIARY PROPULSION SYSTEM MODULE, AFT SKIRT, FORWARD SKIRT, HELIUM SPHERES, AFT INTERSTAGE, FUEL LEVEL SENSORS, CABLE TUNNEL, COLD HELIUM SPHERES, J-2 ENGINE, COMMON BULKHEAD, LIQUID OXYGEN TANK, RETRO MOTORS (4), ULLAGE MOTORS (2), SEPARATION PLANE, THRUST STRUCTURE, SATURN V, MSFC 67 IND 1200-65A

Figure 2-5.

resulted in weight savings of approximately four tons and permitted a 10-foot shorter stage than would have been permitted by separate bulkheads.

A lightweight insulation was also needed for the fuel tank wall exposed to the atmosphere. Therefore, an insulation of a phenolic honeycomb filled with foam was developed. It is about $1\frac{1}{2}$ inches thick and is mounted on the external wall of the tank.

Saturn V Third Stage (S-IVB) (Figure 2-5)

The third stage measures 22 feet in diameter and is 59 feet long. Its single liquid hydrogen and liquid oxygen engine is almost identical to and has the same thrust as each of the second stage engines, namely, 225,000 pounds. The empty stage weighs 26,000 pounds and carries 240,000 pounds of propellants. The third stage structure, of the same material as the second stage, consists of the aft interstage, the aft skirt, a thrust structure, propellant tanks and a forward skirt assembly.

The propellant tank assembly has a similar design to the second stage, i.e., with a common bulkhead separating liquid oxygen and hydrogen tanks. However, the insulation of this third stage tankage differs quite markedly from that of the second stage. The tank wall bears its insulation on the interior surface.

Although the third stage of the Saturn V vehicle is the smallest of the three, it performs the most demanding mission since it not only has to provide propulsion for insertion into a parking orbit (first burn) but it also must power the initial segment of the lunar transfer trajectory (second burn). During the first burn of $2\frac{1}{2}$ minutes, the velocity increases from 22,500 to 25,500 feet per second. For a lunar mission, the second burn of approximately $5\frac{1}{2}$ minutes of the stage places the spacecraft on its way to the Moon by increasing the velocity to 35,000 feet per second.

A unique feature tied to these special requirements is the *continuous vent system* which provides a small thrust force to keep propellants at the aft end of each tank during coast. This force is augmented by ullage motors[9] to be fired prior to engine restart.

---

9 Ullage is defined as the gas volume on top of the liquid in a tank under gravity condition. At O-gravity, the liquid and gas have an undefined position. If a force such as above forces act in flight direction, the propellant masses are seated on the engine inlets and the engine can be started again.

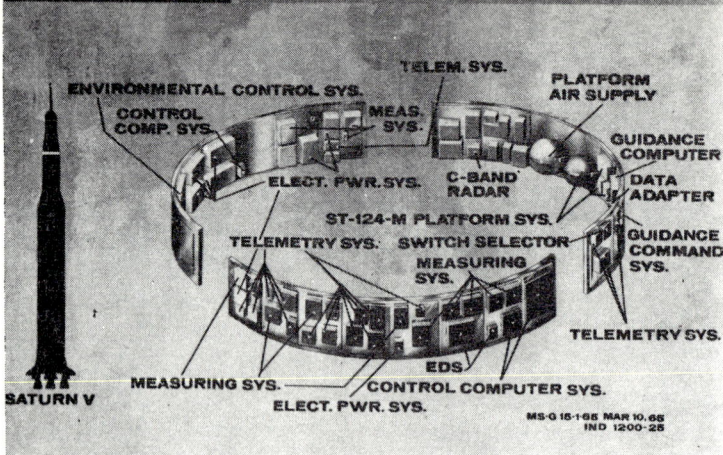

Figure 2-6.

Saturn V Instrument Unit (IU) (Figure 2-6)

The instrument unit is the nerve centre of the Saturn V. It contains the equipment needed for guidance, tracking, communication, environmental control equipment for temperature control, batteries and power supplies to furnish operating power for electronic equipment. The IU is 22 feet in diameter and 3 feet high. As a load-bearing part of the vehicle, it supports the spacecraft. The structure is manufactured in three 120 degree segments, each consisting of thin-wall aluminium honeycomb.

An environmental control system (ECS) cools the electronic equipment in the IU. A water/methanol coolant is circulated through cold plates from a reservoir within the IU. Heat generated by the electronic components is transferred to the coolant by conduction. This heat is finally dissipated through a heat exchanger which cools through the process of sublimation.

The *guidance and control system* of the Saturn V is installed in the IU. Major components include an inertial guidance platform, the launch vehicle digital computer (LVDC), the launch vehicle data adapter (LVDA), an analogue flight control computer, and control and rate gyros. These components have to control the space vehicle for the three main phases of flight: atmospheric-powered flight, boost period after initial entry into space, and the coasting period. During the atmospheric phase, the guidance and control

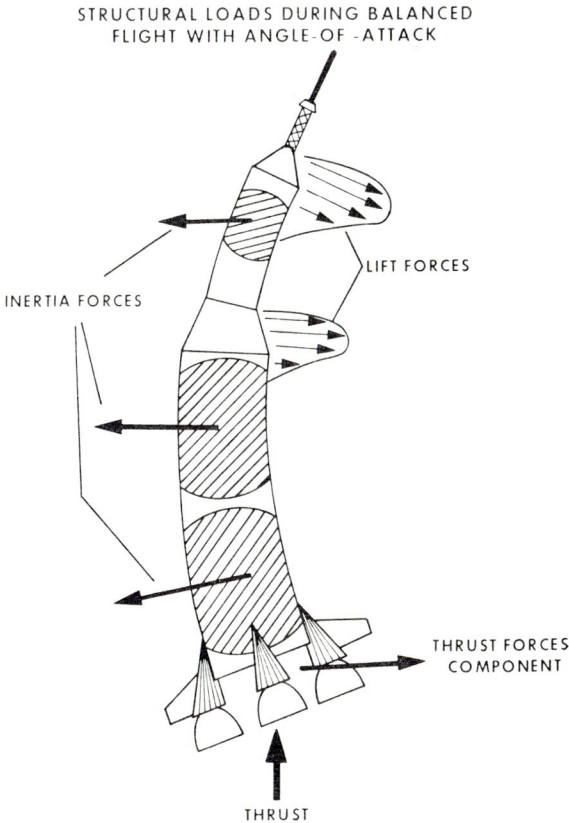

STRUCTURAL LOADS DURING BALANCED
FLIGHT WITH ANGLE-OF-ATTACK

LIFT FORCES

INERTIA FORCES

THRUST FORCES
COMPONENT

THRUST

Figure 2-7.

system is programmed to steer the vehicle along a trajectory which minimizes[10] flight loads on the vehicle. Engine cut-offs and the stage separations are commanded when the IU gets a signal from propellant level sensors that the level has reached a predetermined point. During the second stage powered flight, the launch vehicle

10 This is to say that extreme corrections are avoided to preclude introduction of intolerable dynamic loads on the structure. (See *Figure 2-7.*)

digital computer guides the vehicle along an optimum flight path. During coast, the navigation and guidance information in the digital computer can be updated by data transmission from ground stations through the IU radio command system.

The IU has an extensive instrumentation system which is composed of the measuring system, the telemetry system and the radio frequency system associated with tracking, command and the antenna.

The instrument unit also contains vital parts of the *emergency detection system* which is designed to detect malfunctions. This equipment checks engine thrust, monitors guidance computer status, attitude rates, angle of attack and abort indicators. This information is routed to the emergency detection distributor in the electrical system. The distributor, an interconnector and switching point, has logic circuits which automatically match the input data against predetermined parameters to determine the extent of emergency, if any. If emergency conditions are revealed, the equipment energizes a light, readily seen by the astronaut pilot. If the spacecraft abort selector switch is in the automatic abort position, the abort will take place without further crew participation. Under these conditions the action cannot be vetoed by the astronauts. However, if the selector switch is in the manual position, the crew, consulting with NASA flight controllers, decides when to abort a mission.

E. **The Apollo Spacecraft** (*Figure 2-8*)

As described in a previous paragraph of this chapter, the spacecraft comprises three main systems: command module (CM), service module (SM) and lunar module (LM) formerly also called the lunar excursion module. In addition to these primary systems, there is a launch escape system (LES) and a spacecraft lunar module adapter (SLA).

Command Module (CM) (*Figure 2-9*). The command module or, as it is called on aeroplanes, the cockpit, houses the three-man flight crew and necessary equipment as described later. The primary structure is a pressurized compartment with an access hatch. This compartment is surrounded by a metal structure carrying the heat shield and forming a conically shaped exterior.

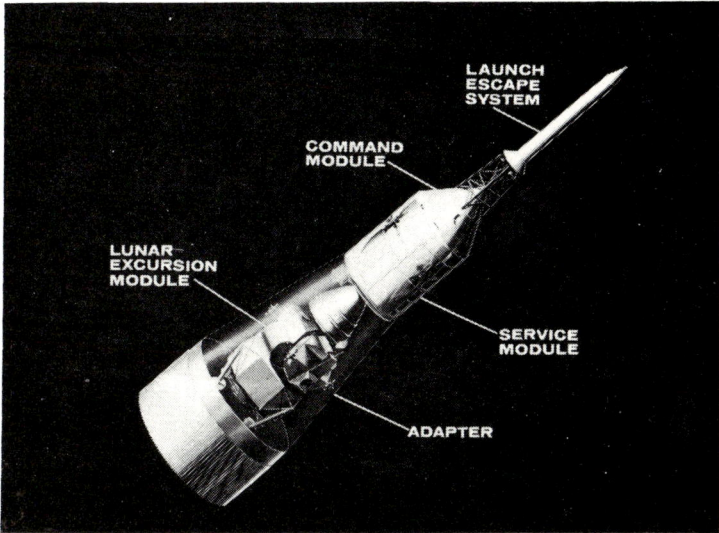

Figure 2-8. The Apollo spacecraft.

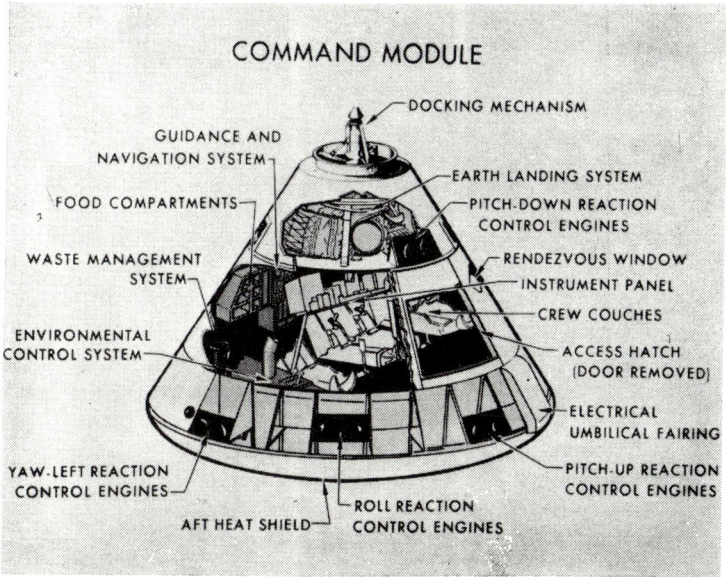

Figure 2-9.

In addition to this heat shield, there are two other structures, namely, the forward heat shield and the aft heat shield. All these structures are coated with ablative material which protects the structure from aerodynamic heating caused by the air friction upon re-entry into the Earth's atmosphere with a velocity of approximately 36,000 ft/sec. This protection process works in the following ways:[11]

1. Heat is absorbed in the decomposition of the outer layer of the ablative material, which then is carried away by the air stream and thus exposes new and cold material.
2. The gases given off by the decomposition are injected into the aerodynamic boundary layer to reduce the heat input.
3. The hot outer layer of charred material reduces the heat input by reducing the temperature difference between the boundary-layer air and the surface.
4. The same hot outer layer radiates heat away from the surface.
5. The undecomposed virgin material below the surface acts as an insulator to absorb heat and slow down the rate at which heat is transferred to the underlying structure.

The relative positions of the heat shields with respect to the primary structure divide the command module into three sections: forward compartment, crew compartment and aft compartment.

The *forward compartment* is an area between the apex of the forward heat shield and the upper side of the forward bulkhead. The centre portion is occupied by a forward tunnel which permits crew members to transfer to the lunar module and return to the crew compartment during the performance of lunar mission tasks. The perimeter of the forward compartment, also called the "upper deck", is divided into four 90-degree segments which contain the recovery equipment, two reaction control motors and the heat shield separating (jettisoning) mechanism. The major portion of this area contains the active components of the Earth landing systems consisting of three main parachutes, three pilot parachutes, two drogue parachutes and drogue and pilot parachute mortars. Four thruster-ejectors are installed in the forward compartment to eject the heat

---

11 A measure of the effectiveness of the system is given by the temperature gradient between heat shield outer surface and spacecraft interior. This gradient lies between 5000°F and 70°F.

shield during landing operations.   The thrusters operate in conjunction with the heat shield release mechanism to produce a rapid, positive release of the heat shield, preventing parachute damage.

The *crew compartment* is a pressurized, three-man cabin with pressurization maintained by the environmental control system. The crew compartment contains spacecraft controls and displays including guidance and navigation equipment, electrical and electronic equipment, observation windows, access hatches, food, water, sanitation and survival equipment.   The compartment incorporates windows and equipment bays as part of the primary structure.

The *aft compartment* is an area encompassed by the lower portion of the crew compartment heat shield, aft heat shield and lower portion of the primary structure.   This compartment contains ten reaction control motors, impact attenuation structure,[12] instrumentation, electrical power, and storage tanks for water, fuel, oxidizer and gaseous helium.

Service Module (SM) (Figure 2-10). The main function of the service module is to give rocket thrust impulse for the following manoeuvres:

1. Braking the velocity of the spacecraft into lunar circular velocity upon approaching the Moon.
2. Escaping the Moon toward Earth
3. Midcourse corrections on the way to and from the Moon.
4. Abort in case of emergency after the emergency escape system has been jettisoned.
5. Transferral from one lunar orbit to another.

The number of burns during one lunar mission flight of the SM's propulsion system may be up to 15.

In order to perform this main function, the service module is equipped with a liquid rocket engine of 21,500 pounds thrust using hypergolic propellants,[13] nitrogen tetroxide as oxidizer and a hydrazine compound as fuel.   For feeding the propellants from the tanks to the engine, helium pressure is used in contrast to the pump feeding systems of the launch vehicle stages.

---

12 A crushable honeycomb that absorbs impact loads by deformation.

13 Hypergolic propellants are chemicals which react with each other (ignite) instantaneously without a foreign ignition source.

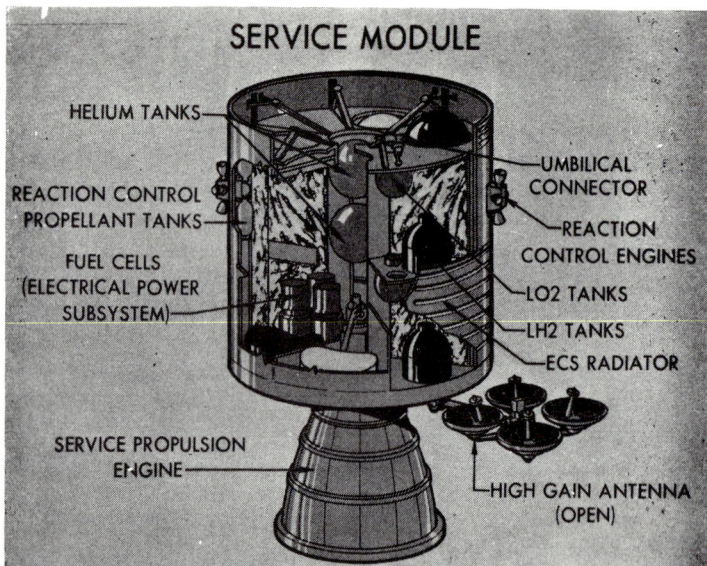

SERVICE MODULE

HELIUM TANKS

REACTION CONTROL
PROPELLANT TANKS

FUEL CELLS
(ELECTRICAL POWER
SUBSYSTEM)

UMBILICAL
CONNECTOR

REACTION
CONTROL ENGINES

LO2 TANKS

LH2 TANKS

ECS RADIATOR

SERVICE PROPULSION
ENGINE

HIGH GAIN ANTENNA
(OPEN)

Figure 2-10.

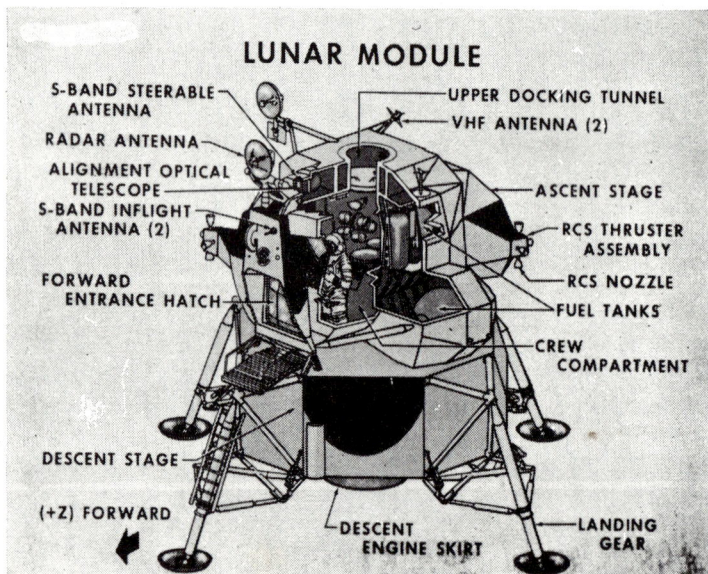

LUNAR MODULE

S-BAND STEERABLE
ANTENNA

RADAR ANTENNA

ALIGNMENT OPTICAL
TELESCOPE

S-BAND INFLIGHT
ANTENNA (2)

FORWARD
ENTRANCE HATCH

DESCENT STAGE

(+Z) FORWARD

UPPER DOCKING TUNNEL

VHF ANTENNA (2)

ASCENT STAGE

RCS THRUSTER
ASSEMBLY

RCS NOZZLE

FUEL TANKS

CREW
COMPARTMENT

DESCENT
ENGINE SKIRT

LANDING
GEAR

Figure 2-11.

In addition to this main propulsion system, the SM carries re-action control systems for attitude control and special small impulse manoeuvres, the fuel cells for electrical power generation, radiators for heat dissipation from CM, etc.

For long-range communication between spacecraft and Earth, the SM contains two antenna systems representing redundant links for this important operation. The one system is a cluster of 2000 Mhz high gain antenna dishes being deployed after separation from S-IVB and the other consists of two 2000 Mhz VHF omni antennas. Transmitted over these systems are mainly the voices of the astronauts and telemetry data.

Lunar Module (LM) (Figure 2-11). The lunar module consists of two stages: the descent stage and the ascent stage. They are mechanically and electrically connected with each other. These connections are separated at the launching of the ascent stage from the surface of the Moon.

The Earth launch weight of the LM is approximately 32,000 pounds.

The *descent stage (Figure 2-12)* is the unmanned portion of the lunar module. It consists only of that equipment necessary for descending from the lunar orbit, for landing on the lunar surface, for serving as a platform for launching the ascent stage after com-pletion of the lunar stay. In addition to the descent engine and related components, the stage houses the descent control instrumen-tation; scientific equipment; and tanks for water, oxygen and hydro-gen. The landing gear is attached externally to the descent stage.

The stage is constructed of aluminium alloy. The descent engine, located in the centre compartment of this stage, can be throttled from a high thrust of 10,000 pounds to approximately 1000 pounds in order to accommodate the decreasing weight during descent. The propulsion system is pressure fed and operates with hypergols (see SM).

The cantilever-type landing gear consists of four equally spaced legs with footpads connected to outriggers that extend from the ends of the descent stage structural beams.

At launch the landing gear is stowed in a retracted position and remains retracted until shortly after separation from the third stage.

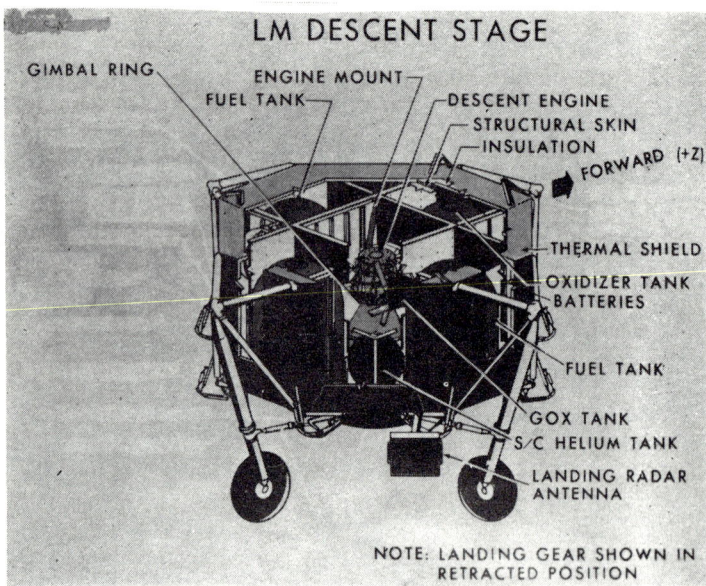

Figure 2-12.

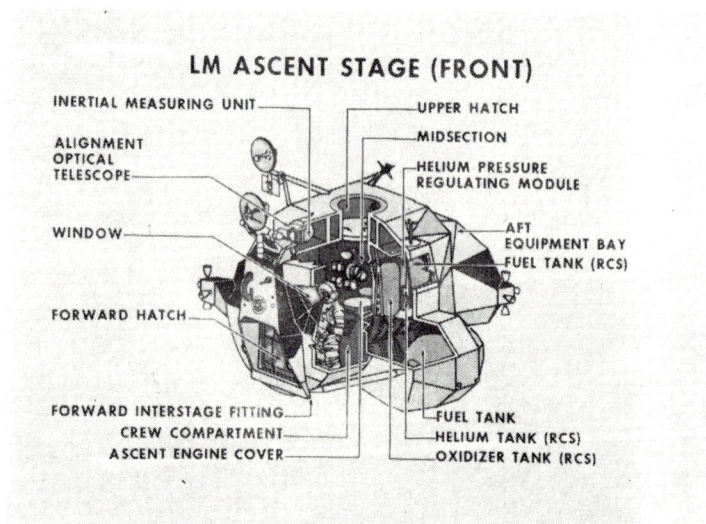

Figure 2-13.

Next, landing gear locks are pyrotechnically released and springs in the drive-out mechanism extend the landing gear. The landing gear is then locked in place and, with the landing gear in position, the lunar module is ready for the touchdown. In order to absorb the impact loads upon landing, all legs are designed with crushable attenuator inserts.

The *ascent stage* (*Figure 2-13*) comprises the propulsion system, the guidance and navigation system, the environmental control system and the electrical power system. It consists of the crew compartment, the midsection, the aft equipment bay and the tanking sections. The two-man crew is housed in the crew compartment, from which they control the flight. This compartment is also used as the operations centre for the crew during the lunar stay. It contains a forward hatch and tunnel, controls and indicators and items necessary for crew comfort and support. The environmental control system located in the midsection keeps the crew compartment under approximately 5 psi pure oxygen atmosphere and a temperature of approximately 75°C.

The upper docking tunnel, at the top centreline of the midsection, is used during the transposition manoeuvre and for transfer of the two crew members from CM to LM, and return. The forward hatch at the lower front of the crew compartment is used for leaving and entering the lunar module while on the lunar surface.

Two triangular windows in the front face of the crew compartment provide visibility during the descent transfer orbit, lunar landing and the rendezvous and docking phases of the mission. Both windows are canted down and to the side to permit adequate side and downward view.

Launch Escape System (LES) (*Figure 2-14*). As indicated before, the launch escape system provides a means for removing the command module from the space vehicle should there be a pad abort or suborbital flight abort. The need is apparent. Should the launch vehicle fail structurally, the propellants would combine to form an explosive mixture and the resulting blast would destroy the spacecraft.

NOSE CONE
AND Q-BALL
ASSEMBLY

BALLAST
ENCLOSURE

PITCH CONTROL
MOTOR
COMPARTMENT

CANARDS

LAUNCH
ESCAPE
MOTOR

TOWER
JETTISON
MOTOR

STRUCTURAL
SKIRT

$+X_C$

$+Y_C$ $+Z_C$

TOWER
STRUCTURE

BOOST
PROTECTIVE
COVER

TOWER
SEPARATION
BOLTS

COMMAND
MODULE

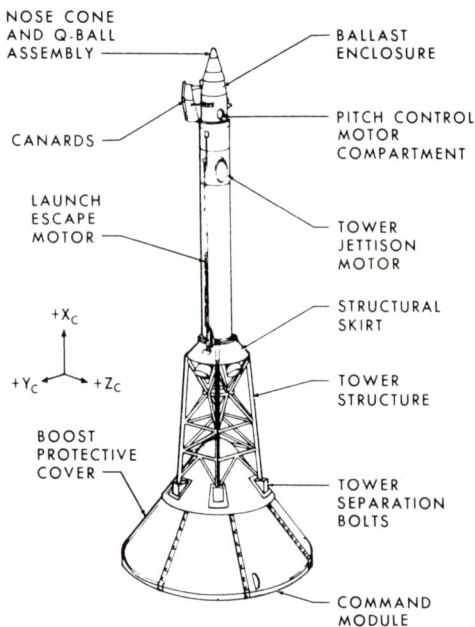

Figure 2-14. Apollo Launch Escape System.

The LES consists of a nose cone with angle of attack meter (Q-Ball), a pitch control motor, a canard system,[14] a 150,000 pounds thrust solid propellant motor, an open-frame tower and a boost protective cover. The boost protective cover which shields the command module exterior during the launch and boost is fastened to the lower end of the tower. Four explosive bolts, one in each tower leg well, secure the tower to the command module structure. When operative at either jettison or abort mode initiation, explosive squibs fracture the bolts and free the tower, together with the boost protective cover. The rocket motors,

---

14 Generally in an aerodynamic vehicle: a system wherein horizontal surfaces used for trim and control are forward of the main lift surface. In this case: an aerodynamic device to control the abort trajectory.

canard and explosive squibs are activated by electronic sequencing devices within the launch escape system.

The launch escape system is jettisoned after first stage burnout when the vehicle is above the sensible atmosphere and the blast danger is thus greatly reduced. This jettisoning reduces the weight that the upper stages must carry into Earth orbit and to the Moon.

Spacecraft Lunar Module Adapter. The spacecraft lunar module adapter represents a structural interstage between the launch vehicle and the spacecraft. It houses the service module engine expansion nozzle and the lunar module.

The adapter is a structure composed of aluminium honeycomb panels with linear shaped-charge[15] explosive cutters installed at panel junctions. At the time of service module/adapter separation, the charges are fired, the explosive force cuts through the adapter structure and four panels fold back exposing the lunar module.

F. Launch Vehicle — Spacecraft — Launch Site Interfaces

In paragraph **C** of this chapter the importance of the interplay between launch vehicle and spacecraft and between the entire space vehicle and ground equipment was emphasized. The significance of the matching of the characteristics of these systems was pointed out under the term "systems engineering".

Following are some few typical examples of interface problems which have to be resolved before the manned Apollo flight series can be initiated.

a. Environmental conditions to which the spacecraft will be subjected by the operations of the launch vehicle and for which design criteria have to be established. Here we are talking mainly about acceleration and vibration loads, vibration patterns, acoustic noise, dynamic loads during high aerodynamic pressure region, bending modes and general dynamic behaviour of the launch vehicle-spacecraft combination. The means of establishing these conditions are unmanned flights equipped with measuring sensors and telemetry transponders, dynamic tests with certain vehicle configurations on test stands, static firings of engines and stages, theoretical analyses with computer programmes, etc.

---

15 A shaped charge is an explosive device so configured that its energy can be controlled in direction. In this case, control is linear.

b. The trajectory during powered flight. The detail design and shaping of optimum trajectories involves a multitude of considerations such as wind profiles and wind statistics during the months of the year, dynamic characteristics of the vehicle combination, minimum propellant consumption, optimum vehicle performance, cut-off characteristics of the stages, guidance modes, minimum structural loads, etc.

c. Emergency detection and launch abort considerations as mentioned in previous paragraph. In this area of interface, the significance lies in the timely sensing of dangers in the launch vehicle and its reporting up to the display panels in the command module for manual or automatic abort.

d. Other interfaces in the launch vehicle-spacecraft area are of mechanical and electrical nature where detail design co-ordination is mandatory. These are, for instance, mechanical fitting of the various stages and electrical plugs, overall network considerations, vehicle alignment, matching of co-ordinate systems launch vehicle-spacecraft, etc.

e. The interfaces between space vehicle and launch facilities and the integration into the entire launch operation comprise a multitude of aspects and problems. Examples:

1. Onboard computers' input and output has to match the ground computers'.

2. Swing arms carrying tanking lines, electrical cables, high pressure gas lines, etc., have to fit the corresponding inlet connections on the vehicle skin. The uppermost swing arm is the access route of the astronauts to the command module. It is equipped with a "White Room" providing spacecraft inside condition. This interface has to be worked out particularly carefully because of, e.g., contamination control, atmosphere, accessibility, etc.

3. Measuring, tracking and command transponders on board have to respond to pertaining ground equipment.

The technical and organizational co-ordination, with all necessary trade-offs and engineering compromises in the area of interfaces, is one of the most difficult tasks for the managers and leaders of such an extraordinary and complex programme as the manned landing on the Moon.

G. The Manned Space Flight Communications and Tracking Network

Centred at the Mission Control Centre in Houston, Texas, and fanning out across the world, is the elaborate communications and tracking network so important to the success of the mission *(Figure 2-15)*. Through the satellite stations comprising the network, the Mission Control Centre, aided by a data refinement centre at Goddard Space Flight Center on the East Coast of the United States, maintains voice communication and tracking contact with the spacecraft.

Communications can be classified under three general categories. The first of these pertains to monitoring of launch vehicle, spacecraft and crew.

It is obvious, of course, that a reliable voice link is required between crew and ground. Then, since the crew cannot possibly

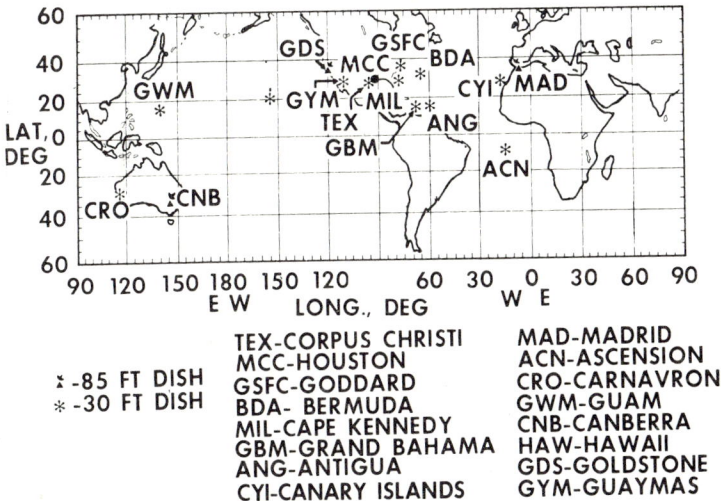

☆ -85 FT DISH	TEX-CORPUS CHRISTI	MAD-MADRID
* -30 FT DISH	MCC-HOUSTON	ACN-ASCENSION
	GSFC-GODDARD	CRO-CARNAVRON
	BDA- BERMUDA	GWM-GUAM
	MIL-CAPE KENNEDY	CNB-CANBERRA
	GBM-GRAND BAHAMA	HAW-HAWAII
	ANG-ANTIGUA	GDS-GOLDSTONE
	CYI-CANARY ISLANDS	GYM-GUAYMAS

Figure 2-15. Manned Space Flight Network.

BJ

take and report the thousands of data readings indicating vehicle performance, an automatic or remote metering system is required. This is furnished through the medium of telemetry which transmits data continuously to ground station recorders. In those cases where the transmission is screened out by virtue of spacecraft position, the data is fed into tape recorders for playback automatically to ground stations when craft position favours communication.

The second category of communications involves trajectory measurements. Both the trajectories of the launch vehicle and of the spacecraft are closely observed and measured from ground stations in a process called "tracking", a function to be discussed in more detail later.

The third communications category includes all the information, other than voice and tracking interrogation, transmitted from the ground to the spacecraft. Since that information usually involves commands to be executed by remote control, this communication channel is called the "command uplink". It is a means of allowing the ground crews to do with the spacecraft what the pilots cannot do, either because they haven't the time, the capability, the equipment, or because they are incapacitated. In coverage, the command uplink system is about comparable to both the voice and the telemetry systems.

How are these functions of tracking, voice communications, telemetering and remote commanding conducted from Earth stations?

The major problem area is tracking. Depending on the space vehicle distance, there are three basic systems for tracking: one based on use of optical equipment and two based on electronic equipment.

As long as the space vehicle is still visible from ground, during the early phases of boost-to-orbit, high precision optical instruments track the vehicle's position and aid in determining velocity. On the periphery of the launch pad at Cape Kennedy there are usually located several dozen documentary cameras including, as an example, an intercept ground optical recorder. This is a heavy 18-inch aperture reflecting telescope with a 35 or 70-mm film-size camera attached to its ocular. Controlled in azimuth and elevation, it has automatic focus and exposure control with a variable focal length up to 500 inches.

For distances beyond the range of visibility, electronic gear becomes necessary. Basically this is radar equipment using either pulsed radio signals or continuous radio energy waves.

These radar systems track the spacecraft on its path. The pulse radar emits short radio pulses which hit the vehicle in flight and are bounced back either by simple reflection from its metal skin (called "skin tracking") or, strengthened in intensity, by a radio beacon or transponder located inside the vehicle ("beacon tracking"). By receiving the returning pulses, the ground station can determine not only the distance to the vehicle but also the line-of-sight direction to it, defined in terms of azimuth, referred to local North, and elevation above the horizon.

The continuous wave or CW-radar systems are based on measuring the phase shift of the radio beam due to the vehicle's speed, otherwise known as the Doppler effect. If a continuous signal of well-stabilized frequency is sent after the vehicle, it is received by it at a somewhat lower frequency because it travels away from the transmitting station. As the received signal is then radioed back to the ground to a second station, the ship is still travelling away from it and an additional Doppler shift occurs in the frequency, changing the "pitch" of the signal. By measuring this shift, the velocity of the vehicle can be determined very accurately and, from it by integration, its position in space.

For the Earth-orbital phases of the Apollo mission and for the near-Earth portions of the voyage to and from the Moon, tracking radars are available from the Mercury and Gemini programmes' tracking networks. Since the latter were concerned with Earth-orbital flights, these radars have a maximum range of only about 930 miles. Obviously, this will cover only a very short portion of the distance to the Moon. Also, the Mercury and Gemini radio stations for communication and telemetry are limited to near-Earth regions. These existing stations, in addition, use a considerable number of different frequency ranges for the functions of tracking, voice, television, telemetry and command. Each link has an individual frequency assigned to it and uses individual equipment. The only common part of the communication equipment is the on-board spacecraft antenna system. There is C-Band radar, in the 5400 to 5900 megaherz range, S-Band radar in the 2700 to 2900 megaherz region and communication networks using Very High Frequency (VHF) and Ultra High Frequency (UHF) systems. These systems, and some others, would be required to meet the specific needs of Apollo and each system would require its own separate ground equipment.

For that reason, it was decided to build the Apollo system, or Manned Space Flight Network (MSFN), using only a single frequency range. Over the lunar distance, the usual radar distance-measuring technique, successfully employed for orbital missions, does not suffice. Therefore, a new ranging system had to be developed, based on pseudorandom codes.[16] Rather than using pulses as before, in this ranging system a long code is transmitted and repeated by the vehicle, allowing accurate range measurement. Because of the wide band width required for this system, the S-Band[17] was selected both for the ranging systems and the communication systems.

The Unified S-Band (USB) system is used for the entire mission from lift-off to landing. It can accommodate all communications between the ground and the spacecraft with one amplifier and one antenna on board the Apollo spacecraft, and identical antennas and support equipment at the ground stations.

Most of the ground stations are clustered around the North Atlantic, above which will occur the launch and insertion into parking orbit. There are, however, important stations in Australia at Carnarvon and Canberra. The tracking station at Carnarvon is part of a network of stations assigned to near-Earth regions in space utilizing 30-foot diameter antennas. The Australian stations are unusually important to the network since, being located about half-way around the Earth from the launch site, they provide the first extremely accurate measure of how well the required flight orbit was attained. Other S-Band stations with 30-foot dish antennas are at Cape Kennedy, Florida, and in Hawaii, Guam, Ascension, Bermuda, Texas, Guaymas, Canary Island, Antigua and Grand Bahama Island.

---

16 A pseudo code is an arbitrary code not directly understandable by a computer. However, by employing an interpreter circuit, the instruction is translated into an instruction understandable by the computer.

17 From 1·55 to 5·2 kilomegaherz.

One interesting feature of the Carnarvon station is the so-called Solar Particle Alert Network (SPAN) which provides data for predicting future solar proton eruptions which would result in radiation hazards to the astronauts. As indicated, the SPAN facility uses a solar radio telescope and a solar optical telescope. With current techniques and knowledge, the SPAN network can provide radiation warnings to the astronauts from one to four hours in advance. Two other stations are located at Grand Canary Island and Lima, Peru, so that the Sun can be monitored continuously for 24 hours.

The Unified S-Band stations with 30-foot dishes are good for distances up to about 12,000 nautical miles. To provide tracking and communication at lunar distances, larger antennas are necessary. For this purpose a Deep Space Network was developed. This network consists of three stations with 85-foot dish antennas, of course also designed for S-Band frequencies. One of these is located at Canberra, Australia. Together with the other two stations (one at Goldstone, California, and the other at Madrid, Spain), the network forms an essentially equilateral triangle around Earth because, if seen from the North Pole, the three stations are spaced approximately 120 degrees apart. Therefore, at least one of these deep-space stations will always keep the spacecraft in its antenna beam which, at the Moon, covers an area about 1400 miles in diameter. Thus, for Apollo mission control on Earth, the space ship never descends below the horizon.

To complete the total coverage of the Apollo network, a fleet of eight four-engine jet aircraft will supplement the ships and land-based stations. Flying at altitudes around 36,000 feet, they will provide two-way relay of voice communications between the spacecraft and surface stations, and reception, recording and transmission of telemetry signals to the ground. Communication links with the ground include a high frequency teletype system.

All data to and from this complex network of stations around the world are brought into a Communications Centre at the NASA Goddard Space Flight Center on the East Coast of the United States. There the information is sorted, somewhat condensed and routed to the Mission Control Center at Houston, Texas.

In conclusion, the total Manned Space Flight Network (MSFN) will provide all necessary contact with the Apollo space vehicle after lift-off. It makes it possible for us to monitor and control the entire trip to and from the Moon from the ground where we have at our command incomparably more resources in terms of computing facilities, readily accessible electronic libraries of systems characteristics and teams of systems experts for decision making than have the astronauts aboard the spacecraft. For the Apollo/ Saturn V mission, the conduct and control of the launch will be the responsibility of Launch Operations Center at Cape Kennedy, Florida, which will rely confidently on the tracking and communication stations of the network. After lift-off and during the flight to and from the Moon, the Apollo mission will be controlled from the Mission Control Center at Houston, Texas. The "eyes", "ears" and the long "reach" of this control centre will again be provided by the world-wide stations of the Space Flight Network. Thus, in a sense, the whole world will have a part in Project Apollo, man's step across space to the Moon.

CHAPTER 3

The Apollo Mission

Introduction

Immediately after the Apollo Programme was established, the first question to be answered was: what would be the most expedient, economical and the safest way to land on the Moon and return?

From a number of possibilities, finally three schemes evolved; namely:

a. Direct route Earth-Moon — landing and direct return from surface of Moon to Earth.

b. Ascent to Earth orbit and retanking from an orbital tanker — Moon landing and direct return from surface of Moon to Earth.

c. Ascent to Earth orbit — second burn of third stage of launch vehicle — transposition with LM and CSM — lunar orbit — decoupling and LM descent to lunar surface with only two astronauts — LM ascent to lunar orbit and docking with CSM — return from lunar orbit to Earth.

Although, at the first glance, mode *c.* looks like the most difficult and cumbersome one, this scheme was selected for a variety of reasons. These reasons involved all kinds of considerations in the area of time schedule and development time, available technology, necessary facilities, money, landing possibilities and manoeuvrability on the Moon, weight and vehicle performance, etc. For the sake of brevity, only one reason for selection of method *c.* may be cited.

Due to lack of knowledge of conditions on a landing site, such as bearing strength of the lunar ground, slopes, rocks or soft soil and other surface features, illumination during landing, etc., it was considered prudent to plan for a lunar landing vehicle as light and small as possible. Since propellants always constitute the highest weight factor in rocketry, it was decided to leave the propellants for the return impulse to Earth in lunar orbit. This also saved propellants for the descent to the Moon and ascent from the Moon. The decision to leave one astronaut in lunar orbit, which also brought other advantages, helped to decrease dimensions and weight of the LM. Expressed in simple terms, by leaving a sizeable mass in lunar orbit; namely, the service module with its return propellants and one astronaut in the command module, only a smaller mass, the lunar module, has to be decelerated for descent and has to overcome gravity of the Moon during landing manoeuvres and ascent. Therefore, a smaller launch vehicle is also needed for method *c.*, especially in comparison with method *a.* For method *b.* two launchings within a period of a few hours would have been necessary.

With these considerations in mind, then, we will turn to a discussion of the mission and how it is to be fulfilled. It must be understood, of course, that, in reality, the mission begins long before the vehicle arrives at the launch site.

A. Launch Preparations and Launch Facilities

Before arrival at Cape Kennedy, each launch vehicle stage is assembled and the propulsive stages give a live firing test — called a "static firing". After static firing, a further period of verification is required to assure that the stage was not damaged or contaminated by the test. The stage is then readied for shipment to the launch site. For reasons not to be explained here, the spacecraft modules (SM and LM stages) are not statically fired.

While the hardware is being designed, developed and ground tested, a very complex non-hardware system is also being designed and prepared. That system comprises the computer programmes which are needed to test, launch and fly the mission. These computer programmes, usually referred to as "software", are in the form of magnetic tape for the ground systems and various forms of core memory units for the onboard flight computers. After the software has been tested and verified, it is sent to the Houston Mission Control Center and to Cape Kennedy for loading into the many computers at these locations for pre-launch tests and the mission. While the computers are being readied, the flight hardware is arriving at Cape Kennedy from many places by sea and air transportation.

Figure 3-1.

Figure 3-2.

Several unique facilities at the Kennedy Space Center (*Figure 3-1*) are used to assist the process of integrating the stages and modules into the Apollo/Saturn vehicle.

The Vehicle Assembly Building (*Figure 3-2*) provides facilities for the inspection, installation, assembly and integrated checkout of the space vehicle. The Vehicle Assembly Building has a total enclosed volume of 129,482,000 cubic feet. It is considered by some to be the largest in the world.

While vehicle assembly proceeds in the Vehicle Assembly Building, several miles away, in the Manned Spacecraft Operations Building, final premating tests are performed on the spacecraft modules. This building houses all the astronaut quarters and their medical clinic. From control rooms in this building, the launch checkout tests for the spacecraft are monitored and controlled. On completion of these tests, the spacecraft modules are moved to the Vehicle Assembly Building, mated to the vehicle, and a series of tests performed verifying that spacecraft and vehicle are fully integrated.

At this point, the fully assembled vehicle is poised on a mobile launcher. The launcher, consisting of a structure 160 feet long, 135 feet wide and 25 feet deep, with a tower extending 380 feet above the top surface, also serves as the platform on which the vehicle is erected and carried from the Vehicle Assembly Building to the launch site. Beneath the tower, within the two-storey platform, there are rooms for a computer and checkout and support equipment. The tower itself supports the service arms which carry electrical lines, pneumatic lines, environmental control lines, fuelling lines, etc., to the eight vehicle umbilicals. The astronauts also gain access to the spacecraft from the tower.

The mobile launcher, with the space vehicle in place, is lifted hydraulically by the crawler transporter and moved from the Vehicle Assembly Building to the launch pad, a distance of three miles, at a speed of one mile per hour. There the mobile launcher is placed on permanent supports and the crawler transporter removed.

The mobile service structure, a tower 402 feet tall, is placed near the vehicle at the launch pad. The structure has movable platforms that encircle the space vehicle and provide access for the launch crew, and will remain in position until about six hours before launch. The structure, weighing about nine million pounds, is moved to and from the launch area by the crawler transporter.

The pre-launch activities now focus on the Launch Control Centre, located beside the Vehicle Assembly Building. The Launch Control Centre is connected to the vehicle, the mobile launcher, and the pad by a combination of wires and radio links. The centre itself contains about 450 consoles and several computers from which operating personnel exercise control of the vehicle and obtain information allowing them to evaluate status and test results. In general, the consoles are divided into related groups that are concerned with individual launch vehicle stages or related subsystems such as propulsion and guidance. In this facility, spacecraft checkout equipment is at a minimum, being primarily status displays, since the comparable spacecraft checkout consoles are located in the Manned Spacecraft Operations Building. Control of the vehicle is exercised either by switches or by a typewriter-like keyboard input to the computers. Test status and results are displayed by lamps, chart recorders, printers, or cathode ray tube displays.

Earlier a series of tests had been run to determine that all vehicle/spacecraft systems, now assembled together, were operational and that no incompatibility existed. During these tests, the vehicle is exercised through all the steps expected to occur during a launch countdown. Simulation was provided where needed. For instance, as the vehicle is not loaded with propellants inside the Vehicle Assembly Building, stimulated signals are transmitted from the propellant level sensors. In other tests, all flight systems are caused to operate as they would during a flight. To do this, special signals are provided to the vehicle in simulation of conditions expected to occur during flight. In this manner, maximum confidence is acquired that both the flight and ground support systems are operational before the vehicle is moved to the launch area.

During a later test — the Countdown Demonstration Test — the vehicle is fully tanked and an actual countdown is conducted down to T-9·0 seconds from expected lift-off. At T-9·0 seconds, the ignition command is given to the Saturn's first stage, but

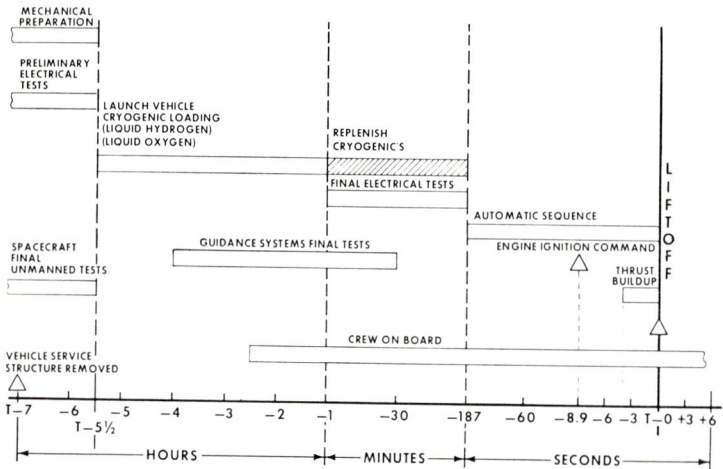

MECHANICAL PREPARATION

PRELIMINARY ELECTRICAL TESTS

LAUNCH VEHICLE CRYOGENIC LOADING (LIQUID HYDROGEN) (LIQUID OXYGEN)

REPLENISH CRYOGENICS

FINAL ELECTRICAL TESTS

AUTOMATIC SEQUENCE

ENGINE IGNITION COMMAND

SPACECRAFT FINAL UNMANNED TESTS

GUIDANCE SYSTEMS FINAL TESTS

THRUST BUILDUP

CREW ON BOARD

VEHICLE SERVICE STRUCTURE REMOVED

LIFT OFF

T−7   −6   −5   −4   −3   −2   −1   −30   −1   87   −60   −8.9 −6 −3 T−0 +3 +6
      T−5½

HOURS ———————————— MINUTES ——— SECONDS ————————

Figure 3-3. Countdown Sequence Outline.

ignition, of course, is not initiated and the test is completed at this point.   The Countdown Demonstration Test is a full-scale dress rehearsal for the hardware and the operating personnel. On completion of the test, propellants, with the exception of the first stage fuel (kerosene), are drained from the launch vehicle, which is put in a stand-by status, while test results are evaluated. If evaluation shows all activities normal, preparation for the actual launch countdown begins.

While the preparation is almost continuous, a strictly-controlled effort for a well-planned final series of tests and preparations starts at about 12 hours before expected lift-off.   At this time — and during the period from T-12 to T-5 hours — preliminary checkout tests are conducted to verify the electrical systems.   The mechanical system is made ready for cryogenic tanking operations.   The vehicle access doors are closed.   The access platforms are removed. After completion of other tests and preparations, the mobile service structure is moved away from the vehicle at T-7 hours.   From that time on (*Figure 3-3*), all planned operations are conducted remotely from the Launch Control Centre with no personnel remaining at the pad area.

Cryogenic propellant loading starts at T-5½ hours. All stages of the launch vehicle must be loaded with liquid oxygen and the second and third stages with liquid hydrogen. The propellant loading activity requires about 4½ hours. During the final hour and until 187 seconds before launch the cryogenic propellant levels are maintained by replenishing these supercold liquids as they boil away.

The terminal count begins at T-60 minutes. As the propellant levels are maintained, all space vehicle systems are brought to their final flight-ready conditions.

The launch conductor in the Launch Control Centre has the job of co-ordinating all activities and ensuring that the countdown sequence proceeds according to plan. Operating personnel report through a well-defined chain of command so that he is aware of the initiation and completion of each required event.

Checkout proceeds to T-187 seconds, at which point a new phase is reached in the countdown. The automatic sequencer and computer now take over, automatically giving all the required commands. The system is so designed that, if any critical system deviates from the normal, the sequence stops without human intervention. At the same time, operating personnel monitor other system parameters from visual displays. These parameters have previously-established limits or "red-line" values associated with them. If a red-line value is exceeded, the person responsible for that parameter requests that the test conductor stop the sequence. If the sequence is stopped, the cause of the discrepancy must be determined, the situation corrected, and the count recycled to a predetermined point — usually about T-22 minutes — and another attempt is made.

During the final 187 seconds, the software is almost totally in charge of the launch. The programmes furnished to the computers in the mobile launch platform and the Launch Control Centre make it possible for these two computers to relay messages to each other during the terminal count. In many instances, it is desirable — and after T-187 seconds it is mandatory — to have an automatic sequence of events take place and to evaluate results of these events automatically. The two computers do that. The volume of data transmitted to and from the vehicle is of such magnitude that the concept of one wire for one signal is impossible, especially when that wire must be over three miles long. Therefore, for data flow and systems control, digital coded messages are transmitted by coax cable and by radio signal between the two computers.

The many switches in the Launch Control Centre for the vehicle are connected by wire to the control centre computer input. There is a corresponding number of devices to be activated on the vehicle. Each of these is connected to the mobile launcher computer output. Since a coded digital message is sent as a result of a switch activation, many different messages (different words) can be sent between the computers over one coax cable. They must be sent in sequence but can be sent very rapidly. In this manner, the commands from many switches to many remote devices can be accomplished by having a large number of wires for the short distance between the switches and Launch Control Centre computer, but only one long coax cable between the two computers.

The reverse of this process is to transmit indicating signals from the vehicle device to the control centre panel lamps. This is, in fact, the normal method of operation in this system. Thus, the two computers "talk" to each other during the pre-launch tests. The system is capable of accomplishing several hundred actions in one second.

The presence of computers in the system offers other advantages. One of these is the ability to gather and display large amounts of data in a very easily readable form. The television-like display can present any combination of letters and numbers, and, in addition, can display graphic forms. The computers can do the computation necessary to change voltage levels to the true engineering units that the voltage represents (such as pressure in pounds per square inch) and to display that information to the operator. The computers are also exploited during automatic sequences, as their speed and accuracy far exceed human ability to perform faultlessly an extended series of routine actions.

The spacecraft checkout system also relies heavily on a computer system. The Manned Spacecraft Operations Building, the equivalent of the Launch Control Centre, is used for spacecraft checkout. There, operators control the spacecraft in a manner similar to that for the launch vehicle. One primary difference is that a computer is not used in the vicinity of the mobile launcher — the only computer is in the Manned Spacecraft Operations Building. Again, as with the launch vehicle, signals (commands) are transmitted to the spacecraft and information is received from it by coded digital messages. Special purpose electronic equipment is used to decode the command messages and give the proper signals to the spacecraft. Computers are vital in bringing the system to T-9·0 seconds.

B. The Launch

At T-9·0 seconds the ignition command is sent by the computers to the launch vehicle. Ignition is confirmed on the closed circuit television screens and by the computer. The centre engine ignites first; a quarter of a second later two opposite outer engines ignite; finally a quarter of a second after that, the final two engines ignite. This thrust buildup in per cent over the time is shown on *Figure 3-4.*

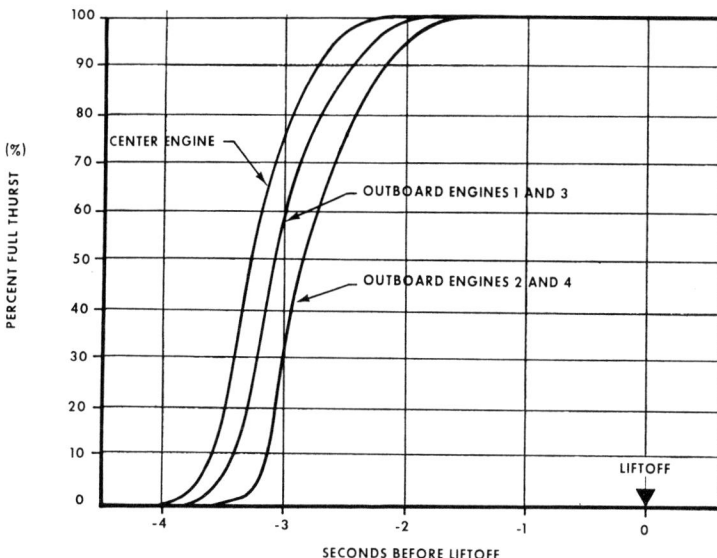

Figure 3-4. Typical First Stage Thrust Buildup.

When engine thrust is at full scale and all systems are operating satisfactorily, the vehicle hold-down arms retract and the Apollo/Saturn vehicle begins its mission to the Moon. The entire period from ignition to lift-off has taken four seconds, and about 46 tons of fuel and oxidizer have been consumed.

As the vehicle rises beside the tower, it is stabilized in the vertical position by the four control engines. At this time the total launch vehicle, together with the spacecraft, has a mass of over 6·2 million pounds. In order to lift this weight, the total thrust force of all the rocket engines has to be higher. In the case of the Saturn V, this force is 7·5 million pounds which results in an initial acceleration of approximately ·12g, indicating that the space vehicle rises very slowly.

With propellants being consumed at a rate of 29,000 pounds per second, the vehicle rapidly becomes lighter. During the same time, the thrust force of the five engines, amounting to over 7·5 million pounds at lift-off, starts to increase with altitude, as explained in an earlier lecture.

After 12 seconds of vertical rise, Apollo/Saturn clears the launch umbilical tower. Instead of continuing a vertical path, the guidance system is programmed to start tilting the vehicle slowly in an easterly direction. Since we want to reach Earth orbit, the rocket will eventually have to assume a horizontal attitude. The sooner this manoeuvre begins, the less we have to fight against the full component of the gravity. Of course, more time is spent in the denser atmosphere this way, but with aerodynamically-shaped space vehicles, atmospheric drag can be tolerated. Extensive study, of course, is required to achieve the desired trajectory. A launch in an easterly direction gives the additional benefit of the Earth's rotation, adding 1000 feet per second or more as a bonus to our flight velocity; the exact velocity value depends on the latitude of the launch base and the exact launch heading.

After about 84 seconds, the vehicle has reached an altitude of 43,000 feet. At this point, the atmospheric forces exerted on the vehicle reach a maximum. These forces are proportioned to the so-called dynamic pressure, which is a function of the atmospheric density and the square of the vehicle velocity relative to the ambient air. The forces cause the space vehicle to undergo structural loads which are a continuously changing combination of steady-state loads from dynamic pressure, wind loads, buffet loads caused by pressure oscillations around the vehicle, panel flutter, fuel sloshing in the tanks and the acoustic loads due to the high level of noise bombarding the space vehicle.

The terms "trajectory" and "orbit" both refer to the path of a body in space. "Trajectory", commonly used in connection with projectiles, missiles and launch vehicles, is often associated with paths of limited extent — that is, paths having clearly identified beginning and end points. "Orbit", commonly used in connection with natural bodies and satellites, is often associated with paths that are more or less infinitely extended or of a repetitive character, such as the "orbit" of the Moon around the Earth. Thus, we speak of "trajectories from the Earth to the Moon" and of "satellite orbits around the Earth".

As the vehicle flight continues, the atmosphere is soon traversed. The 4·4 million pounds of propellants in the first stage are almost totally consumed. About 155 seconds after lift-off, the centre or "inboard" engine of the first stage is automatically shut off. Four seconds later follows cut-off of the four "outboard" engines. This point is approximately 37 miles high, the velocity is 8900 feet/sec. and the trajectory has already tilted down until it is about 25 degrees above the horizontal. At this time there is also maximum acceleration on the three astronauts who will experience about four-and-one-half times their normal weight before the engines shut down.

As shown in *Figure 3-5*, an explosive charge separates the empty hulk of the S-IC stage from the remainder of the space vehicle. Retro-rockets push the first stage backward so that the second stage has room enough to ignite its own five engines in safety. The first stage is jettisoned because there is no reason to carry its now useless mass into orbit. The stage will re-enter the atmosphere, breaking into many small pieces which burn due to air friction.

The second stage is now almost out of the atmosphere. The major part of the gravity pull has been successfully overcome by the first stage by tilting the vehicle nearly horizontal. But more velocity is required to achieve an orbit and, therefore, the five

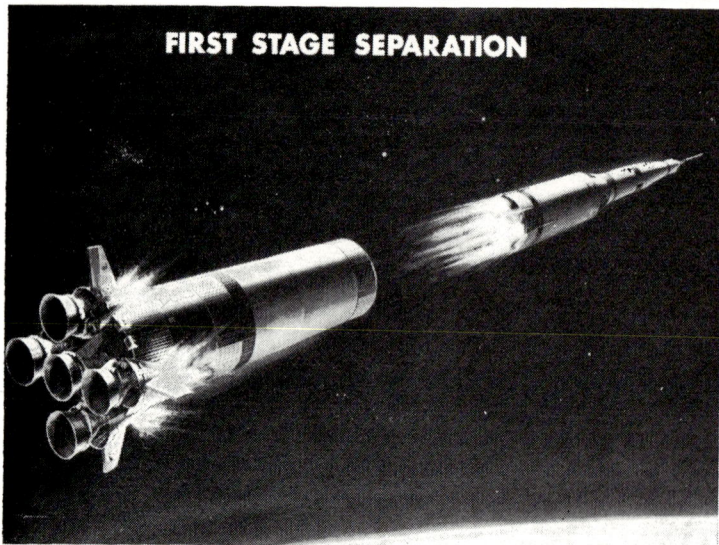

Figure 3-5

Figure 3-6.

engines are not so much designed to produce large thrust, to counteract gravity, but to operate more efficiently and obtain more velocity per pound of propellant.

Approximately 26 ·seconds after S-II ignition, the S-IC/S-II interstage structure is dropped off, followed a few seconds later by the jettisoning of the Launch Escape Tower. The engines of the second stage burn for a total duration of about 375 seconds — more than six minutes. During this time, the launch vehicle is no longer following the natural pull of gravity, bending down its trajectory as during first stage operation, but is steered upward by its guidance, navigation and control system.

From each location on the trajectory, there is a whole multitude of possible flight continuations for the vehicle by which it could reach its desired goal, some higher, some lower in altitude. Only *one* of these many possible trajectories causes the vehicle to reach its end point in a minimum amount of time or — what is the same, in view of the constant burning rate — with a minimum expenditure of propellants. This is what is called the "optimum trajectory". The onboard guidance system is designed to bring the vehicle automatically along such an optimum trajectory to its end point, regardless of where the vehicle may find itself at any time. Simultaneously, the onboard guidance must account for such disturbing forces as mass unbalances. Wind forces and air density variations become gradually negligible during second stage operations.

Finally, at propellant depletion and at a velocity of 22,500 ft./sec., which occurs at an altitude of about 98 miles, the second stage engines shut down automatically, the stage is separated, and the operation of the third stage begins. (See *Figure 3-6*.) At this point the vehicle has almost reached the desired orbital altitude and now flies an essentially horizontal trajectory. But the velocity is still less than required by about 3000 feet per second — a deficit which is made up by the first burn of the third stage. To maintain approximately constant altitude during subsequent stage operations, the control computer swivels the engine on its universal joint into the direction necessary to counteract forces trying to pull the vehicle to other altitudes; thus, to counteract a centrifugal force

SATURN/APOLLO EARTH TO ORBIT FLIGHT PROFILE

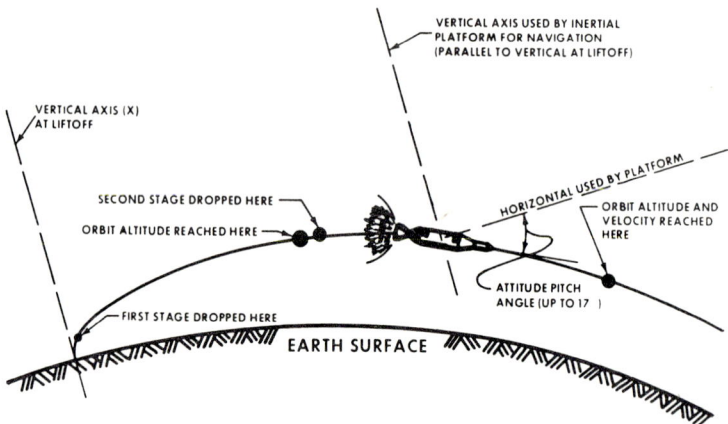

VERTICAL AXIS USED BY INERTIAL
PLATFORM FOR NAVIGATION
(PARALLEL TO VERTICAL AT LIFTOFF)

VERTICAL AXIS (X)
AT LIFTOFF

SECOND STAGE DROPPED HERE

HORIZONTAL USED BY PLATFORM

ORBIT ALTITUDE REACHED HERE

ORBIT ALTITUDE AND
VELOCITY REACHED
HERE

ATTITUDE PITCH
ANGLE (UP TO 17)

FIRST STAGE DROPPED HERE

EARTH SURFACE

Figure 3-7.

trying to lift it to higher altitudes, the engine actually thrusts into a downward direction.   (*Figure 3-7.*)

To impart the missing 3000 ft/sec to the vehicle requires a burn of about 2½ minutes, the engine thrusting at about 225,000 pounds of force.   At the conclusion of this burn, the spacecraft is at 100 nautical miles altitude and has travelled another 600 nautical miles downrange for a total distance, during the boost phase, of almost 1500 nautical miles.   The entire ascent to orbit has taken about 11½ minutes.   The third stage remains attached to the spacecraft to provide, with the remaining propellant, the required velocity for injection into a trajectory to the Moon.

During the orbit insertion burn (first burn) of the third stage, the space vehicle has passed beyond the horizon of the land-based tracking stations.   But a tracking ship located in the Atlantic Ocean beyond the Lesser Antilles, and its tracking radars, provide enough measurements to determine whether the orbit has been achieved.   Only about three minutes of tracking are necessary to provide these data.

Following confirmation of a good orbit, the Apollo spacecraft, with the third stage attached, will remain in parking orbit for at least one or two revolutions of the Earth, but no more than three (about 90 minutes for each orbit). During this time, the cryogenic tanks of the third stage are vented to the outside releasing internal pressures caused by vaporizing of the propellants. Following a brief onboard checkout of spacecraft systems and instruments, the CSM pilot, who is also the navigator, will go to his equipment bay and set up the gyroscopes. These are designed to maintain the inertial reference attitude of the spacecraft using optical sightings of stars and angular measurements between them and landmarks on Earth as reference.

Meanwhile, the other two crew members — the command pilot and the LM pilot — are checking out other systems. Data transmission by telemetry channels and voice communication are maintained as the spacecraft passes over each ground station. Also, the spacecraft is rolled, as necessary, around its longitudinal axis, so that the navigator may conduct star sightings without hindrance by the Sun.

C. **The Translunar Profile**

After the checkout of the space vehicle has been successfully concluded, it is ready for departure for the Moon. But because the Moon moves around the Earth on its own orbit, similar to but slower than the spacecraft, the astronauts cannot command the start of the third stage engine whenever they wish. There are certain limitations imposed on them by the heavenly geometries involved.

Figure 3-8 gives a simplified schematic view of the problem. The figure shows the Earth with its companion, the Moon, circling in an orbital plane according to Kepler's laws. The Moon's orbital plane is inclined to the Earth's equator by about $28 \cdot 5$ degrees in April, 1969 (see also Chapter 4). On the other hand, the spacecraft ascent trajectory and its orbit around the Earth (commonly referred to as its "waiting" or "parking" orbit) are part of another plane which is at an angle to the Moon's orbital plane. The spacecraft plane will also be more or less identical with the plane containing the translunar trajectory because it would require additional propellant to steer from the parking orbit into a new

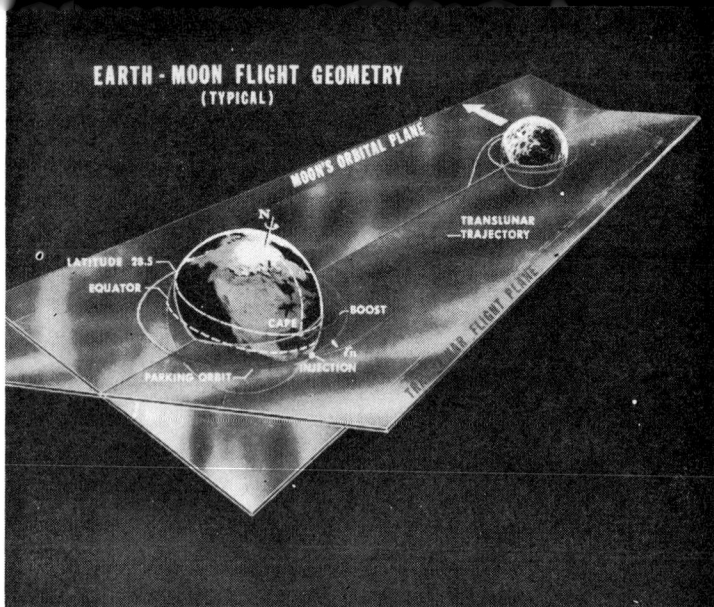

Figure 3-8.

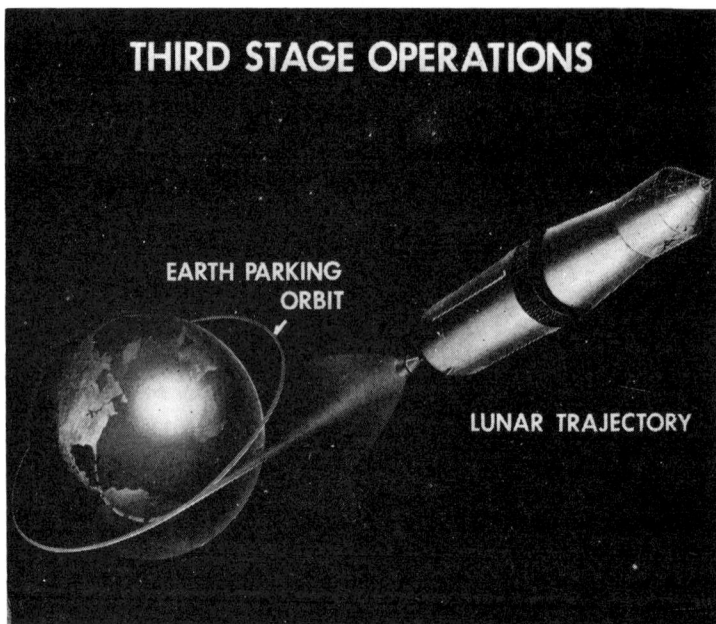

Figure 3-9.

translunar plane.   (Later, however, some minor plane changes must usually be performed.)

For the example shown in *Figure 3-8*, the translunar trajectory begins by passing "over" the lunar orbit plane, while later it will penetrate the lunar plane and reach the back side of the Moon "under" it.   If you consider the direction of the Earth's polar axis shown in the figure, you will see that the translunar trajectory plane requires a launch from parking orbit in a northerly direction and an injection toward the Moon somewhere over the Pacific Ocean.   If you mentally move the Moon (shown in the figure) to the other side of its orbit so that it travels now "downward" in front of the Earth, you can see — by following the turning direction of the parking orbit and its plane — that to reach the Moon during this time of the month one would have to launch in a southerly direction, from "behind" the Earth, over the Atlantic Ocean. These relationships will be examined more closely in Chapter 4.

As mentioned above, the third stage is able to conduct a small plane-change manoeuvre at the time of the insertion firing.

The magnitude of that manoeuvre depends on the particular flight geometry involved — and it could be as much as three degrees.

Approximately seven minutes before the spacecraft reaches the predetermined point of injection in its orbit, a computer in the vehicle instrument unit begins issuing directions to prepare the third stage for another firing.   The stage vent valves are closed and the pressure inside the tanks begins to build up.   Then, small solid propellant rockets — called ullage rockets — are fired to provide a slight forward acceleration.   You could call it "artificial gravity".   It has the purpose of forcing the liquid oxygen and hydrogen, which are floating freely in the tanks, down to the engine intake lines at the bottom of the tanks.   Finally, the engine ignites and thrusts the spacecraft out of orbit into a translunar trajectory.   (See *Figure 3-9*.)

The third stage thrusts for approximately $5\frac{1}{2}$ minutes, adding more than 10,000 feet per second to the orbital velocity of the spacecraft and boosting the speed to about 35,000 feet per second. (During this time, voice communications and data transmission will, for the most part, be maintained by the relay aircraft of the tracking network.)

After this velocity has been reached by the space vehicle, the stage is shut down and again vented to low pressure. The spacecraft is now on trajectory, essentially an elliptical track.

According to Kepler's First Law, an elliptic orbit in space has an attracting body at one focus. In this case, this would be the Earth. The attraction of the Moon, the Sun and of the other planets is also there, but near Earth they are so small as to be negligible. However, farther out they have to be accounted for. To compute the trajectory with the high precision required by the Apollo navigation and guidance system, the trajectory is considered not as an ellipse on a plane but as a three-dimensional curve whose shape is influenced by gravitational forces of Earth, Moon, Sun, Jupiter and other effects.

The translunar flight takes about 70 to 72 hours. It could be performed in less time although more injection velocity would be required and, in consequence, more propellants. The flight could also be planned for longer than 72 hours; however, while that would require less fuel, more weight would have to be allocated for oxygen, food, water, batteries and other time-dependent on-board resources. (Additional restrictions will be discussed in Chapter 4.)

During the coasting time toward the Moon, the crew has been preparing the spacecraft for the lunar activities. First, the command and service modules have been separated from the third stage. Using the reaction control system of their spacecraft, the astronauts turn the spacecraft and manoeuvre it, nose forward, back to the rocket stage. The lunar module is still attached to the third stage. The command/service modules and the lunar module are then docked. When they are fastened together, the lunar module is separated from the adapter and the third stage by explosives. The lunar module is then withdrawn. The entire operations, called "Apollo Transposition and Docking", is described in greater detail in Chapter 4.

The third stage is discarded and the spacecraft continues on its course in its new configuration. At this point the spacecraft is about one hour past the translunar injection point and has reached an altitude of about 9000 nautical miles above the Earth. During the hour, the velocity has already decreased by about 13,000 feet per second according to Kepler's Second Law. In about two hours the astronauts will perform the first midcourse correction.

The spacecraft is tracked by one of NASA's three deep-space tracking stations and two of the other stations described in Chapter 2. The trajectory is accurately determined on the ground by triangulation and Doppler measurements and its deviations from the desired trajectory computed. The correction data are transmitted to the spacecraft where they are combined with computations by its onboard computers. These computations provide the data required to place the spacecraft on the proper path. The spacecraft is then manoeuvred by the astronauts to point it in the required direction and the engine on the service module is ignited for a short burn (typically three seconds) to correct the trajectory. A second correction will not be required until the spacecraft nears the Moon about 2½ days later.

One of the major activities during this 2½-day period of flight is concerned with the Sun's heat and radiation. If we would let the spacecraft coast toward the Moon in a fixed attitude, one

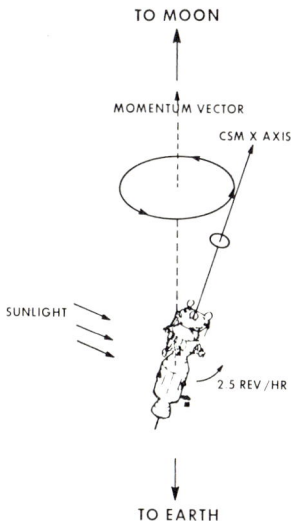

TO MOON

MOMENTUM VECTOR

CSM X AXIS

SUNLIGHT

2.5 REV/HR

TO EARTH

Figure 3-10. Passive Thermal Control Scheme.

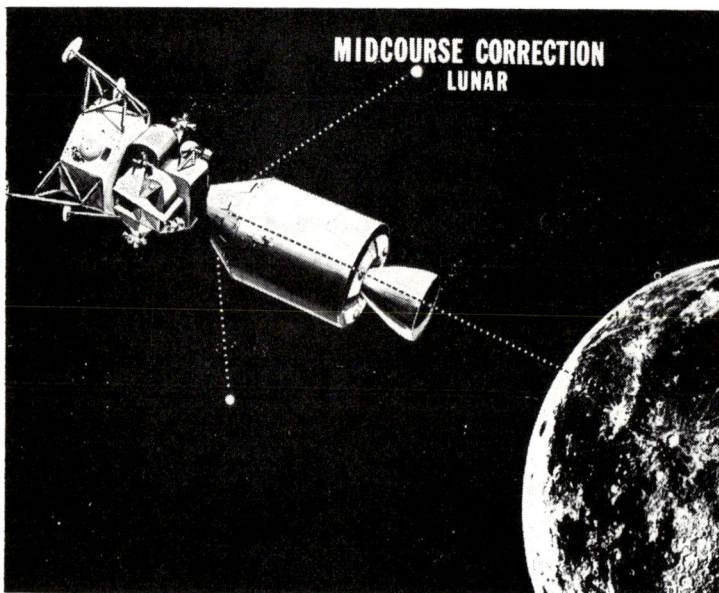

Figure 3-11. Midcourse correction during lunar trajectory.

part of it would look directly at the Sun for the next two days while another part of it would look directly away into space. As a result, it would receive severe heating on one side and severe chilling on the other causing critical components to experience unusual thermal strain. A spacecraft could be built with refrigeration on the hot side and electrical heating on the other side to balance temperatures but that would mean carrying much unnecessary weight along on the flight. There is a better way of establishing the desired temperature balance — passive thermal control, also referred to as "toasting" (*Figure 3-10*).

The spacecraft is manoeuvred until its longitudinal axis is perpendicular to the vehicle-Sun line. Then a slow rotation about the line-of-flight direction is established with the reaction controls. This roll manoeuvre is so designed that it does not interfere with the transmission cone of the high-gain antenna back to Earth.

By this time the Moon's gravitational forces are having an ever increasing effect on the spacecraft flight path. The spacecraft

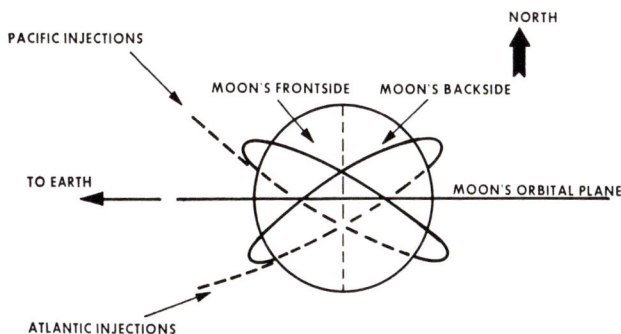

Figure 3-12.   Possible Apollo lunar orbits.

enters what is called the Moon's "sphere of influence", a point approximately 32,400 miles from the Moon's centre.   About one hour after reaching that point, the astronauts perform the second midcourse correction if that is necessary.

For this manoeuvre the exact position of the spacecraft has to be known.   The navigator measures angles of typical stars and known landmarks on the Moon and feeds this information into the onboard computer which calculates the true position.   With the position of the spacecraft known, the second midcourse correction is executed with the service module engine in the same manner as the first correction.   (*Figure 3-11.*)

D. **Lunar Orbit and Landing**

As discussed earlier, the spacecraft will approach the Moon from "above" the lunar orbit plane, penetrate that plane, and travel behind it "below" the plane, if the injection were in a northerly direction over the Pacific.   Conversely, if the injection from parking orbit were over the Atlantic, in a southerly direction, the spacecraft will approach the Moon "under" the orbital plane and rise through it around the Moon (*Figure 3-12*).   Thus, a Pacific injection will bring the spacecraft to northern latitudes on the front side of the Moon while an Atlantic injection carries the spacecraft to the southern latitudes on the Moon's front side.

From this it can be shown that the desired landing spot on the Moon more or less determines the orientation of the lunar parking orbit and even the point of injection back at Earth. If there is only a single landing spot it would place such extremely narrow constraints on the trajectory that there would be very little flexibility to take care of unforeseen deviations. It is even quite possible that the lunar landing attempt would be unsuccessful if there were such an inflexible schedule. Therefore, the Apollo mission is based not on one single landing site but on at least five landing sites from which can be selected the one which is most easily reached by the particular flight.

As the spacecraft approaches the Moon, sunset (on the Moon) occurs. Shortly thereafter the ship travels behind the Moon and out of line of sight to the Earth. At an altitude of about 150 nautical miles, the service module engine is ignited the third time, this time to brake the spacecraft into an orbit around the Moon (*Figure 3-13*). To achieve the desired reduction in velocity of about 3200 feet per second, the engine burns at 21,500 pounds of thrust for about six minutes. During this time it consumes about 24,000 pounds of propellants. The vector of this thrust may point out of the plane of the trajectory. The result is a parking orbit with a new plane which passes over the intended landing site. This manoeuvre provides a certain additional flexibility in the selection of the landing site.

After shutdown of the engine, the spacecraft coasts in a circular orbit of about 60 nautical miles altitude around the Moon. Periodically it will appear in the region of sight of the Earth tracking and telemetry stations which measure its position and velocity during each pass over the front side of the Moon. After the third pass, $5\frac{1}{2}$ hours after the braking manoeuvre, there will be enough measurements to compute accurately the lunar orbit of the Apollo spacecraft.

Finally, the crew will be ready to proceed to the lunar landing. Two astronauts will pull themselves through the connecting tunnel and an access hatch into the pilot compartment of the lunar module (*Figure 3-14*). There they will activate and check systems required for the subsequent landing, align the gyro-stabilized platform of the lunar module attitude reference unit, transfer all necessary information from the command module computer to the lunar module computer, close the hatches and prepare for separation.

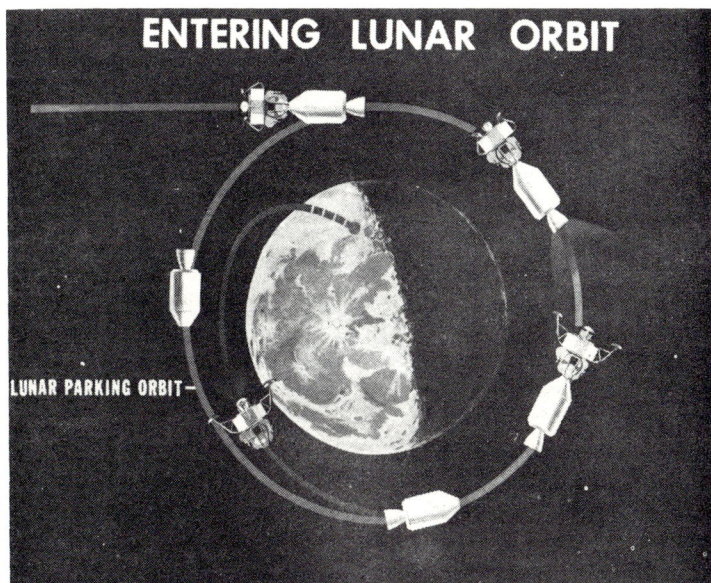

Figure 3-13.

Figure 3-14. Transfer to lunar module.

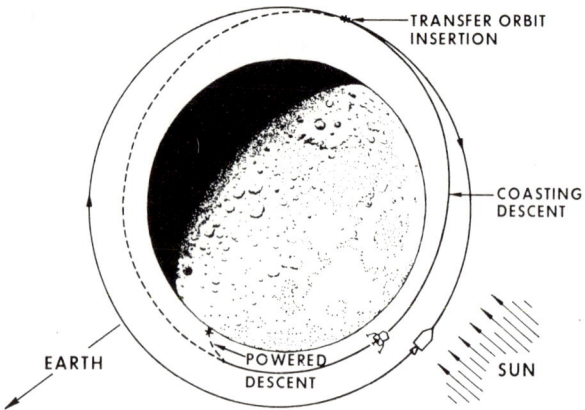

Figure 3-15. Lunar descent.

The docking mechanism is unlocked and the reaction control system of the landing module burns for five seconds to push the craft away from the command module which, with one crewman aboard, will remain in lunar orbit until the lunar module returns.

Because the lunar module is to land at a pre-selected spot on the front side of the Moon, it must start its descent at an accurately specified point in orbit. Since the lunar module will travel through a central angle of about 200 degrees, the descent manoeuvre must begin on the back side of the Moon, out of sight of Earth (*Figure 3-15.*)

The descent stage engine is ignited firing in the direction of flight for about 32 seconds... The engine is throttleable meaning that its thrust level can be changed smoothly almost like an aircraft engine. After the orbital velocity of the lunar module is reduced by about 100 feet per second (using 330 pounds of fuel), it is shut down and the vehicle continues its flight coasting freely. Its trajectory is now a portion of an ellipse reaching around the Moon with the Moon's centre at one focus. The lowest point of this ellipse, the "pericynthion", is 50,000 feet above the lunar surface, safely out of reach of even the highest mountain peaks (*Figure 3-16*). During this coast phase the crew will track the still orbiting Apollo command/service modules with rendezvous

Figure 3-16.

Figure 3-17.

radar and determine their own descent trajectory from its stored knowledge of the command/service modules' orbital data. Then the surface features of the approaching land areas are surveyed through the windows. If, for some reason, the astronauts decide not to go ahead with the landing they can continue the coast. The landing ellipse is so designed that it will carry the lunar module back to its original altitude of 60 miles, where it can rendezvous with the other spacecraft or be rescued by it.

If everything goes well landing will be initiated. To do this the reaction controls first fire a five-second burst to settle the propellants in their tanks; then the descent engine is ignited. This manoeuvre takes place at about 50,000 feet, the lowest point reached before the spacecraft orbit rises to higher altitudes (*Figure 3-17*). During the manoeuvre the lunar module is flying almost horizontally with the rocket firing horizontally forward. Braking takes almost eight minutes while covering 200 to 250 miles and carrying the vehicle down to about 8600 feet altitude under auto-pilot control. This flight phase is depicted in the upper portion of *Figure 3-17*. Mission planning based on terrain photographs (discussed in Chapter 1) from Lunar Orbiter and Surveyor space-craft, as well as the survey by the crew prior to pericynthion, has assured that the approach path is free of such dangerous obstructions as high mountain ridges. Thus, at 8600 feet altitude the ship is turned upright so that the crew can better observe the landing area, still eight miles away from them. After about 1½ minutes of flight, still under autopilot control, the vehicle's position is 500 feet above and 1200 feet away from the landing site. The trajectory has turned increasingly downward. During the final phase of the landing, descent is vertical. Control is now by the pilot who will be able to select a suitable landing spot and manoeuvre the module to it. The final landing phase is shown on the lower part of *Figure 3-17*.

The lunar module comes to rest settling down on its shock absorbers. Wide footpads on the four legs sink a few inches into the lunar soil which has about the same consistency as wet beach sand according to information from Surveyor lunar landings. After making sure that the vehicle has safely landed, the astronauts begin post-landing checkout. The descent engine is disarmed being no longer needed. The descent propellant tanks are vented to make sure that they cannot burst. All systems not required during the stay are turned off.

E. Lunar Exploration and Experiments

For about $1\frac{3}{4}$ hours, communication with Earth is checked and exercised. Each crewman puts on his space suit — or Extra-vehicular Mobility Unit. *Figure 3-18* shows the two major parts of this suit — the actual pressure suit on the left and the protective overgarment with backpack on the right. The space suit consists of seven major components: a liquid-cooled undergarment close to the skin, the pressure garment assembly, the portable life support system worn on the back, the thermal-meteoroid garment, lunar boots, lunar gloves, and a protective over-visor on the helmet. Each space suit has its own communications, electrical power and air conditioning system. The suit is almost a small spacecraft in itself operating completely independent of other vehicles while on the lunar surface.

To explore the Moon and conduct scientific experiments on its surface, about 250 pounds of tools, instruments, containers and other equipment — the Apollo Lunar Surface Experiments Package — have been brought from the Earth. About 80 pounds of soil and rock samples from the lunar surface will also be returned to Earth. The samples will be stored within a two cubic feet volume in the ascent stage.

Figure 3-19 shows the astronauts gathering lunar soil and rock samples. The standing astronaut carries a sample return container in his left hand and holds a calibrated staff with a terrain camera in the right hand. Since the length of the staff is known, those who will later evaluate the terrain photographs will be able to determine accurately from what height above the surface the pictures have been taken and they will be able to make geometric measurements from the pictures.

The kneeling astronaut is filling plastic bags with soil samples and placing them in a sample return container. During this work he is being photographed by his companion so that evaluators can later identify the exact source of the particular samples. (Behind the sample return container is illustrated a tool carrier to which

Figure 3-18.

Figure 3-19.

are attached the necessary tools for surface exploration.) Later, on return to the spaceship, the filled containers will be weighed and hermetically sealed.

Another crew activity will be a thorough external inspection of their vehicle. The Apollo Lunar Surface Experiments Package (ALSEP) will be unloaded from its storage bay in the descent stage, deployed and emplaced on the surface. The experiments package is actually a small automatic research laboratory powered entirely by nuclear power. It is designed to remain active long after the Apollo crew has departed for Earth. A data system will process commands from Earth and will transmit back scientific and engineering data which describes the lunar environmental conditions. The data system will use the Manned Space Flight Network. Some of the following seven major experiments are onboard this flight:

A Passive Seismic Experiment will establish the Moon's natural seismicity and will provide data on the physical properties of the lunar interior. Additional seismic stations will permit location of quake epicentres and correlation of seismic events with observed surface features. The 25-pound package will be emplaced on the lunar surface and levelled by one astronaut. It contains two three-axis seismometers. One of these measures long period motions; the other monitors high frequency seismicity signals. If the Moon's seismic activity is similar to the Earth's, about five events per day would be recorded.

An Active Seismic Experiment will measure properties of the lunar sub-surface material to depths in excess of 500 feet. Unlike the passive experiment, explosive devices will impart seismic signals which will provide information on the structure, thickness, physical properties and elasticity of the Moon's top layer. Two energy sources create artificial seismic waves. The astronaut will detonate 21 explosive squibs electronically and "thump" the lunar surface at varying distances from the detectors. The second energy source will be deployed by the astronaut but will be armed and operated by Earth command. That source consists of four self-propelled grenades which will be fired to various points on the lunar surface up to 5000 feet apart. The astronaut has placed geophones, connected by a 300-foot cable, to record the sound

waves travelling from the energy source through the lunar sub-
surface to the detectors.   The packaged experiment weighs nearly
25 pounds.

A Magnetometer Experiment will measure the magnitude and
direction of the total magnetic field during the lunar day and the
interior or residual field during the lunar night.   Changes to the
interior field brought about by the interaction with the solar
magnetic field will permit investigators to infer the electrical conductivity and magnetic permeability of the Moon's interior.   This
experiment contains three fluxgate magnetometers mounted on
booms and placed about five feet apart.   The experiment weighs
less than 20 pounds and must be deployed about 50 feet from the
other experiments to avoid interferences.   Its range is adjustable
to a maximum field value of 400 gammas.

A Solar Wind Experiment will measure the flux, energy and
direction of medium energy electrons, protons and other charged
particles contained in the solar wind and impinging upon the lunar
surface.   The experiment, weighing less than 10 pounds, will
provide an opportunity to study the interaction of the solar wind
with the low magnetic field of the Moon.

A Suprathermal Ion Detector package weighs about 12 pounds
and contains instrumentation to study density and temperature of
ions which are thought to exist in an ionosphere near the lunar
surface.   The density is measured by counting the number of
ions in a given time interval and the temperature by measurement
of the particle velocity and energy.   The instrument can detect
and resolve particles ranging from one to four AMU and can,
therefore, determine the percentages of hydrogen, helium, neon
and other ions that might be present in the ionosphere.   A cold
cathode gauge can measure the pressure of the lunar atmosphere
down to 10^{-12} torr.

A Heat Flow Experiment will measure the net flux of heat
energy originally in the Moon and measurable at the lunar surface.
The instrumentation consists of two probes each capable of making
two thermal gradient measurements and four thermal conductivity
measurements to permit calculation of net heat flow.   The probes
have to be placed in two holes each one inch in diameter and
three metres deep drilled into the lunar sub-surface by the

astronaut using the Apollo lunar surface drill. This is a battery powdered rotary-percussive coring tool capable of drilling three metres into the surface. The process of setting up, drilling and emplacing the heat flow probes is estimated to require one-half hour per hole. Several months will be required for the heat flow experiment to reach a steady state condition, following the perturbing influences of drilling and emplacing the probes. The experiment, together with the drill, will weigh approximately 30 pounds. The drilled-out cores will be returned to Earth for thermal conductivity measurements and other analyses at the Lunar Receiving Laboratory.

A Charged Particle Experiment will measure proton and electron fluxes at the lunar surface and will study their energy distribution and time variations. Bombardment by these particles may result from a variety of phenomena, including the solar wind. Of particular interest are the measurements the experiment will record as the Moon passes through the tail of the Earth's magnetic field. The instrument will measure separately electrons and protons over an energy range of 50 to 150,000 electron volts. The instrumentation weighs five pounds and requires five watts of power for operation. The instruments will transmit to Earth during their expected operational life of one to two years.

The lunar surface stay time on the first Apollo mission is planned to be approximately 18 hours, including a six-hour sleep period for the astronauts. Both will sleep at the same time because, as discovered during the Gemini programme, the activities of the non-sleeping crewman disturbs the other.

The astronauts will bed down in the confines of the lunar module. One of them will sit on the floor and lean against the wall; the other will recline in a sort of hammock placed above the first astronaut.

After the rest period there will be three more hours of exploration during which more soil samples are collected. *Figure 3-20* shows schematically the communications network used during these excursions. While on the lunar surface, the astronauts will be able to talk to each other by radio. Both of them are also in communication contact with the third astronaut in the orbiting command/service module when it is in sight and all three crewmen

**COMMUNICATIONS NETWORK DURING
LUNAR SURFACE STAY**

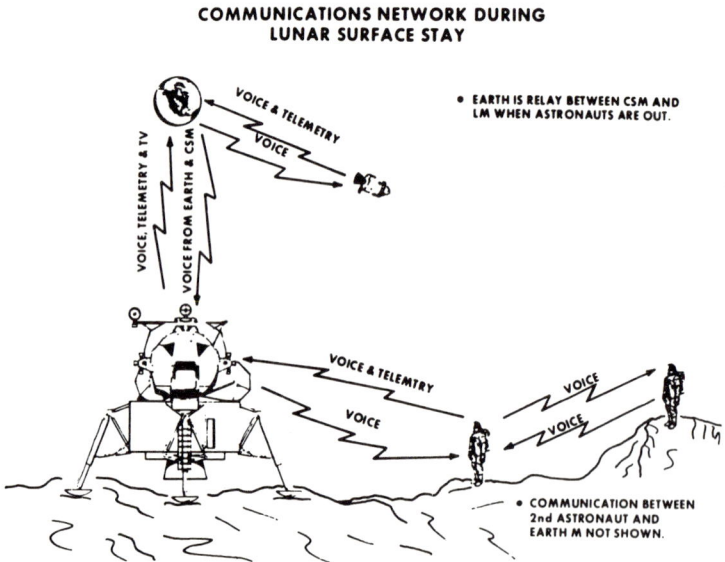

Figure 3-20.

are continuously communicating with the ground stations on Earth. From the two astronauts on the lunar surface, the Earth stations will receive voice reports, biomedical data over the telemetry channels and live pictures over the television link.

The astronauts will place their scientific samples in the ascent stage. When they are ready to depart, that equipment no longer needed including such items as spent batteries, cameras, used containers and waste material, will be stored in the descent stage. By discarding all unessential material a larger weight of scientific material can be returned to Earth for analysis.

F. Lunar Takeoff and Return Profile

A thorough pre-launch checkout of the ascent stage is now performed, all systems being activated, inspected and prepared for launch. The astronauts then wait until the command/service module is in the required position before the launch occurs. They will acquire the orbiting command/service module on the radar.

The parking orbit of the command/service module no longer passes over the landing site since its direction is, aside from perturbations due to gravitational irregularities of the Moon, anchored in space while the Moon slowly rotates under it at a rate of approximately one revolution per month. Therefore, the third astronaut in the command/service module must activate its propulsion system and make a plane change, bringing the orbit back to its position before during the landing of the lunar module. When the command/service module is again in the orbital plane over the landing site the lunar module radar is used to measure the spacecraft's precise orbit and position. These measurements are then fed into the onboard computer and the exact lift-off time is calculated. Once the time of launch is known in advance the astronauts can split the waiting period in specific operational steps and conduct a regular countdown.

Because the problem is to rendezvous with the orbiting command/service module in space, the lunar module launch has to be timed according to the command/service module position on its orbit. This is shown in *Figure 3-21* which depicts the

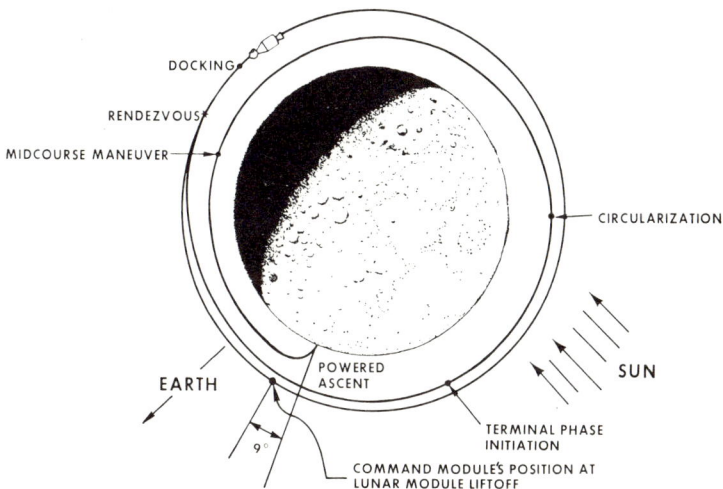

Figure 3-21. Lunar ascent flight plan.

location of the landing site on the Moon in relation to the position
of the command/service module in its orbit and the manoeuvre
necessary for the ascent stage to catch up with the spacecraft.
The orbiting craft leads the ascending ship by about a nine-degree
central angle on take-off, and the lunar module goes through an
intermediate catch-up orbit or "phasing orbit", which allows some
flexibility in the launch timing. As a result, the lunar module
crew will not be forced to launch on a split second of time but will
have about $5\frac{1}{2}$ minutes in which to initiate lift-off.

When the lunar module computer has ticked off the last seconds
it gives the signal to ignite the ascent engine. The ascent stage
separates from the descent stage and rises from it, leaving it behind
on the lunar surface (*Figure 3-22*). The climb will be vertical
for about 12 seconds; then the guidance system steers it into an
optimum trajectory slanting upward to about 50,000 feet altitude.
From there an intermediate orbit is established. (See *Figure 3-23*).
The lift-off time (within the $5\frac{1}{2}$ minutes available) determines the
amount of time in which to achieve the final rendezvous. The
intermediate orbit is established so that it accommodates the lunar
module for exactly this period of time and simultaneously carries
the lunar module to the command/service module in its higher
orbit. During the final ascent flight the lunar module onboard
radar tracks the spacecraft and determines the exact distance to it.
When this distance is reduced to about 30 nautical miles, reaction
control motors are fired on the lunar module and the final
approach begins. This manoeuvre is called the "Terminal Phase
Initiation" (*Figure 3-23*). As the two orbiting craft approach
each other, the astronauts in the lunar module fire their engine —
not to drive them toward the command/service module but
apparently to drive themselves away. The explanation of that
apparent paradox lies in a fundamental property of orbits of
celestial bodies. The closer these bodies are to the centre of
attraction around which they move, the faster they travel in
their orbits (Kepler's Second Law). Therefore, the lunar module
is actually travelling faster than the command/service module.
When the lunar module gets close to the command/service module,
the velocity of the lunar module must be decreased to achieve
final rendezvous. This then explains why the astronauts point the

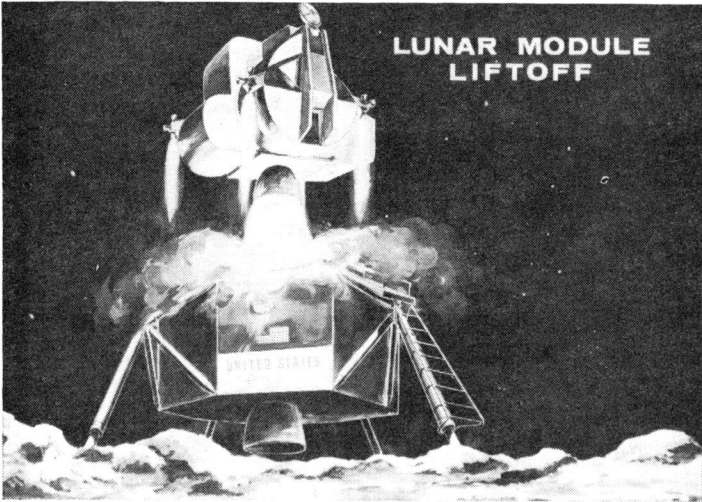

Figure 3-22.

Figure 3-23.

lunar module rocket engines straight along the line of sight toward the target vehicle. Firing the engines and thus braking the velocity will "expand" the lunar module's orbit and it will climb, finally meeting the orbiting spacecraft (*Figure 3-23*).

The two astronauts manually manoeuvre their vehicle closer to the target — a manoeuvre already rehearsed several times during the Gemini programme. Finally, contact is made between the two vehicles, docking clamps are latched and locked and the connecting tunnel is sealed and pressurized.

The two crewmen in the lunar module turn off all switches and deactivate the ascent stage. Now it will be separated from the command/service module and remain in orbit around the Moon. Eventually its orbit will deteriorate, even in the high vacuum of space, due to a complex combination of gravitational irregularities caused by the Moon's irregular shape and the dis· turbing pulls of the Earth, the Sun and some of the other planets.

After the transfer of their equipment and scientific samples, the two crewmen join their colleague in the command/service module, close the tunnel hatch and jettison the ascent stage.

For the return flight to Earth, certain geometrical restrictions must be observed which are similar to the translunar injection discussed earlier. One difference between the two flight phases lies in the fact that the return trip is to commence from an orbiting body to a stationary target while during the Earth-Moon transfer the situation was reversed. In lunar orbit the command/service module, together with the Moon, is moving around the Earth at a speed of over 2200 miles per hour while the Earth is at rest with respect to the Moon. Therefore, the spacecraft must actually reduce its inertial velocity in order to return to Earth. This is illustrated by the velocity vectors in *Figure 3-24*.

If V_M is the orbital velocity of the Moon of about 2200 mph, and if V_{re} is the velocity needed to establish a proper return ellipse with the Earth as focus, then the velocity V_{RM} must be applied by the spacecraft's engine to achieve V_{re}. It must be remembered, however, that the return velocity V_{re}, with which the proper ellipse is established, assumes that the spacecraft has already escaped the Moon's gravitational attraction. In other words, at S we are

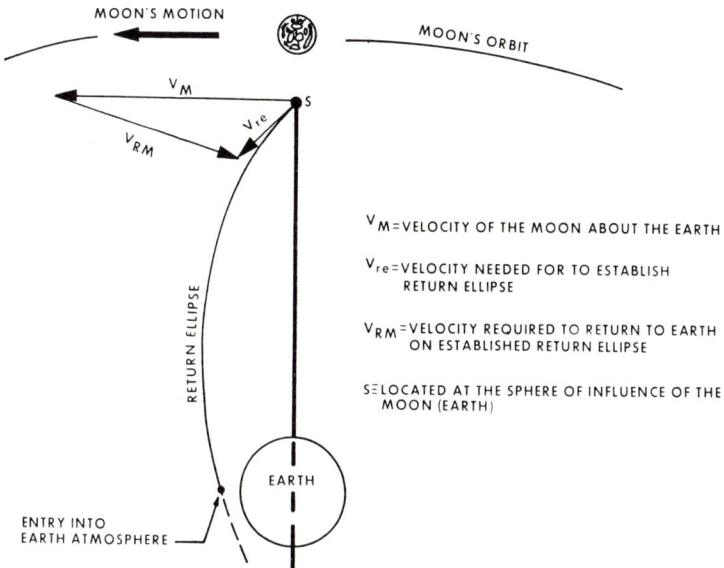

V_M = VELOCITY OF THE MOON ABOUT THE EARTH

V_{re} = VELOCITY NEEDED FOR TO ESTABLISH RETURN ELLIPSE

V_{RM} = VELOCITY REQUIRED TO RETURN TO EARTH ON ESTABLISHED RETURN ELLIPSE

S ≡ LOCATED AT THE SPHERE OF INFLUENCE OF THE MOON (EARTH)

Figure 3-24. Earth return diagram.

already at the point where the Earth's gravity starts overriding the Moon's sphere of influence. To reach this point the velocity of the spacecraft must be increased relative to Moon to escape from it. The escape trajectory was a hyperbola leaving the circular orbit. The problem of timing the trans-earth injection properly now consists simply of firing the rocket propulsion in such a direction and at such a magnitude that, on reaching the sphere of influence (point S on *Figure 3-24*), the spacecraft's velocity corresponds exactly to V_{re} in magnitude and direction. Thus, V_{RM} must have been part of the injection firing.

Figure 3-25 shows the injection of the spacecraft onto its correct return trajectory. The flight time will be between 86 and 110 hours. The trajectory was chosen with a flight time calculated to have Earth rotate the geographical region, where the recovery ships are waiting, into the return flight path at the time when the spacecraft arrives there. The ignition of the rocket engine in the service module and the injection occur actually on the back side of the Moon. The entire firing period is no longer than about two minutes, using up about 8000 pounds of propellants.

Figure 3-25.

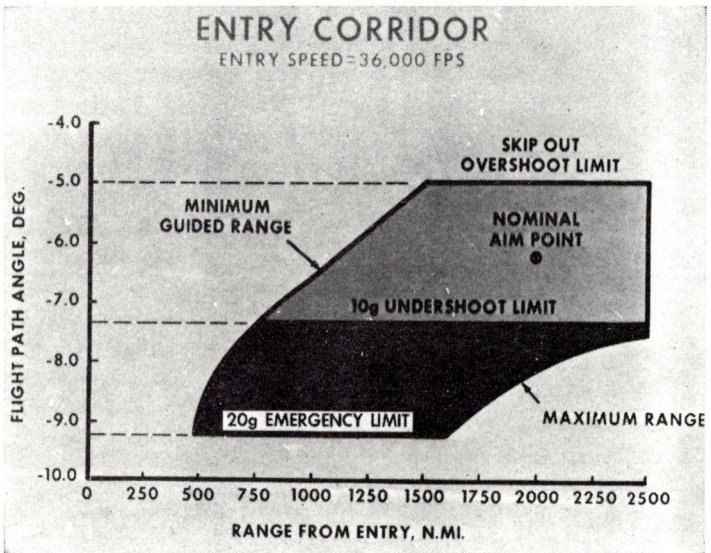

Figure 3-26.

Soon after shutdown of the engine, the crew will experience "Earthrise"; that is, the Earth coming above the Moon's horizon. From that point on, Earth stations will continuously track the returning spacecraft, making sure it is exactly on course. The astronauts, on the other hand, initiate another passive thermal control manoeuvre as before and then will get some sleep.

About 10 hours later in flight, the crew will perform a midcourse correction manoeuvre. This is followed two days later by a second correction. These manoeuvres will adjust the trajectory so that the entry into the Earth's atmosphere will occur exactly under the desired conditions.

As the vehicle approaches the Earth, it has a velocity of about 36,000 feet per second. This tremendous velocity and therefore the associated spacecraft energy must be dissipated in a manner that will not destroy the vehicle or hurt the crew.

If all of the vehicle's energy were converted into heat within the body itself during entry it would be more than sufficient to vaporize the entire vehicle and passengers. However, the survival of many natural meteorites which enter the atmosphere at far higher speeds obviously indicates that not all energy goes into the body. Actually the initial energy is transformed through aerodynamic drag into thermal energy or heat in the air around the body and is rapidly carried away by it. During this extremely violent energy transformation, the chief effects on the body are heating which cannot be completely prevented, and deceleration which will be definitely felt by the astronauts.

Both deceleration and heating are most severe when there is a combination of high atmospheric density and high vehicle velocity, i.e., when vehicles fly at high speeds down to low altitudes.

A shallow entry will have very little deceleration but if the entry is too shallow there will not be enough deceleration to pull the vehicle in for a landing. Instead, it will go past the planet and if the vehicle is so built that it generates lifting force like an aircraft it will even skip out of the atmosphere like a flat stone on a water surface.

Figure 3-26 shows how the combination of all these factors result in the so-called entry corridor. If the spacecraft enters the atmosphere at 36,000 feet per second and the flight condition is within the bounded region, everything will work out successfully. In the figure the flight path angle is shown against the horizontal. For example, a zero flight path angle would mean that the module skims the atmosphere tangentially without entering it. To come down, negative flight path angles are needed but, as shown in the figure, if the flight path angle is smaller than about 5 degrees downward, the module would skip out again. Thus, that comprises the upper corridor limit. The lower limit, the steepness mentioned before, is caused by the deceleration loads the astronauts are able to withstand. At slightly above 7 degrees they would experience about ten times their normal weight which they can bear well. The lift borderline keeps them from getting into an area where the flight duration in the atmosphere, or the range covered, would be too short to be handled by the onboard guidance system. The right boundary limits the flight duration of the module as to exceed those limits means that the ablative heat shield would be melted away. Thus, the entry corridor is only approximately 2 degrees wide. To hit this corridor on return from the Moon is rather like threading a needle's eye. The midcourse corrections are designed to achieve it.

Entry into the Earth atmosphere begins at about 400,000 feet or about 120 kilometers. Fifteen minutes before reaching this point, when the spacecraft is still about 2500 nautical miles away, the service module is separated from the command module and pushed off. Then the command module is turned by its reaction control system and oriented in the entry attitude, blunt heat shield facing forward.

Soon the first effects of the friction with the air become apparent (*Figure 3-27*). A shock wave forms around the forward part of the body and begins to glow as the air molecules and the particles from the heat shield begin to consume and re-emit the energy drawn off from the spacecraft. Deceleration makes itself felt as a steadily increasing retardation. After a few minutes the astronauts' contour couches are sagging deep under the braking loads. These couches are especially constructed with springs and shock absorbers to absorb these loads as safely and comfortably as possible.

The heating around the spacecraft becomes so intense that an electrically conducting sheath of ions is formed. As this cannot be

APOLLO REENTRY

Figure 3-27.

APOLLO SEA RECOVERY

Figure 3-28.

penetrated by the communication system, the command module communications are blacked out for many seconds. After the vehicle emerges from the "blackout" at an altitude of about 24,000 feet, two orientation and stabilization parachutes, called "drogue chutes", deploy on the command of a computer. The parachutes reduce the velocity from about 400 ft/sec to about 200 ft/sec. This velocity is reached at 10,000 feet. Three small pilot chutes deploy and pull out the three main parachutes while the drogue chutes are disconnected.

The vehicle splashes into the ocean at a velocity of about 25 ft/sec (*Figure 3-28*). The Apollo voyage is now over. The spacecraft has returned from the longest expedition in human history and its three astronauts, two of whom have actually walked on another celestial body, are back on Earth.

In this chapter the extensive effort culminating in the Apollo mission has been discussed. In Chapter 4 a few of the technical problems briefly mentioned here will be reviewed more fully. With the accomplishment of the Apollo flight a new era will have opened — an era of flights into near-Earth and cis-lunar space. And it will move the portal for manned exploration from the surface of the Earth to beyond the Moon. In reality man will have ventured 240,000 miles across "this new ocean" in less than 10 years.

CHAPTER 4

Selected Technological Aspects of the Apollo Programme

Introduction

As perhaps in no other endeavour, the attempts to solve the mysteries of space have brought together a great array of technologies representing just about all of the disciplines from science through production. This rather huge effort culminated in the Saturn Apollo programme. For its execution and achievement, a concentration of some of the best talents in America and an application of impressive amounts of resources has been necessary. The technologies involved have been close to, and in many cases beyond, the state of the art and much research and development has had to be undertaken.

For this series of lectures a selection of technological aspects from the many confronting the manager, the scientist and the engineer has been most difficult. Those few chosen and discussed here are representative of only a small slice of interesting problems of a multitude in the areas of space vehicle navigation, guidance and control, automatic attitude control in space, rendezvous and docking of spacecraft, high velocity atmospheric re-entry, stage separation, restart of high energy rocket engines in space and a host of others.

Since "quality assurance and reliability" play a predominant role in such a technological venture, this subject was selected as representative of technical and operational aspects. In the purely operational field, the ensuing discussion will touch briefly on the problem of the "launch window" which is one of the most typical aspects of space flight. In the purely technical area, the highly complex and interesting "spacecraft transposition and docking manoeuvre" was chosen. Finally, I have felt a short narrative on "crew environment" would be quite intriguing to you.

RELIABILITY AND QUALITY ASSURANCE DISCIPLINES

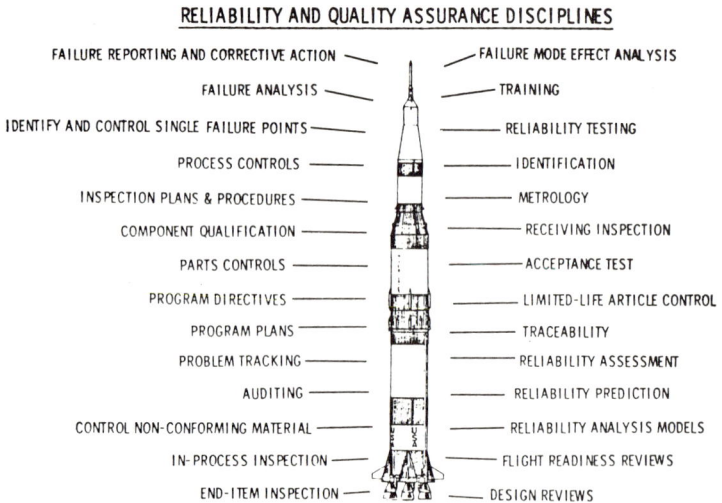

FAILURE REPORTING AND CORRECTIVE ACTION	FAILURE MODE EFFECT ANALYSIS
FAILURE ANALYSIS	TRAINING
IDENTIFY AND CONTROL SINGLE FAILURE POINTS	RELIABILITY TESTING
PROCESS CONTROLS	IDENTIFICATION
INSPECTION PLANS & PROCEDURES	METROLOGY
COMPONENT QUALIFICATION	RECEIVING INSPECTION
PARTS CONTROLS	ACCEPTANCE TEST
PROGRAM DIRECTIVES	LIMITED-LIFE ARTICLE CONTROL
PROGRAM PLANS	TRACEABILITY
PROBLEM TRACKING	RELIABILITY ASSESSMENT
AUDITING	RELIABILITY PREDICTION
CONTROL NON-CONFORMING MATERIAL	RELIABILITY ANALYSIS MODELS
IN-PROCESS INSPECTION	FLIGHT READINESS REVIEWS
END-ITEM INSPECTION	DESIGN REVIEWS

Figure 4-1.

A. **Quality Assurance and Reliability**

It is critically important in the space programme to achieve a high reliability very early in the design and development process. This need is clearly dictated by the relatively small number of items produced, the high cost of each and, last but not least, by the fact that human life is at stake. Every flight and its mission has to have highest probability of success. Therefore, also clearly dictated is the need for comprehensive systems analysis, increased and more severe component testing and meticulous, detailed analysis of each failure to detect cause, effects and to derive therefrom corrective measures. In actual application, these general considerations represent the sum of many highly technical and managerial disciplines, some of which are shown in *Figure 4-1*.

The term "reliability" is defined as the probability of successful operation of a component or assembly over a given time period under a specified operating environment. Reliability must be proven by tests and analyses with a reliability goal being assigned for each component. If the component fails to meet the goal it must be redesigned until that goal is met.

Reliability, then, has become a design requirement on an equal basis with volume, weight, power and other physical and performance requirements. As discussed in Chapter 2, the Apollo hardware is an assembly of hundreds of thousands of components. The extended mission duration, the execution of precise manoeuvres in space, the critical sequencing of events, the large energies released and the severe environmental requirements demand both an operational capability and a reliability beyond anything previously contemplated in order that there may be no compromise with safety.

During the design phase, in the normal approach to reliability, the system concept is reduced to its simplest functional form. A reliability analysis is then performed to obtain a system-predicted reliability. If the predicted reliability is satisfactory, the system is developed by selecting and using the most reliable parts preferably those with proven flight experience. After assembly, the system is tested and analysed to see if the predicted reliability goal can be materialized. If the tests are successful the concept is released for final design and manufacturing. If failures occur redesigns and redundancies are considered in meeting the reliability requirements. Extensive testing is again performed to demonstrate the capabilities of the system.

The point to be emphasized here is that reliability is *designed* into the system during the design phase and *proven* during the prototype operational phase. This method has been used extensively on major programmes and has been shown to be logical and rewarding.

In those cases where system performance falls short of the reliability goals, redesign or redundant applications are used where feasible. The instrument unit for the launch vehicle is a good example. The reliability goal for the instrument unit is 0·992 for the total period of the prelaunch tests and a subsequent flight of 6·8 hours. Analysis showed that, even with conservative practice, only an 0·98 reliability goal was possible. Therefore, extraordinary redundancy schemes to cover critical failure modes were applied to the components of the system with the result that the reliability goal was met.

As a particularly appropriate example, the navigation, guidance and control subsystem located in the instrument unit represents one of the most extensive applications of redundancy existing in a flight system. Some of the basic redundancy approaches (or schemes) employed in this system are Duplex, Triple Modular (TM), Prime-Reference-Standby (PRS), Multiple Parallel Element (MPE), and Quad Redundancy. These techniques are used to augment basic reliability which may be calculated from the product rule:

$$P = R_1 \times R_2 \times R_3$$

where P is the reliability of a three part system and R_1, R_2, and R_3 refer to the reliability numbers of three subsystems. For example, assume a reliability assigned to each of the three above mentioned subsystems as $R_1 = 0 \cdot 999$, $R_2 = 0 \cdot 998$, and $R_3 = 0 \cdot 997$. Then, using the product rule, the reliability of the assembly is $P = 0 \cdot 994$. Actually, these high reliabilities of each of the three subsystems could never be reached by applying the product rule, even if every

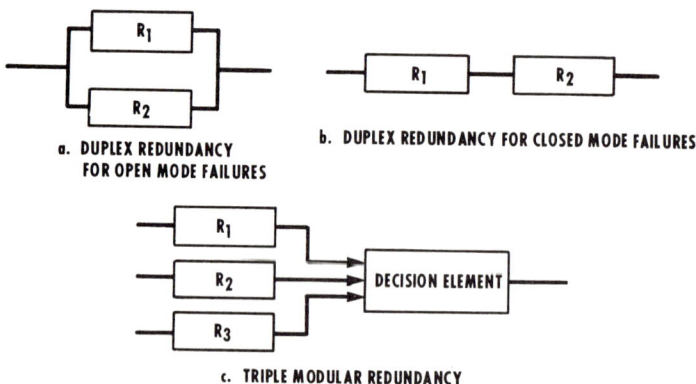

a. DUPLEX REDUNDANCY
FOR OPEN MODE FAILURES

b. DUPLEX REDUNDANCY FOR CLOSED MODE FAILURES

c. TRIPLE MODULAR REDUNDANCY

R_1=RELIABILITY OF COMPONENT 1
R_2=RELIABILITY OF COMPONENT 2
R_3=RELIABILITY OF COMPONENT 3

REDUNDANCY SCHEMES FOR RELIABILITY

Figure 4-2.

one of these would have a proven reliability of $0 \cdot 9999$, because each subsystem is composed of thousands of components and parts.

Some basics of how redundancy schemes improve reliability can be illustrated by the duplex and the triple modular redundancy schemes. The duplex arrangement is the simplest form of redundancy. If a component fails primarily in the open mode, then an identical part is added in parallel (*Figure 4-2a*). If the part fails in the closed mode, then a part is added in series (*Figure 4-2b*). Assuming identical units, the reliability, P_D, for the duplex scheme may be expressed as:

$$P_D = 2R - R^2$$

If a predominant failure mode cannot be assumed, a decision element must be added to determine which channel is operating correctly. The triple modular redundancy (TMR)[1] (*Figure 4-2c*) arrangement is an extension of the duplex method with a decision element that reacts to the majority inputs from three components. Assuming identical components and a decision element reliability of $1 \cdot 000$, the redundant reliability can be expressed as:

$$P_{TMR} = 3R^2 - 2R^3$$

There are numerous analytical methods by which a complex assembly of components like the instrument unit may be evaluated to determine if it will fulfil its intended purpose. One method employed on the Apollo hardware has been the systems design analysis (Failure Mode, Effect and Criticality Analysis). The purpose of this analysis is to determine the possible ways a given component can fail; the effect of that failure on the subsystem, stage, vehicle and mission; and how the failure compares in severity with other possible failures.

The Failure Mode and Effect Analysis is of value for compiling the critical components lists and establishing their relative criticality ranking which serves as a tool for evaluating comparable components and for identifying parts of the system requiring additional engineering attention and tighter quality assurance disciplines.

Flight data have confirmed that the considerable effort devoted to reliability is well worth the time required. For example, the

---

1 The triple modular redundant arrangement is frequently referred to as the voting circuit. Any two of the three elements override the third.

Saturn I, described in Chapter 2, proved several reliability concepts including the cluster design concept of rocket engines. In addition, it was proved that a flight could be made with one engine of the cluster inoperative. On the Saturn's fourth flight an inboard engine on the first stage was intentionally cut off early to demonstrate the reliability of the vehicle after a major component failure. In spite of that cut-off, all flight objectives were achieved. On the sixth flight a first stage inboard engine failed, causing large deviations in the trajectory that the guidance system had to and did overcome to achieve final orbit. Prior to the launch this occurrence and its result were considered and assessed to lie within the required reliability envelope.

On the spacecraft, redundance is used to the extreme in order to achieve maximum safety for the astronauts and highest probability of achieving the mission. This is especially evident by the redundant number of thrusters in the control system or the clusters of parachutes, or the number of redundant components and circuits in the environmental conditioning system in the spacecraft.

B. Launch Window to the Moon

As for many space missions and especially planetary flights, the launch of Saturn V/Apollo to the Moon can only occur at specific times during the year, during the month and even during the day. The time span in which the space vehicle can be launched for a specific mission is called the "launch window". The duration of this time span creates a problem that must be considered in the design of the mission. For example, if, for operational reasons, the vehicle can be launched only during daylight hours, then "launch window" would define the time between sunrise and sunset at the launch site. If an additional constraint requires that the Sun be behind some optical tracking equipment covering the easterly sky, then the launch window would be narrowed down to the time span between noon and sunset. For the Apollo mission, there are not only one or two but an entire list of conditions to be fulfilled simultaneously if the mission is to succeed. Therefore, the job of determining the launch window, that is, the time when it "opens" and how long it remains "open", becomes a complex one.

The launch window narrows as the number of conditions to be fulfilled increases. And with a smaller launch window, the chances of getting the vehicle off the pad are also shrinking. In the previous chapter it was shown that to meet the launch window the highly complex countdown must be so timed as to permit completion in time to accommodate the restrictions of the window.

It is, then, necessary to examine some of the conditions and parameters which shape the flight plan to the Moon and which, in combination, result in the launch window or launch windows since it is necessary to deal with a monthly launch window and a daily launch window.

A beginning consideration in discussing the launch window centres around the overall geometry of the Earth-Moon relationship as schematically depicted in *Figure 4-3* which shows the Earth and the Moon as they might be seen from a position high up in the northern celestial hemisphere.

On this geometry is drawn a schematic of the spacecraft flight trajectory discussed in Chapter 3. This trajectory is, for practical purposes, a portion of an ellipse until the spacecraft reaches the sphere of influence of the Moon's gravitational field. There the path is disturbed as shown in the figure.

To inject the spacecraft into an elliptical trajectory, it is obvious that energy must be spent and it is important that such an expenditure be kept to a minimum. This minimum is achieved through observance of one of the fundamental trajectory considerations of astronautics published by Walter Hohmann in 1925 which states that the injection energy would be a minimum if the spacecraft is at the point of closest approach to the attracting body of the focus. This point, called "pericynthion" for the case of the Moon, is called "perigee" for the Earth.

Applying the Hohmann trajectory then, it can be concluded that the spacecraft should be injected at perigee. Further, the perigee should be on the antipode[2] so that the high point (or "apogee") of

---

2 Consider that the Earth and Moon are connected by a line and that this line emerges from the Earth at a point diametrically opposed to the point of entrance. The point of emergence is the Moon antipode. Similarly, the Earth antipode lies at a corresponding point on the back side of the Moon.

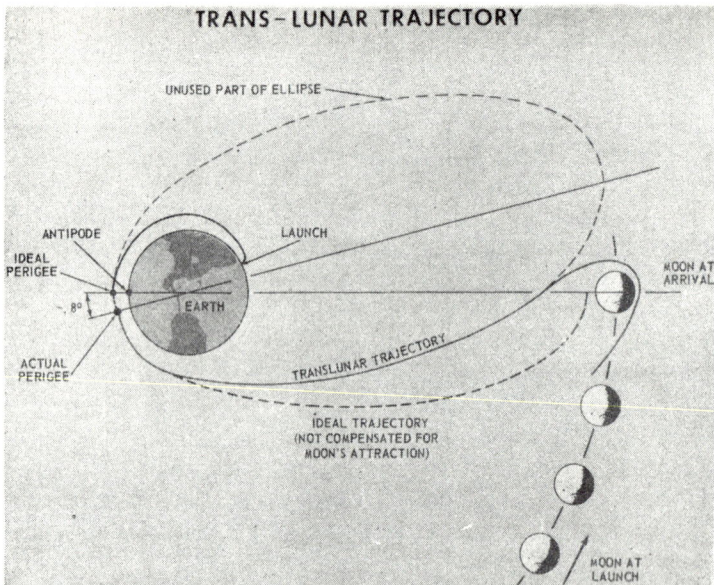

Figure 4-3.

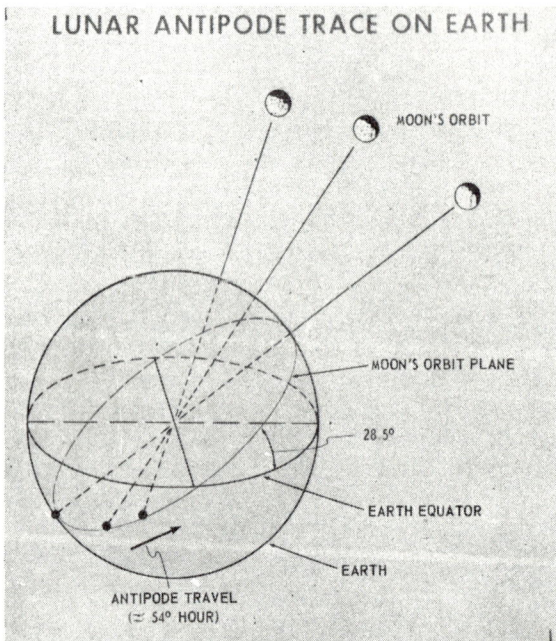

Figure 4-4.

our ellipse is roughly in the vicinity of the Moon at the time of spacecraft arrival. Since the Moon perturbs the trajectory away from the ideal trajectory (shown in broken line), these two points will in reality not coincide and the perigee will lead the antipode (by approximately 8 degrees) to compensate for this perturbation. In general, though, the injection point should always be close to the antipode (this is also the case for Earth's antipode for the return flight). In fact, for the purposes of the ensuing discussion, the injection point and the antipode will be assumed to coincide. The problem then will be to get from the launch pad at Cape Kennedy to Moon's antipode.

Since the antipode is always diametrically opposed to the Moon, it moves as the Moon travels in its orbit. As shown in *Figure 4-4*, the Moon orbital plane cuts through the Earth globe, forming a circle along its surface called a "great circle". Obviously, in one lunar cycle — about 28 days — the antipode moves once around the Earth along this circle, first into the northern hemisphere up to about 28·5° latitude, then into the southern hemisphere down to about 28·5°, and then up again.

In addition to the movement of the Moon and antipode, the launch pad also rotates with Earth from west to east. Combining the motion of the Moon and the motion of Earth, it can be shown that the antipode actually sweeps daily along the Earth's surface from east to west, while simultaneously moving slowly up and down to 28·5° northern and southern latitudes in a period of 28 days.

In principle, all that must be done now is to launch the space vehicle in such a direction that it intersects the Moon orbital plane at the exact time that the antipode arrives at the point of intersection. (This is shown schematically in *Figure 4-5*). As depicted in the figure, the direction of launch is usually measured from a line going toward the North Pole. The measured angle is the azimuth. Obviously, to meet the antipode at Point 1 the vehicle is launched with azimuth A; to meet it at Point 2 the launch must take place under azimuth B, and so on.

In *Figure 4-6* the same situation on a nonrotating Earth surface is depicted. As seen from this surface, the great circle of the Moon orbital plane and the antipode travelling along it move rapidly over

LAUNCH AZIMUTHS FOR LUNAR MISSION

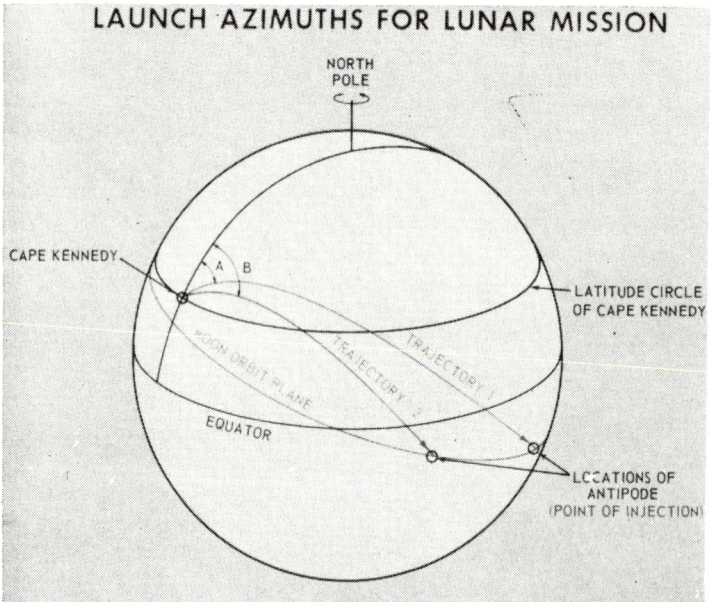

NORTH POLE

CAPE KENNEDY

A B

LATITUDE CIRCLE OF CAPE KENNEDY

MOON ORBIT PLANE

TRAJECTORY 2

TRAJECTORY 1

EQUATOR

LOCATIONS OF ANTIPODE (POINT OF INJECTION)

Figure 4-5.

LAUNCH AZIMUTH AND LUNAR ANTIPODE RELATIONSHIP ON EARTH MAP

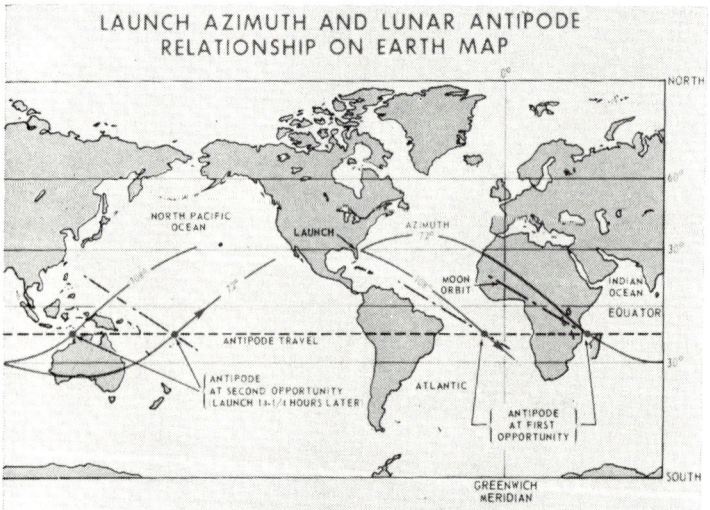

NORTH

60°

NORTH PACIFIC OCEAN

LAUNCH

AZIMUTH 72°

MOON ORBIT

INDIAN OCEAN

EQUATOR

72°

ANTIPODE TRAVEL

ANTIPODE AT SECOND OPPORTUNITY (LAUNCH 1¼-1¼ HOURS LATER)

ATLANTIC

30°

ANTIPODE AT FIRST OPPORTUNITY

GREENWICH MERIDIAN

SOUTH

Figure 4-6.

the surface of the Earth from east to west.   Again, the launch vehicle is launched at a specific azimuth and intercepts the antipode on crossing the Moon orbital plane.   If the launch vehicle trajectory is extended around the Earth, reappearing on the left side of the figure, it would again intersect the Moon's plane — but at a location shifted to the west because of the rotation of the Earth.   If the space vehicle would have been launched to *this* interception point about 14¼ hours after the launch time associated with the first interception, the antipode would have been also at this point of intersection.   Thus, a second opportunity of launch to the Moon results approximately 14¼ hours after the first.   However, while the first opportunity was in a southerly direction over the Atlantic Ocean, this second opportunity is obviously in a north-easterly direction over the Pacific Ocean, as shown in the figure.

Having two opportunities does not mean that if the astronauts are unable to utilize the first they can switch to the second.   Once the mission planners have decided on whether to inject over the Atlantic or over the Pacific, tracking ships will be sent to their appropriate locations and the spacecraft will use only the chosen injection point.   In summary, while two launch windows theoretically exist for each Earth rotation for Apollo, only one of them shall be considered for operational reasons.   In the following only the opportunity over the Pacific Ocean will be considered.

In *Figure 4-6* it was shown that interception of the antipode on the Moon's orbital plane is possible with any selected launch direction or azimuth if it occurs at a certain time, i.e., when the antipode nears the intended intersection point.   Therefore, each time of launch is rigidly associated with a certain specific azimuth. If the vehicle is launched at a certain time, it must be done under a particular azimuth or it won't meet the antipode.   Of course, the space vehicle must be launched somewhat ahead of time, *before* the antipode reaches the intersection point, because the flight from pad to the crossing with the Moon orbital plane takes a certain time. The longer the flight prior to the crossing becomes, the earlier the vehicle must be launched.   If the flight plan includes a coast period or even a parking orbit or two before injection at the antipode, this waiting time also would have to be allowed for in the timing.

In reality, this will happen, for the reason that the exact time of launch from Cape Kennedy would have to be extremely precise to meet the antipode and fly directly to the Moon without a parking orbit. By including the Earth parking orbit, the requirement for a split second timing from the Cape launch pad is relaxed. The capability of performing a small plane change manoeuvre at injection is gained as mentioned earlier in Chapter 3. With this capability, the space vehicle does not have to be injected on a split second timing either, but the engine can be ignited a little earlier or later and used for manoeuvre into the desired plane.

As just described, the launch azimuth for the flight to the Moon is dependent on the time of day. The later in the day, the larger the azimuth. Unfortunately, there is not a free choice of the azimuths throughout the day and thus the choice of the time of day is also limited. The range of permissible launch azimuths is limited by three considerations: range safety, booster performance and insertion tracking coverage.

The primary concern of range safety is to keep the ascending space vehicle over geographic areas where a malfunction could do no harm to anyone. These areas are called the "range", and largely consist of open sea.

A second consideration involves booster performance. In the previous chapter it was pointed out that a launch due east would gain the full benefit of the Earth's rotation. As the azimuth shifts away from due east (or 90°), this advantage decreases and more performance is required from the booster vehicle.

The third consideration, insertion tracking, is based on the fact that the last part of the trajectory (including insertion into parking orbit) must be monitored by a tracking ship in the middle of the North Atlantic. If the azimuth changes, the ship must move laterally south once the trajectory shifted out of the cone of accessibility of its radar antenna. However, the azimuth moves much more rapidly than the ship. Therefore, the ship remains where it is and the azimuths are restricted to the tracking cone offered by the ship. Combining all these requirements, the permissible range is between the azimuths of 72° and 108°, with 90° indicating a due east launch.

The appropriate launch circumstances then devolve finally into a range of azimuths that are obviously time-coupled. Within the bounds of the time-azimuth range, the vehicle may be launched into a parking orbit from which it may subsequently reach the antipode over the Pacific. Termed the "daily launch window", the time-azimuth span represents a $2\frac{1}{2}$ hour period.

Let us now turn to the other end of the mission — the Moon. Since the Moon reappears over the same point of the Earth's surface every 28 days, any restrictions which it places on the flight geometry must also have a monthly cycle. Thus, there is the "monthly launch window".

Two considerations which are really inseparable in their effects on the launch window are the lighting conditions at landing on the Moon and the location of the lunar landing site.

As discussed in Chapter 3, two astronauts will land on the Moon in the lunar module (LM). To assure that they have the best possible visibility during landing, it has been determined that the Sun must be behind them, neither too high nor too low over the horizon. (That is the eastern horizon, as the LM comes in for a landing from east to west on the Moon's front side.) For best visibility, the irregularities of the lunar surface should throw shadows. If the Sun is lower than about 5° to 7° over the horizon, there would be too many shadows and not enough light. On the other hand, if the Sun is higher than about 20° over the horizon, then for some portions of the approach path the landscape would not have enough shadows and would appear "washed out". Therefore, a landing region has been selected where the Sun is between 7° and 20° above the eastern horizon.

Figure 4-7 illustrates this condition. The landing must occur within the area enclosed by the dotted lines, close to the terminator, or day/night line, where the Sun elevation angle, of course, would be 0°. Since this terminator, which heralds sunrise, moves across the face of the Moon from east to west, the region of acceptable lighting moves also across the Moon's front side. The motion of the Moon along its orbit is exactly known and tabulated in large ephemeris[3] tables; thus, we can determine the days in the month

───────────

3   A periodical publication tabulating the predicted positions of celestial bodies at regular intervals.

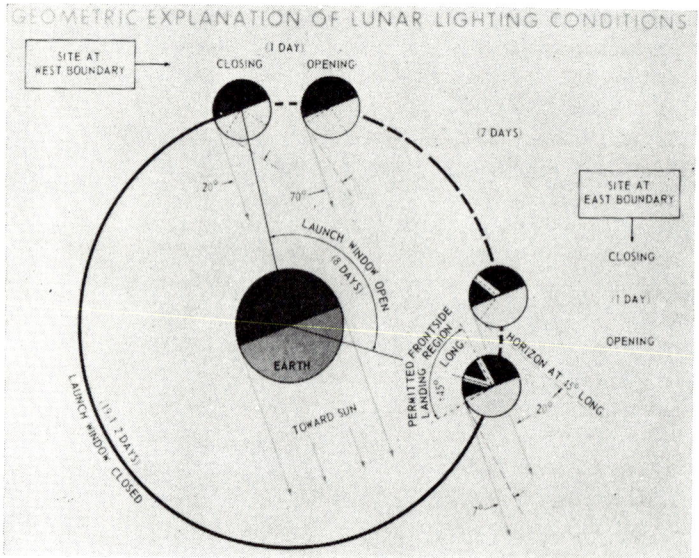

Figure 4-7.

when the permissible landing area is on the right edge of the Moon's face, in the centre of it, or on the left side.

Figure 4-8 shows how the lighting constraint would limit the launch in two typical months in the first quarter of 1968. Even if the LM were free to land anywhere between the longitudes 45° west and 45° east, it could not do so for about 60% of the month due to the lighting restriction. However, as an additional problem, we are not even free in the landing longitude. There are three reasons why the LM can land only at a few carefully selected spots. An explanation of this will be helpful before the discussion continues on the monthly launch window.

One important requirement which the lunar flight trajectory must fulfil is that it must be a free-return trajectory. This means a trajectory which curves around the Moon and returns to Earth, if the spacecraft keeps following it, instead of injecting itself into the lunar parking orbit. *Figure 4-9* shows such a free-return trajectory. This condition has been imposed on the selection of a

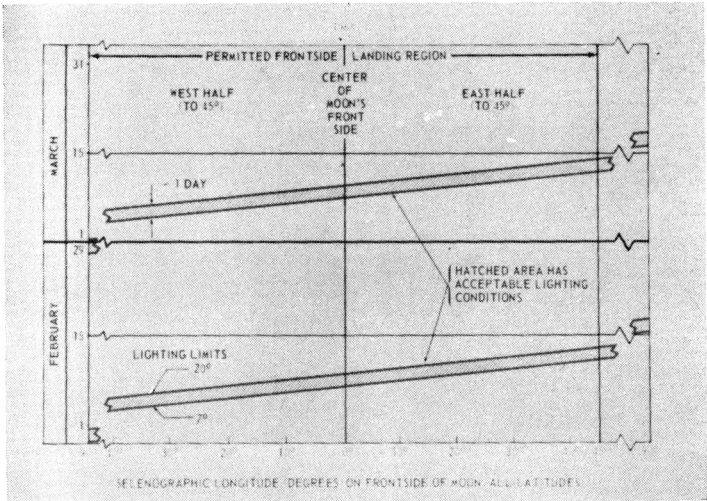

Figure 4-8. Launch window for February and March, 1968.

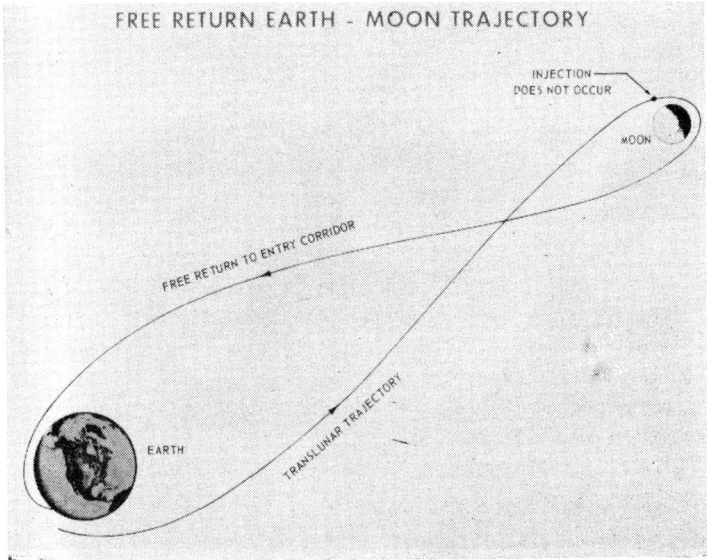

FREE RETURN EARTH - MOON TRAJECTORY

Figure 4-9.

lunar trajectory to make sure that the Apollo crew can return safely to Earth in the event that something goes wrong in the space ship and they are unable to use their propulsion systems. Unfortunately, free-return trajectories are not easy to obtain. All these have a very low inclination to the Moon orbital plane — less than 11°. Also, all have rather inflexible flight times — between 60 and 80 hours — which is the main reason why a flight time of around 70 hours was chosen. As a result of these two features, any lunar orbit resulting from such an approach trajectory would be of a very low inclination to the Moon's equator, i.e., very close to its plane. The second constraint is that the spacecraft will have only a very limited amount of propellants to conduct a plane change manoeuvre at orbit insertion (as was discussed earlier in Chapter 3) and, therefore, will be restricted to orbits close to the equator. Consequently, the landing sites will have to be along the equator while the northern and southern halves of the Moon's face are inaccessible to us. (*Figure 3-12* in Chapter 3 illustrated this point.)

Even this narrow equatorial band is not fully accessible because of a third constraint: terrain features. With Lunar Orbiter and Surveyor spacecraft, the Moon's surface was examined and most of this region found to be too mountainous and generally too dangerous for the spacecraft landing. Therefore, only a few landing sites remain. Two of these are on the right side of the Moon's face, one or two are around the centre and those remaining are on the left side.

If all the conditions discussed earlier are now combined, we will obtain the actual launch window during the year, month, and day, which includes both daily and monthly launch restrictions.

To illustrate an actual launch opportunity, see *Figure 4-10*, which shows how the Earth, the Moon and the Sun combine under the discussed constraints to give a realistic flight plan for, say, the year 1969. The figure shows that there is a five-day window in February from the 19th to the 24th. Similar windows would be available in 1969 throughout most of the year.

If for some reason the launch does not occur on the 19th, another attempt is not possible until two days later. (That much time will be needed to "recycle" the space vehicle, i.e., to prepare

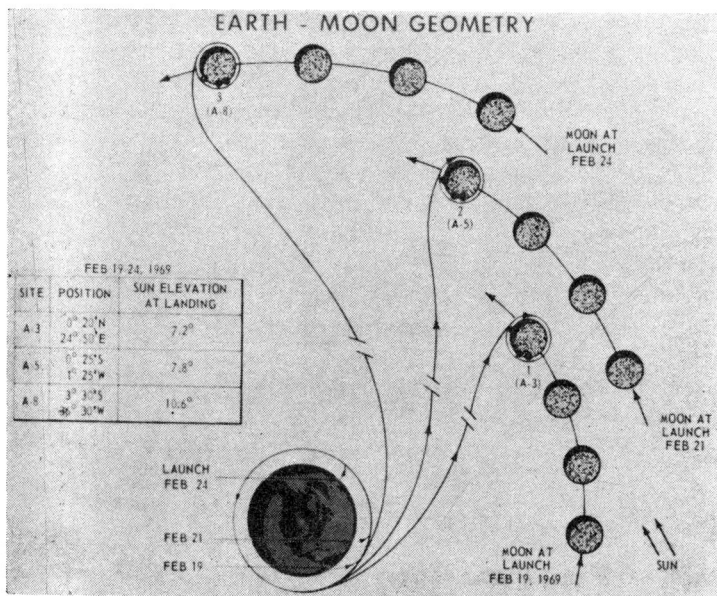

Figure 4-10.

it for a second attempt, involving for instance crew fatigue, retanking, exchange of batteries, etc.) Thus, the second launch attempt could come on the 21st. If we again have to hold or scrub,[4] the launch window allows a third attempt on the 24th. On each of these three days there is a daily launch window of about $2\frac{1}{2}$ hours and an injection point over the Pacific Ocean. It is estimated that the probability of getting the vehicle successfully off the pad is about 96 out of 100 if there are three "chances". Three chances, however, amount to at least a five-day window and, therefore, launch windows of this minimum duration are preferable.

As shown in the figure, each launch opportunity is associated with its own specific lunar landing site, as was discussed earlier. These sites are called A-3, A-5 and A-8, and their positions on the Moon's face are given. If the launch occurs on February 19th, the

---

4 To "scrub" has become an American idiom for cancelling a launching for the day it was to occur.

spacecraft would reach the Moon just as A-3, in the eastern half, is moving into the permissible lighting region close to the morning terminator. Two days later A-5 would arrive there, to be replaced two-and-a-half days later by A-8 in the western half. Thus, by launching in the darkness of the early morning hours of February 24th, the astronauts would step from the LM onto the mare Oceanus Procellarun not far from the crater family called Wichmann.

C. Apollo Transposition and Docking

During the Apollo mission the transposition and docking manoeuvre is the first of a number of activities of the Apollo spacecraft which are directly and completely controlled by the astronauts. The purpose of this manoeuvre is twofold: *a.* to attach the Apollo spacecraft to the lunar module to obtain the configuration required for the flight to the lunar orbit and for preparation for landing on the Moon, and *b.* to separate the spacecraft-lunar module combination from the instrument unit and the third stage of the Saturn vehicle which is then discarded. The manoeuvre takes place shortly after injection of the Apollo system into the trajectory to the Moon by the third stage.

Basically, in performing the transposition and docking manoeuvre it is necessary to close in on the passive vehicle (lunar module) with the active vehicle (command and service module) and to achieve the proper orientation of one to the other before the docking mechanism is engaged. This orientation involves two distinct aspects: the alignment of the two longitudinal axes through the components of the docking system on each vehicle so that the actual closure is as "straight" as possible: and the "indexing" of one vehicle to the other, i.e., the proper roll alignment between the vehicles around the longitudinal axis, since docking usually is not omni-directional in roll, due to mechanical limitations.

For the Apollo mission, however, there are three additional conditions which must be fulfilled before the manoeuvre can be executed by the astronauts. First, both the Apollo spacecraft and the third stage must maintain individual radio contact with Earth stations during the entire manoeuvre. Second, lateral (yaw) motions of the stage must not exceed angles larger than $\pm 45°$, to keep the gimbals of the inertial platform in the Saturn instrument unit from "locking" (contacting their stops) and thus losing inertial attitude in yaw. Third, the manoeuvre must be conducted so that the docking mechanism on the lunar module is illuminated by the Sun.

The resulting manoeuvre sequence is shown in *Figure 4-11* through *4-16*. Before initiation of the manoeuvre, the instrument unit maintains the third stage/spacecraft combination in the original inertial attitude existing at termination of injection thrust. The subsequent orientation manoeuvre, which precedes the actual transposition and docking, is executed by the third stage for the entire system. First, to reach an orientation where the Sun falls on the docking tunnel of the lunar module (still shrouded within the spacecraft-launch vehicle adapter), the third stage is pitched up by its auxiliary propulsion system through an angle of 60°. Both the spacecraft and stage are communicating with Earth over their respective S-Band omni-directional antennas. However, a short time later automatic switch selectors will switch the stage communications to directional antennas. To maintain the communication link with Earth, even with the relatively narrow beam, the vehicle must be rolled around its longitudinal axis by about 180°, till it points "heads down", so that the directional antenna will be in the proper position for transmission when the switchover is made. Actually, the pitch-up and roll-down manoeuvres will be done simultaneously and very slowly to conserve reaction control fuel, in a manner not conflicting with the gimbal-lock constraint.

During this period of reorientation, the vehicle is being tracked from the ground. It is expected that after about 10 minutes of tracking the vehicle's trajectory can be accurately determined. If the flight path is satisfactory, the astronauts receive approval to proceed with transposition and docking.

This manoeuvre is executed in a number of clearly defined, carefully designed steps. First, the combined command and service modules separate from their position on the adapter atop the stage and are moved away under the power of the service module reaction control system. The astronauts are now no longer connected to the stage by electrical wiring. Any further manoeuvres of the empty

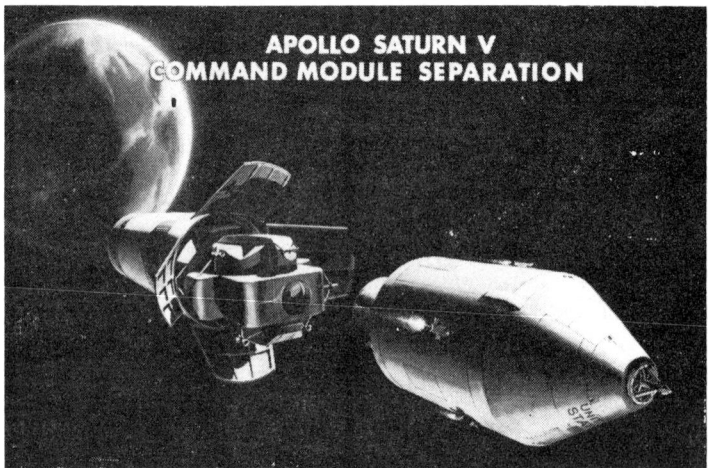

Figure 4-11.

stage must occur at ground command. However, the attitude selected before the spacecraft separation will normally not require any additional adjustments. The instrument unit on top of the stage will merely keep the inertial attitude of the stage fixed for the duration of the manoeuvre. During this time the directional antenna of the stage has been switched to a higher gain (a narrower beam) which, at least for an hour or so, will still be able to "see" the ground stations if the vehicle attitude in space is kept unchanged.

As shown in *Figure 4-11*, the four panels of the spacecraft/launch vehicle adapter are now automatically deployed like the petals of a flower. They open out to about 45° of angle with the longitudinal axis. If folded back completely, they would cover the stage antennas located around the periphery of the instrument unit and make communications impossible.

Using the service module reaction control system, the spacecraft is moved about 100 feet away from the stage. It is then rolled around its longitudinal axis to the proper indexing for docking (offset by 60° in roll from the lunar module axis) and is finally pitched around through 180° to point back at the lunar module (*Figure 4-12* and *4-13*). It is rolled *before* turning around to keep

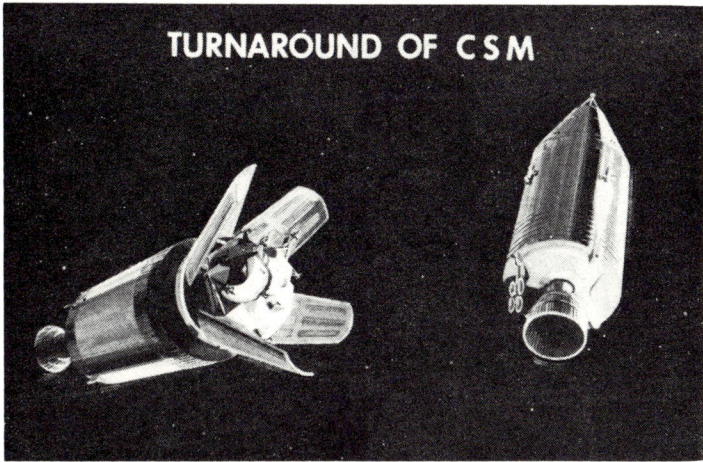

Figure 4-12.

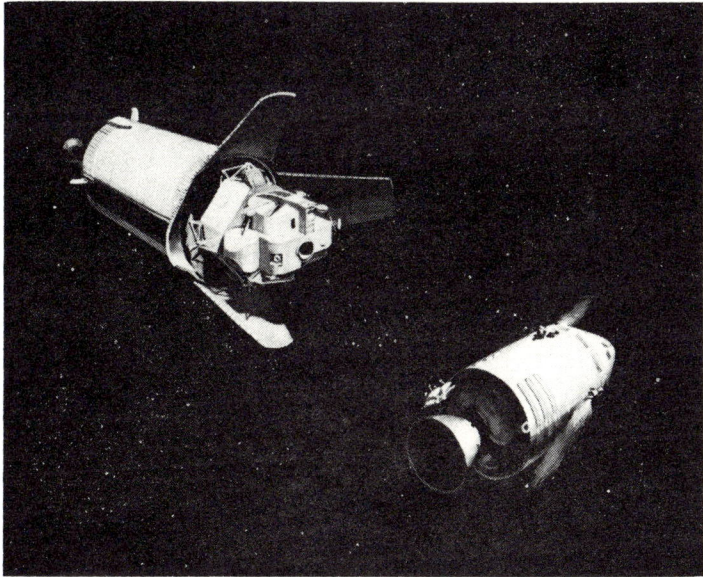

Figure 4-13.   Turnaround manoeuvre.

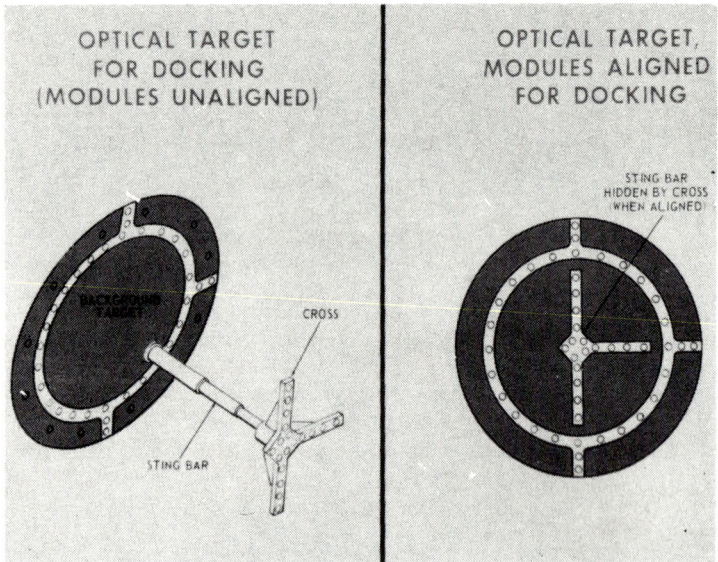

Figure 4-14.

the spacecraft omni-directional antenna from losing contact with the Earth stations during the pitch-over. As in the case of the stage, yaw manoeuvres must be restricted to avoid gimbal-lock of the inertial platform (to less than $\pm$ 70°). Having turned around, the astronaut will now deploy the high gain steerable antenna of the spacecraft and orient it to Earth before closing on the lunar module for docking.

The docking operation continues under the manual control of the pilot as the final translation is made and the command and service module (CSM) slowly nears the top of the lunar module.

To allow the crew the best view of the approaching stage and lunar module, their couches are moved around pivot points and shifted to docking position so that the faces of the command pilot and pilot are close to the rendezvous and docking windows in the command module.

The pilot conducts the rendezvous and docking manoeuvre by observing a target on the lunar module through an optical sighting

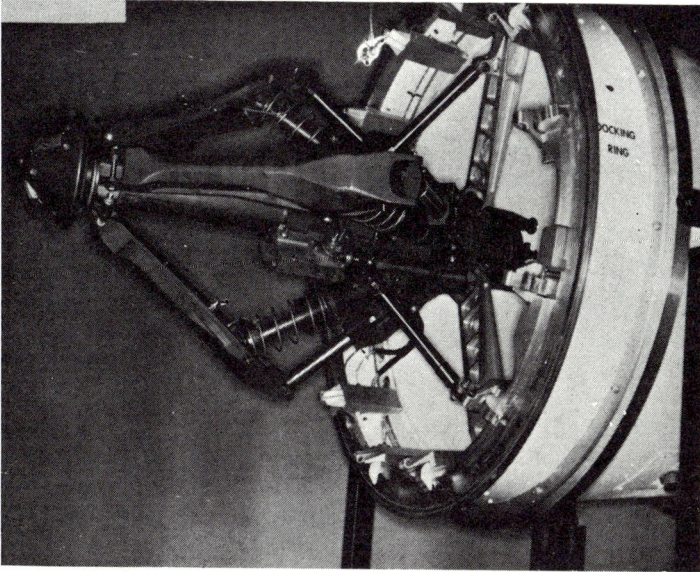

Figure 4-15. Apollo docking probe.

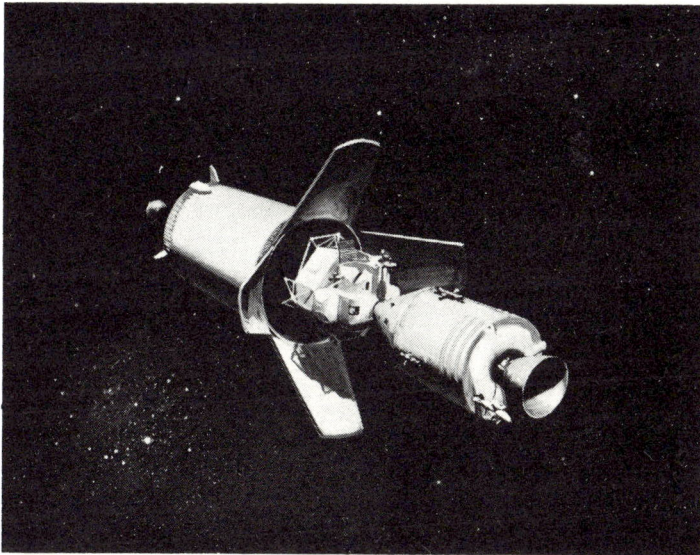

Figure 4-16. Apollo command module — lunar module engagement.

device much like a gun sight. The target on the lunar module top is a cross with a diamond in its centre *(Figure 4-14)*. It is mounted about 14 inches away from the lunar module top surface on a sting, at the foot of which is drawn and painted a target in red. By observing the white cross through the sighting device, which projects it two-dimensionally into its background, the pilot steers the CSM such that the diamond on top of the sting remains within the confines of the red target, as indicated in *Figure 4-14*. As long as this is the case, the docking is being performed within the capture range of the docking mechanism.

As shown in *Figure 4-15*, the docking mechanism consists of a probe on the spacecraft apex and of a conical receptacle, called a "drogue," at the lunar module docking hatch. The lunar module target with its T-bars provides a means for the pilot to determine if the two vehicles are properly indexed with respect to each other before the probe and drogue make contact. By applying closing thrust from the reaction control system of the service module, the pilot brings the probe into a drogue receptacle which is a funnel-type device on the LM. *(Figure 4-16)*. Capture latches are activated automatically; a panel light in the command module will indicate if they are properly engaged. Docking contact will be indicated by a disturbance in motion and by a small but sensible deceleration. Then the pilot will terminate closing thrust. He must be careful to apply thrust up until less than one second of achieving contact or else the energy stored in the springs of the docking system from first contact will, in some contact conditions, generate a rebound that will cause the probe to back out of the drogue.

The next step is initial seal latching in which the probe mechanism is actuated by high-pressure gas from pressure bottles in the gas-operated probe retract system to pull the two docked vehicles firmly together the final few inches. On pilot command, four semi-automatic docking latches are actuated to provide initial sealing at the docking interface. The forceful alignment and pulling together by the retract mechanism may cause the combined vehicles to start shifting their attitude in space. The pilot's task is to monitor the attitude rate indicators on his display board and stabilize the configuration, if necessary.

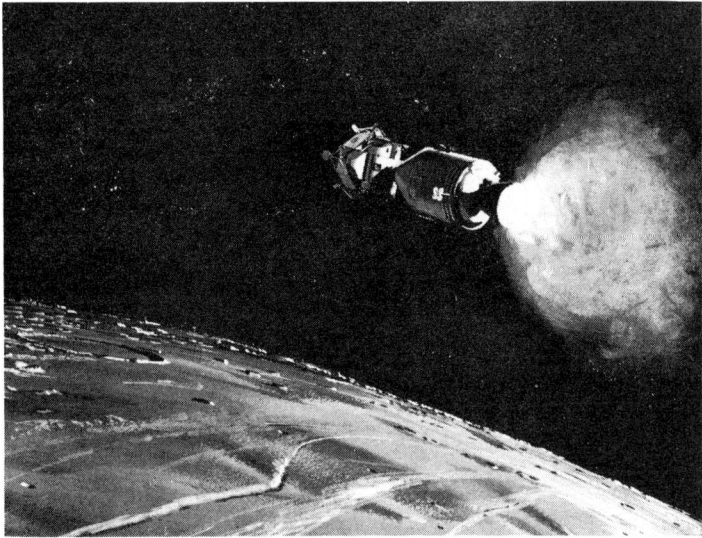

Figure 4-17.   Command lunar module flight configuration.

After the docking has been successfully achieved, the next step will be to make the tunnel between the command module and the lunar module accessible to the astronauts.   To do so, it must be pressurized with breathing oxygen at 5 psi.

The centre astronaut ("CSM pilot" and "navigator") removes a pressure hatch and a thermal hatch from the apex of the conical command module.   Then eight additional latches are hand-set by him to achieve a hard dock and structural integrity.   Redundant electrical cables are connected to the lunar module to provide for the electrical current required during the translunar flight (e.g., to feed heaters in the lunar module inertial measurement unit), and to provide power for the pyrotechnical devices which will release the lunar module from the rocket stage adapter.

After the connecting of the umbilicals, the centre astronaut restores the docking mechanisms in the tunnel which had to be removed earlier because they were blocking the tunnel.

On completion of this operation, the crew activates the pyrotechnic devices and withdraws the lunar module from the adapter (*as shown in the picture on the jacket of this book*).   This operation is supported by compression springs on the LM-adapter mounting points.   The service module reaction control system engines are used with short intermittent bursts to reduce the amount of flame impingement of their nozzles on the thermal coatings of the lunar module.

A separation velocity of about 3 feet per second is established. Then the space ship is oriented for passive thermal control and the flight is continued without the stage, the IU, and the four petals of the adapter.   (*Figure 4-17.*)

D. **The Crew Environment**

In man's manifest drive to probe deep space, he is forced to confront a deadly hostile environment comprised of components each inimical to life, especially to life of higher order like mammals or humans. A moment's unprotected exposure to any one of these perils would be fatal. It is, then, appropriate to explore certain of these hazards in greater detail and the measures taken to circumvent them or protect against them.

Best known of the hazards of space is the absence of oxygen. Above an altitude of about three miles, man must have special protection to survive against oxygen starvation, sometimes referred to as hypoxia.

At higher levels, other critical regions are reached in rapid succession.   Thus, when total atmospheric pressure reaches the point where it equals the combined partial pressures of water vapour and carbon dioxide, a condition of anoxia[5] sets in.   Moreover, at the low total pressure above the 24,000 foot level, nitrogen gas normally dissolved in the blood begins to bubble out clogging arteries. This condition is termed aeroembolism, otherwise known as the "bends".

At approximately 52,000 feet and above the phenomenon of ebullism or body fluid boiling is observed. Ambient pressures are lower than the pressure of the gases dissolved in the various body fluids such as, for example, saliva.   Çonsequently, these gases bubble out creating conditions akin to bends.

---

5 Anoxia — a complete lack of oxygen available for physiological use within the body.

The Apollo command module is, of course, designed to accommodate these and other rigours of flight that confront the astronaut. It is air and vacuum tight. To assure the integrity of the seal and to provide protection against dynamic loads, extreme cold and heat, noise, vibration, radiation, meteoroid impacts, etc., the module is encased in an outer shell. The atmosphere or climate of the crew compartment within the module is carefully regulated by the life support system which consists of a versatile apparatus that checks and adjusts temperature, pressure, purity and humidity both of the spacecraft environment and of the space suit when attached.   This environmental control system supplies oxygen at 5 psi, holds the temperature at 75°F, maintains a humidity index of between 40 and 70%.   It also provides portable water and power.   To continue citing the functions performed by this unique system, it reclaims oxygen from the spent atmosphere at the same time removing contaminants and odours and it provides for waste disposal and hygiene.

Similar functions are performed by the basic space suit life support system contained in a back pack that weighs but 65 pounds fully charged. Used by the astronauts as they venture from the lunar module to the surface of the Moon, this ingenious pack, called the "portable life support system", supplies the astronauts' needs for four hours after which it is easily recharged on board the lunar module.

In the absence of the normal evaporative cooling process, body heat generated by exertions of the astronauts is controlled by means of a very special undergarment.   Cool water is circulated at a rate of four pounds per minute through tubular passages in the garment.   Since the garment is in direct contact with the body, heat conduction to the water is easily effected.   The astronauts don this garment only before they go out on the surface of the Moon.

Yet another strange element of the ominous environment faced by the astronauts is weightlessness, otherwise referred to as "zero gravity", "zero-g", the "gravity free state" or the "agravic state".

Until quite recently, little was known about the effects of weightlessness on the body. What is more, studies were severely hampered by the inability to simulate the condition for appreciable periods. One method was suggested in 1950 by the brothers Heinz and Fritz Haber, who showed that zero-g could be obtained in an aircraft if the plane follows a certain parobolic or Keplerian trajectory. Using this technique, weightlessness periods of 30 to 35 seconds have been achieved in a modified version of specially equipped jet aircraft. The NASA research rocket plane, the X-15, has held to Keplerian trajectories for as long as 300 seconds.

Experiments with these aircraft did much to dispel previous theories regarding the zero-g impact on the body. These experiments showed, for example, that the eyes are so dominant in control as to overwhelm any disorienting that is suffered by the labyrinthian system of the inner ear, or the kinesthetic system consisting of sensory receptions located throughout the body. By aiding the visual apparatus with a horizon indicating instrument, orientation is rather easily controlled.

The real problems of weightlessness appeared in the extra-vehicular activities (EVA) undertaken in the United States Gemini Programme. In EVA, whenever the astronaut tried to perform physical tasks, the lack of bodily restraint forced unaccustomed muscular reactions and body movements. It was learned rather quickly that Keplerian trajectories in aircraft did not last long enough to make the problem evident or to allow the astronaut to learn to overcome the problem. However, it was found that neutral buoyancy conditions as provided in water tanks produced effects much like those of weightlessness. Complete compatibility has not been achieved, however, and the astronaut must therefore learn "on the job" in space.

Physiologically, weightlessness impacts hydrostatic pressures in the blood system, causing changes in the blood distribution which has the net effect of increasing heart activity. Secondary effects include dehydration and a greater-than-normal extraction of calcium mainly from the bones of the body. Upon landing, the fluid balance is restored promptly although some unpleasantries result from inability of the cardiovascular system to readjust promptly to the restored hydrostatic forces. So far 14 days of weightlessness has been experienced by two astronauts during Gemini flight without being harmful to their health. The question is whether this remains so during longer flights.

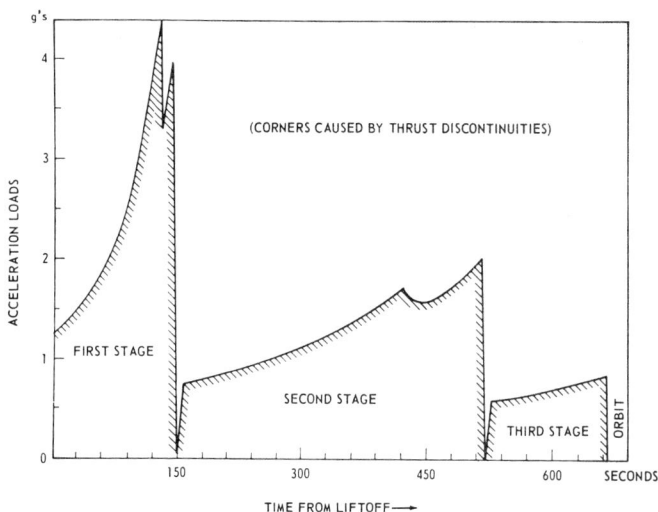

Figure 4-18.   Apollo/Saturn V acceleration loads.

In a sense, acceleration can be said to be the converse of weight-lessness since its effect is felt as an increase in weight. It is apparent that sufficiently high g or acceleration forces could severely limit or incapacitate an astronaut. Fortunately, however, accelerations experienced in normal flight are tolerable although levels reach 4½ g's at approximately 150 seconds after lift-off. *Figure 4-18* shows the profile of acceleration loads during launch vehicle operation. As an indicator, it has been determined that an astronaut can lift his hand and do useful work at an 8 g level.

Later in the Apollo flight, higher accelerations will be reached. As previously mentioned, decelerating forces during re-entry will reach 10 g's for a brief period. Since man's tolerance varies with time and position relative to the loads and since the time involved is brief, investigators devoted considerable attention to the position. As a consequence, couches fitting the form of the body of each individual astronaut have been designed and tested and are cur-

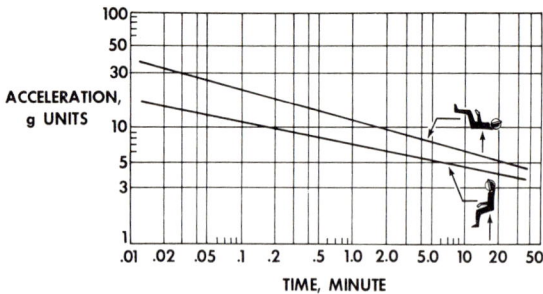

Figure 4-19. Man's tolerance to acceleration.

rently used in space flight programmes.[6] (*Figure 4-19.*) This couch is equipped with crushable honeycomb shock attenuators located at appropriate support points to accommodate hard landing shocks.

Finally, in the selection of particular phenomena, the problem of possible meteoroid[7] penetration arises. In the event of such an occurrence, two possibilities are presented. If the penetration is quite small, proportionally gradual or noncatastrophic decompression can occur. The astronauts will, under this condition, have time- to get into their space suits before unsafe pressures are reached. If accessible, the leak would, of course, be repaired.

In the event of a large meteoroid penetration, however, a serious, explosive decompression occurs during which cabin pressure drops in a matter of seconds with catastrophic results to the occupants. It remains, then, for the spacecraft designer to assure that the likelihood of meteoroid puncture is sufficiently remote to be acceptable in terms of mission reliability. One important step in this

---

6 The attitude of the spacecraft during periods of acceleration and deceleration is controlled in a way that the vector of the forces presses the astronaut with his back into the form-fitted couch.

7 The distinction between meteoroid, meteorite and meteor is frequently clouded. Conventionally, a meteoroid is a solid object moving in interplanetary space. A meteorite is any meteoroid that has reached the Earth's surface without being completely vaporized. A meteor, on the other hand, is the light phenomenon resulting from the entry into the Earth's atmosphere of a solid particle from space.

direction was taken with the introduction of large meteoroid detection satellites mentioned earlier. It may be recalled that this project provided and, indeed, is still providing valuable statistical data pertinent to meteoroid density. Smaller scientific satellites and space probes have also provided information on meteoroid density.

The meteoroid environment can be divided into two classes: sporadic meteoroids and showers. Sporadic meteoroids, as their name implies, approach Earth from all, although not random, directions and at irregular intervals. Showers, on the other hand, are encountered as concentrated, predictable groups. Some of these showers are the Leonids and the Perseids which return to the vicinity of Earth at regular intervals.

Optical and radar measurements, particularly those conducted by Canadian researches, show that meteoroids travel at velocities relative to the Earth that can vary anywhere between 11 km/sec (due to the gravitational attraction of the Earth) and 72 km/sec (if the particle comes from outside the solar system and meets Earth head-on).

To determine if a meteoroid will penetrate a spacecraft wall, one would have to know the velocity of the particle relative to the spacecraft, the mass of the meteoroid and its density. The kinetic energy of the impacting particle dissipates in several forms: *a*. melting, vaporizing and ionizing both meteorite and wall material; *b*. in local deformations of the spacecraft wall; *c*. in material ejected; and *d*. through infrared radiation. The latter, as well as ionization, dissociation and vaporization, can dissipate a much greater amount of energy per unit mass of material than the relatively low-energy process of melting, mechanical displacement and deformation. Thus, one consideration in designing meteoroid shields must be to promote maximum radiation dissipation. *Figure 4-20* depicts such a meteoroid impact.

With respect to the likelihood of meteoroid encounter, probabilities against the event appear quite favourable. As an example, consider an aluminium spacecraft with an exposed surface area of 50 square metres and an effective wall thickness of 0·25 centimetres or about a tenth of an inch. For a lunar mission of 14 days' duration, the equations (as used by NASA in 1964) would

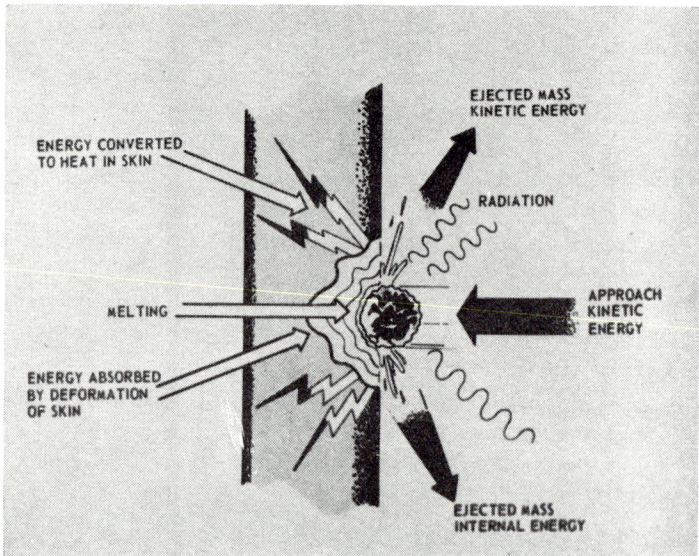

Figure 4-20. Meteoroid impact.

indicate that the probability of no penetration by a meteoroid is about 92·5%, or conversely — in 100 flights to the Moon of 14 days' duration each, only seven to eight flights will have a meteoroid penetration. These penetrating meteoroids must be larger than about 7 millimetres in diameter. By slightly increasing the wall thickness, the chances of a puncture can be decreased considerably, and this can be continued until a point is reached where the remaining probability of penetration is so small that the additional shielding will not be worth the disadvantage of additional structural weight.

In general, we may conclude that though meteoroids may be hazardous to space vehicles, the hazard will certainly not be as severe as formerly anticipated by many people. Present knowledge permits adequate shielding of spacecraft, or space suits, against the more numerous micrometeoroids (as well as against space radiation). The larger meteoroids that would penetrate any shield are relatively rare. Thus, there is justification for setting aside protective measures. As a parting example, it would be necessary to fly about 6000 times to the Moon and back to be struck one time by a meteoroid which is 1 cm across or larger.

One of the severest environmental phenomena the astronauts have to face is radiation. The atmosphere of the Earth protects us all extremely well against radiation. However, in space even short time exposure to the unprotected body would be lethal. Although the protection of the spacecraft shields the astronauts well during Earth orbital flights below the Van Allen Belt (mentioned in an earlier chapter), flights to the Moon and into deep space should go as fast as possible through this zone of trapped radiation. Above this zone the major radiation danger stems from solar flares which are caused by solar eruptions and which shower the solar system with protons, among other types of radiation. For more about solar flares see previous chapters.

CHAPTER 5

The Future

A. Objectives and Goals

Project Mercury, the first manned space flight programme, essentially developed basic transportation and proved that man could exist and work in space. Project Gemini extended the knowledge of the range of man's capabilities in the space flight environment and permitted conduct of a large number of engineering and scientific experiments in which man actively participated.

Project Apollo, with the goal of manned lunar landing and return, will sufficiently extend available technology so that in defining further activity, the planners can ask *why* should the next step be taken and what should its *goals* be.

Achievement of the primary goal envisioned by Project Apollo will provide a basic capability to fulfil many types of manned space missions. While many of these missions could be accomplished by unmanned vehicles, there have been and will be situations where the presence of a trained man on the scene would add immeasurably to the understanding of the data obtained. Certainly, astronomy and space physics, exploration of the Moon and planets, and similar activities require the flexibility of response and the ability to make value judgments which, characteristically, are human capabilities.

In other areas where continuous recording of data is the desired goal, the unmanned satellite provides a high standard of performance. Typical satellites are weather, communications and navigation satellites which obtain and transmit data indirectly. The survey satellite provides still another attractive capability. Studies indicate that there is a high economic benefit through overall surveys from space of such natural resources as crops and crop conditions, timber, water potential from various sources, and possible mineral locations.

Contributions, then, can be made by both manned and unmanned satellites which supplement each other. The issue is not, therefore, one between manned *or* unmanned space flight. The manned satellite provides a high degree of flexibility and selectivity in observation. The unmanned satellite provides continuous coverage of specifically selected data.

These, of course, are extremely generalized guidelines. Other factors strongly influence the final decision as to whether a satellite space system should be manned or unmanned: the level of available technology, the amount of available resources — both in terms of time and funds—and the type of mission to be performed. In determining, for example, a basic approach to satellite maintenance and repair, we need to understand whether it would be economically advantageous to rely on unmanned satellites which must be replaced by another satellite when a malfunction occurs; whether a manned vehicle should be sent up to make repairs or whether man should be stationed aboard the satellite to provide continual repair and maintenance.

Although operational concepts are still being defined, it is possible even at this early time, to identify some of the larger areas of activity towards which the space programme is moving. There are many ways to go and many things to do with the capabilities thus far developed. There are, in fact, so many possibilities that it is necessary to organize and group them so that lines of primary activity can be identified from numbers of smaller competing efforts.

To define these possibilities and secure the greatest possible scientific return in the 1970s, one of the leading space advisory bodies in the United States — the President's Science Advisory Committee — formulated three sets of key questions with which the space programme should concern itself.

1. Does life exist in places other than the Earth and, if so, what is its nature, how did it evolve and what are its probable forms?

2. What is the origin and evolution of the universe and what is its ultimate destiny? What is the place of our Sun and solar system in it? Do natural Earth physical science laws govern the behaviour of every observable part of space?

3. What are the physical conditions of the Moon and other planets in the solar system, and how did the solar system evolve? What dynamic relationships between the Sun and the planets shape the environments of these planets?

After demonstrating the capability of landing man on the Moon and returning him safely, it is intended that man explore the Moon and seek answers to some of these questions. A first step will be to extend the time that man can stay in the environment of space. The technology to support that extension can be developed in Earth orbit.

Planned are orbiting workshops or space laboratories manned by astronauts whose number now includes many scientists. The first workshops will be based on and evolve from an upper stage structure of the uprated Saturn I (Saturn IB) and will provide the environment to advance man's capability for long-term productive activity in space. Workshops may eventually lead to sophisticated space stations and will also provide experience and technology which will be directly applicable to planetary exploration.

Planetary objectives of the early seventies include further flights of unmanned planetary spacecraft. These will gather scientific data concerning planetary environments and the environment of outer space. Plans include more sophisticated Mariner spacecraft which will visit Mars and, perhaps later, Venus, Mercury, Jupiter and other planets. In addition to gathering data for basic scientific knowledge unmanned planetary exploration is extremely desirable before manned expeditions are established as an objective in space. Manned expeditions to Mars are feasible, of course, although they will require from 18 months to two years.

Manned lunar exploration, Earth orbital workshops and unmanned planetary exploration are all ways of seeking answers to the three basic questions proposed by the Science Advisory Committee. That committee has also proposed some objectives which, if pursued, may provide some answers. These objectives can be summarized as follows:

1. A limited but important extension of Apollo in order to exploit the anticipated capability of exploring the Moon.

2. A strongly upgraded programme of early unmanned exploration of the nearby planets on a scale of time and effort consistent with the requirements for planning future manned expeditions.

3. A programme of technology development and of qualification of man for long duration space flight in anticipation of manned planetary exploration.

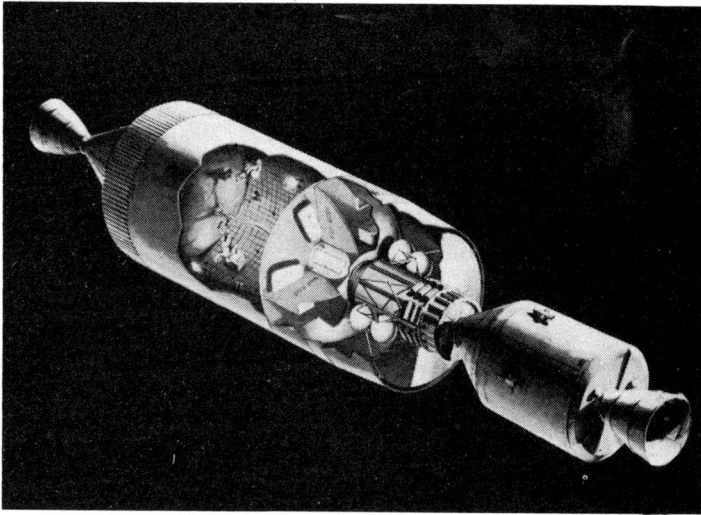

Figure 5-1. Orbital workshop.

4. The vigorous exploitation of space applications for the social and economic well-being of man.

B. Earth Orbital Operations

Two major categories of space flight activity are included in Earth orbital operations planning. The first category primarily includes manned scientific laboratories. The second one describes unmanned applications and resources satellites.

1. Manned Scientific Laboratories

Before real exploration and exploitation of space is possible, long duration space flight of man and system is a must if stated objectives are to be achieved. Significantly longer duration missions must be accomplished if meaningful data and experience is to be compiled in the fields of habitability, bio-medicine, behavioural science, work effectivity and in a host of additional disciplines. This competence and experience can most readily be secured in Earth orbital operations where, if some system or component should go amiss, safe recovery of crew and vehicle

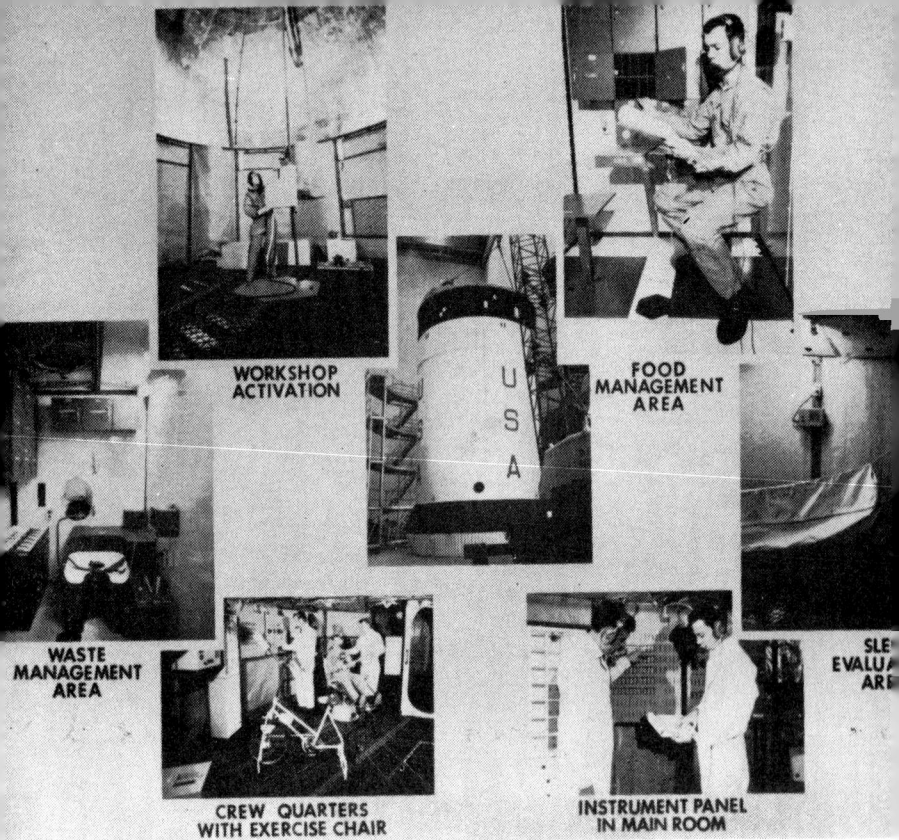

WORKSHOP
ACTIVATION

FOOD
MANAGEMENT
AREA

WASTE
MANAGEMENT
AREA

SLE
EVALU
ARE

CREW QUARTERS
WITH EXERCISE CHAIR

INSTRUMENT PANEL
IN MAIN ROOM

Figure 5-2. Workshop interior equipment.

would be a matter of hours rather than the days involved in recovery from far out in space, if in certain cases, such recovery is possible at all.

In the near future range, there is planned a series of experiments directed toward accomplishment of a limited spectrum of scientific, technological and medical investigations. The vehicle for this effort will be the so-called Orbital Workshop (*Figure 5-1*). The first Saturn IB Orbital Workshop mission could be launched after the successful lunar mission. The primary hardware is drawn from the Apollo Programme and will be modified to meet the special requirements of longer mission duration. Two vehicles will be used: the first is a Saturn IB launch vehicle carrying a manned Apollo command module plus a service module. The second vehicle will be a Saturn IB with its second stage especially designed to serve as the orbital

Figure 5-3. Saturn 1B orbital workshop with airlock and multiple docking adapter.

workshop after it has been utilized as a rocket stage to obtain orbital velocity (*Figure 5-2*). On the upper bulkhead of the second stage there is attached an airlock module and a multiple docking adapter (MDA). This is all generally referred to as the uprated Saturn 1B Orbital Workshop. (*See Figure 5-3.*) The airlock will provide access to and from the workshop and will also provide life support for crewmen in the workshop. The multiple docking adapter will provide a means for docking other payloads to the workshop while in orbit and will carry equipment and experiments that cannot be pre-installed in the workshop. Among these other payloads is the Apollo Telescope Mount (ATM) which, operating from a modified lunar module, will provide the means for solar astronomy experiments. This subject will be developed further later in this chapter.

Subsequent to this orbital workshop which constitutes a first step and is limited in its applicability since it also must function as a stage, current project studies consider the planning of a Saturn V workshop. This is an attractive possibility. The third stage of the Saturn V, in its basic structure, is essentially the same as the second

Figure 5-4.

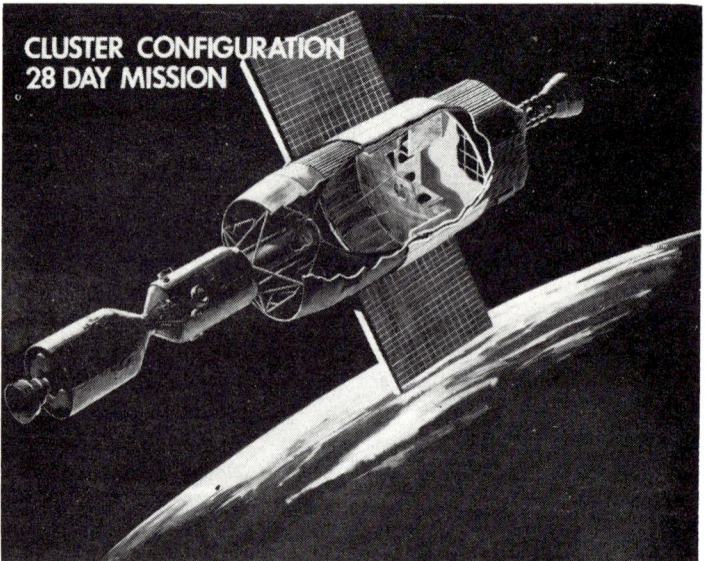

Figure 5-5.

stage of the Saturn IB which, as previously pointed out, will become the Saturn IB workshop. In the Saturn V application, all features of the workshop (or laboratory and house) would be built on the ground. The stage would not be used as a propulsive stage, but would be a sophisticated payload (*Figure 5-4*). This concept calls for use of the experience and designs developed on the Saturn IB workshop. The Saturn V workshop would be placed in orbit by the first two stages of the Saturn V. The command and service modules would also be part of this vehicle (*Figure 5-5*). Manoeuvres similar to those performed during the lunar mission would allow the spacecraft to rendezvous and dock with the workshop once in orbit.

Present configuration studies show that the docking adapter and airlock described for the Saturn IB workshop can be used. The arrangement of five docking ports on the multiple docking adapter allows considerable flexibility in the mission profile. Additional experiments could be developed later and brought up on future logistics flights, docked to the workshop and operated by the workshop crew. To take advantage of the additional payload capability, extra storage for breathing gas, water and food would be provided so that the astronauts would have sufficient supplies for one-and-a-half years.

Studies indicate, furthermore, that many of the designs, other than those mentioned above, from the Saturn IB workshop would be retained. The methods for food preparation and waste management will be suitable without modifications for the Saturn V workshop. The large solar panels for generation of electrical power as well as the electrical distribution system, could be carried over for application. Control panels developed for Saturn IB Apollo Telescope Mount could be mounted within the workshop, enabling the astronauts to operate solar and stellar astronomy experiments from interior control panels.

Concerning the solar experiments, it is appropriate to devote some attention to the Apollo Telescope Mount or ATM. This highly significant solar astronomy experiment package, which includes a modified lunar module, will be used for a variety of studies by the astronauts. These investigations using the ATM which consists of several small telescopes will be directed primarily towards events related to the active period of the Sun (*Figure 5-6*). Power for the experiments on the ATM will be obtained through solar arrays included in the workshop cluster. Reasons underlying this heavy concentration on solar astronomy are many.

The ultimate source of all energy on the Earth is the Sun. The natural conversion of solar energy has provided the Earth with such resources as oil, coal and wood. During the past 100 years it has been estimated that the world used an amount of energy about equal to the quantity consumed during the previous 1800 years. In this century the rate of increase has been tenfold and the consumption is increasing.

Approximately 32,000 times as much energy as the human race is currently using reaches the Earth's surface each year from the Sun. If a sizeable portion of this energy could be efficiently harnessed, it would help to solve the energy source problem on Earth. For this reason a better understanding of the Sun, its activities and its influence upon the Earth is a basic practical application.

The Apollo Telescope Mount with its improved capabilities and techniques will provide major contributions towards a better understanding of these questions by studying solar areas of scientific

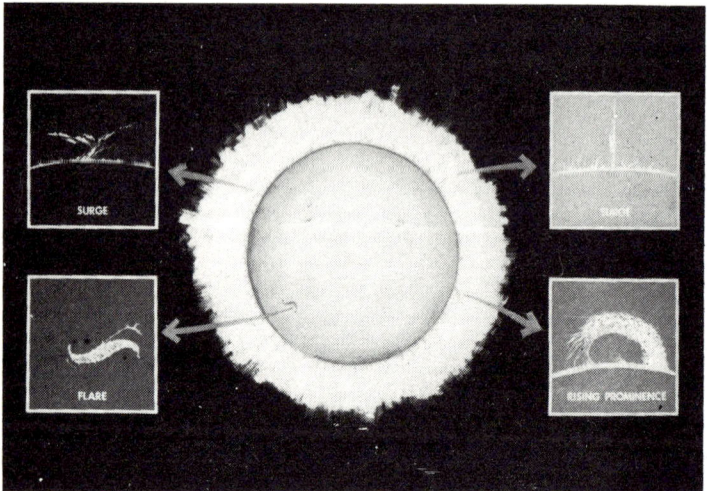

Figure 5-6. Solar activity.

CLUSTER CONFIGURATION-56 DAY MISSION

Figure 5-7.

interest. This should provide fundamental information on which
to base practical application for the benefit of mankind.

The Saturn V workshop, by providing a more extensive experiments programme, more sophisticated experimental hardware and
longer astronaut stay-time in orbit, would add new and meaningful
data to our initial space knowledge (*Figure 5-7*).

In its definition phase, special attention is being paid to the
operational mode, its resupply requirements, further identification
of the experiment programme and the impact of commonality
requirements with planetary and lunar developments. A long-
duration manned orbital station like the Saturn V workshop would
be useful to all space sciences.

2. Unmanned Applications and Resources Satellites

A second category of activity in Earth orbit includes investigations performed with applications and resources satellites.

For the most part, applications satellites would be unmanned
and would be either of the direct service type such as communica-

tions, or of an information gathering type like the Earth resources survey satellites which commonly would bring more indirect benefits.

From space high-resolution cameras can identify such surface environmental features as crop conditions, water potential from mountain snow, probable mineral locations, etc. As discussed earlier in Chapter 1, the pictures from the Gemini flights have already been used for such identification.

In the management and development of world food potential, the use of satellites as a highly sophisticated tool offers significant and attractive possibilities. The need for such information grows daily more pressing. The Earth is presently undergoing a population explosion. Between the birth of Christ and the year 1700, mankind doubled. It doubled again between 1700 and 1900 — an interval of 200 years, not 1700 years. Today the population on Earth is about three billion people and will double again in 33 years. This means there will be about six billion people by the year 2000 and from 12 to 13 billion by the year 2033. Despite this staggering increase, many world leaders believe that the Earth has enough basic resources to support 12 or 13 billion people.

But the problem is deeper than these absolute figures or the rate of growth. The areas of greatest population growth on Earth today are the underdeveloped nations. Modern medicine has begun to make its influence felt in these nations, reducing the infant mortality rate and increasing life expectancy. The birth rate has also been rising. More than 50% of the population in many underdeveloped countries consist of children under 15 years of age. Breadwinning in these areas becomes the sole preoccupation. These nations spend their meagre resources to stay alive, with little or no money left over to improve their lot. Currently it is a hopeless, vicious circle type situation.

One method for at least partially alleviating the consequences of these circumstances is to begin development of a world-wide Earth resources management system since but 9% of the land surface of the Earth is presently used for the production of animal and vegetable food. Much of the land cannot be used because it is too mountainous or too cold. But there remains plenty of land with a high agricultural potential. For example, the jungles of Brazil might be converted into an agricultural area.

How does this relate to the space programme? The photographs of the Earth in Chapter 1 were taken with a simple hand-held camera from the Gemini spacecraft. Nevertheless, the pictures show a tremendous amount of detail. If more sophisticated cameras were used, an important tool could be available for continuous surveys from space of the Earth's resources. The photographs would not only distinguish such individual food crops as rye, barley, soy beans, wheat and rice, but comparison of the photographs would monitor progress of these crops through their growing seasons, evaluating the impact of the weather and assessing the potential crop yield. This would be done by computers.

The computer would report many weeks ahead of time the possibility of famine somewhere in the world so that food shipments could be initiated before disaster hits.

But there are other benefits. In the event that a poor crop yield is identified in certain parts of the world, multi-spectral sensors[1] could be used to help determine the cause, such as too much salinity in the soil, poor fertilization, not enough water, or soil erosion. This information could result in the solution of a chronic agricultural problem. Rather than supplying food after crop failure, fertilizer could be shipped beforehand or consultants could be dispatched to give some on-the-spot advice. If, in addition, the communication satellites are used to educate people directly on steps they can take to help themselves, space applications will have provided another powerful tool for problem solving.

The remote-sensing space system could also be used to prospect for ore and oil deposits and to up-date maps. It has been estimated that it costs about one billion dollars a year to up-date the world's maps. No method is simpler and more effective than up-dating the maps with photography from orbit.

Satellite sensors can also be applied in oceanography, another important aspect of Earth resources, to determine such things as sea state, water temperature, the salinity of water and plankton content. These elements, in combination, directly affect the habits of fish as plankton is a basic food material of the sea. Where there

---

1 A multispectral sensor system would identify thermal radiation properties for ratios such as emittance, reflectance and transmittance at specified wavelengths.

is plankton, little fish are found. And where the little fish are found, the big fish are found. So by keeping an eye on the plankton distribution in the oceans, fishing fleets can be advised where to fish.

The Earth resources programme cannot be inaugurated tomorrow. This is a very major research and development effort. Much work would have to be done in what is called ground truth correlation tests. That is, the conditions that are really found on the ground are checked against what is shown by sensors and photographic equipment flown over the site.

Why go into orbit to make these surveys? Perhaps in the United States, Australia and Japan the necessary information could be collected at ground level. But it would be difficult, if not impossible, in areas like India or Central Africa, or northern Latin America, where large portions of terrain are relatively inaccessible and these data have not been collected and organized. So photography from above is necessary.

Why not take the required pictures from an aeroplane? There are two reasons. First, it would require many more pictures, resulting in poor overall resolution. Second, the cost of aerial surveillance is high and continued aerial surveillance results in very large operating costs. Nevertheless, continued surveillance is necessary — and on a global scale — if coherent records and measurements are to be kept on something that is constantly changing like crops or the movement of sea ice. Placing a space system in orbit may be more costly, but once it is there it stays there, and the longer it stays there the more economical it becomes. For example, a Saturn V payload, after half a year, costs less than a jet airliner; after one year, less than a luxury car; and after a year and a half, an economy sports car; and beyond that it even costs less than a motor scooter, which gets 125 miles per gallon. Whenever there is a need for a system to do a job for a long, long time, such as constantly surveying vast expanses of Earth, the space system is simply more economical.

C. Lunar Surface Transportation

A small scientific base on the Moon will afford man the opportunity of exploring the lunar surface and studying its characteristics in a depth of detail. In support of that exploration, preliminary

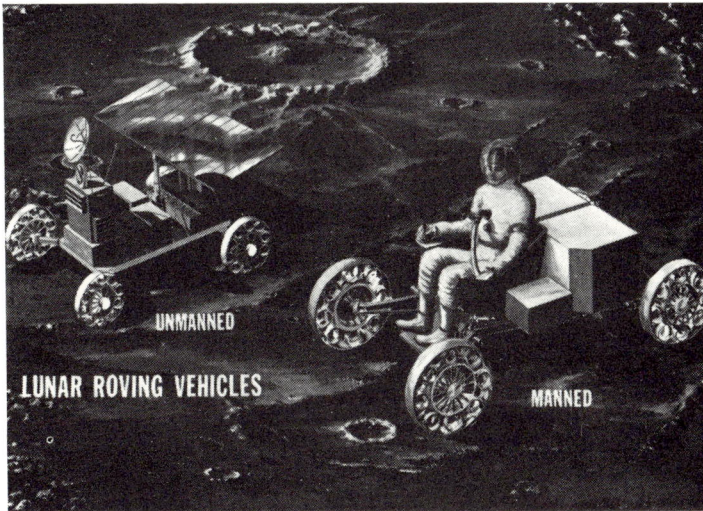

Figure 5-8.

planning, including systems concept and design studies, have begun
in representative areas such as scientific instrumentation, logistics
systems, housing for extended stay-time and transportation while on
the lunar surface.

Studies have identified a design of the lunar roving vehicle (*Figure
5-8*) which could be transported to the lunar surface in one bay of
the lunar module's descent stage. The vehicle would weigh about
500 pounds and be able to carry a suited astronaut plus a spare
portable life support system and 300 pounds of scientific equipment
over a five-mile radius area from the lunar module.

The lunar roving vehicle could be adapted to remote controlled,
unmanned operations. Once so adapted, it could be operated
from Earth and could traverse 620 miles over the lunar surface,
acquiring samples in preparation for a later visit by an astronaut.

This concept keeps open several promising alternative developments. For example, there may be a need to provide transportation
for men and equipment between different sites on the Moon, to
deploy scientific experiments for example, or to perform geological
surveys. As stay time on the Moon increases, the need develops for

Figure 5-9. Local scientific survey module.

larger wheeled vehicles such as the local scientific survey module
(*Figure 5-9*), which could transport more equipment and also carry
two astronauts, if desired. The survey module would be capable
of manoeuvring in rougher terrains. Because of its size and weight
the vehicle, along with other cargo, would be transported to the
lunar surface by one launch while the astronauts would arrive
on a later launch. The dual launch approach permits landing three
to five times the cargo of the single launch approach and increases
astronaut stay time from 3 to 14 days.

Another method of providing mobility on the Moon is through
use of a small rocket-powered flying vehicle (*Figure 5-10*). This
vehicle weighs approximately 150 pounds, has a propellant capacity
of 230 pounds and, in addition to the astronaut, can carry up to
250 pounds of equipment and supplies. The range of this vehicle
varies depending on distance to be traversed and the payload to
be carried. Typically, it could travel about nine miles and return
with one stop, or it could make a four-mile trip with a maximum
payload. Propellants used are the same as for the lunar module:
nitrogen tetroxide and 50-50 mixture of hydrazine and unsymmetrical dimethylhydrazine. The vehicle is refuelled on the Moon by

Figure 5-10. Lunar rocket-powered flying vehicle.

exchanging tanks. Two throttleable 100 lb thrust rocket engines provide the required propulsion. The vehicle is controlled completely by the pilot and could provide rapid transportation for exploration of sites inaccessible to surface travel or could be used to rescue an astronaut stranded away from the base. Mobility systems such as these will significantly extend man's range of activities in the event that exploration of the lunar surface is begun.

D. Planetary Exploration

Earth orbital programme plans have considered the use of orbital stations as a stepping stone to planetary exploration. For the immediate future, however, planetary plans primarily involve unmanned probes for scientific missions.

The planetary activities will develop knowledge and technical competence to keep open a series of alternative choices for whatever activities are later chosen as the most desirable space objectives.

In the immediate future the planetary effort will be directed towards three intermediate objectives: (1) to acquire as much knowledge as possible about the planets themselves; (2) to investigate the physiological and psychological qualifications of man to undertake extended space journeys; and (3) to advance through research in all the areas that technology might be required for in extended space journeys.

The presently planned planetary missions of the United States for the next few years consist of unmanned missions to Mars by Mariner and Pioneer class payloads. Pioneer is an interplanetary satellite of smaller weight whose objective is to gather interplanetary and solar data. Mariner is a satellite especially designed for planetary exploration with larger payload weights. In the 1970s Mariner class orbiters and hard landing probes are being considered for Mars.

The nuclear rocket programme is closely connected to the planetary programme. The more ambitious missions of the 1980s and beyond, such as manned Mars landings, will require a nuclear propulsion system as well as large deep space probes to outer planets such as Neptune and Uranus. Since a nuclear stage is not used to boost the payload into Earth orbit, it would serve as a spacecraft propulsion module.

There are two basic planetary alternatives for development in the future: (1) more advanced, unmanned planetary spacecraft for the next generation after the Mariner probes, and (2) manned flyby missions to Venus and Mars. Both types of mission would enable the collection of considerable data from the science experiments they carry. Before that data can be collected, however, it must first be decided what areas are to be studied and what types of data are desirable.

The approach to the study of a planet must proceed along the lines of inquiry developed from the experience with the Earth by applying the classical geophysics or planetology. This begins by definition of various geophysical entities relating to gross structure such as the solid core, its interior and surface, a hydrosphere, an atmosphere, an ionosphere and a magnetosphere. Each major entity has characteristic compositions and phenomena which, from terrestrial experience, can be studied indefinitely, rewardingly and in ever-increasing detail.

Particular objectives for the future planetary missions include the mapping of Venus and Mars, the determination of the structure of the Martian atmosphere, the composition of the atmosphere and clouds of Venus, determination of a Martian ecology (if any) and the search for extraterrestrial life on any of Earth's planetary neighbours. These same scientific areas will also serve for the probes to the other planets of this solar system. While all these subjects are of academic interest at the moment, it is certain that some results of planetary studies will ultimately be of great significance for man.

E. Space Astronomy

The astronomer has long been limited in his observations of the heavenly bodies by the Earth's atmosphere. This atmospheric filter has not only limited the astronomer's viewing but restricted the extent of electromagnetic spectrum in which he can observe. These limitations can be best understood by considering the objectives of astronomy and the techniques available to fulfil them.

The basic objective of astronomy is to understand the physical universe. In practice the astronomer observes with his instruments the current structure and content of the universe. He then deduces its probable origin and evolution from these observations, his knowledge and various theories. To obtain this information about our universe, the astronomer must intercept and analyse the electromagnetic radiations emitted, reflected or absorbed by the subject of interest. But all of these radiations cannot reach the astronomer's telescopes on the Earth. Therefore, Earth-bound observations will always be limited to the narrow window of the visible and radio wavelengths. The visible spectrum, which has been the most revealing, extends over only one octave. On the other hand, the unobservable ultraviolet and X-ray regions extend beyond 15 octaves — regions of vast potential knowledge (*Figure 1-10*). In addition, adverse Earth-bound observation conditions caused by turbulence, sky brightness and atmospheric scattering result in fuzzy and smeared images on the astronomer's photographic plates.

In the future, space astronomy will provide astronomical observations over the entire electromagnetic spectrum, as well as astronomical observations with maximum spatial resolution and of faint and extended sources.

The ultimate objective of the space astronomy discipline is to place into Earth orbit a large astronomy facility which would possess the same versatility as ground-based observatories. This facility would support the experiments of many astronomers. It would possess optical telescopes comparable in size to our large ground-based telescopes, instruments capable of analysing the energetic X- and gamma-ray emissions from celestial objects, and antennas of sufficient size to monitor the radio emissions of many galactic and extragalactic sources. This large astronomical observatory, because it depends on many other future Earth orbital space activities, will require a decade or more to reach reality.

Meanwhile, space astronomy will undergo the same evolutionary development as other space projects. During the evolutionary period the limitations that the space environment places upon both man, his support equipment and his scientific tools will be defined. The astronomy experiments performed during this period will provide an opportunity to obtain very worthwhile scientific data while simultaneously determining how to develop the equipment required for a space astronomy facility.

These experiments will be directed towards investigations within the following seven major areas:

1. Nature and evolution of the universe.
2. Nature and evolution of the galaxies.
3. Origin and evolution of stars and planetary systems.
4. Origin and evolution of cold bodies in the solar system.
5. Solar physics.
6. Non-thermal energy processes.
7. Existence of environmental conditions favouring the emergence of life.

Experiments in the ultraviolet region will be used to study stellar atmospheres, which includes the detection, the determination of the state of ionization and the composition of a celestial object. By studying the absorption and emission lines, the result of atomic electron transitions, the astronomer will be able to derive the chemical composition of both the stellar and interstellar medium.

Even since the accidental discovery of celestial X-ray sources by R. Giacconi *et al.* in 1962, it has been felt that this field of experimentation will play a very important role in advancing our understanding of the structure of the universe. X-ray astronomy conducted in the past by rocket flights, has been helpful in establishing new fundamental concepts of stellar evolution. In the near future space astronomy programmes will conduct X-ray investigations which will search for additional sources and study the structure of discreet sources and their extent. These investigations will also establish precise source location and correlate with any optical or radio emissions. At the same time astronomers will perform special analyses on discreet sources, look for emission lines and establish the source of the diffuse radiation and the properties of the intervening medium.

Gamma-ray astronomy is in its infancy. For the immediate future experiments are designed to look for sources of these extremely short wavelength emissions. These energetic photons are produced in nuclear reactions and in processes which involve the annihilation of matter. Since gamma-rays have infinite lifetimes and do not possess a charge, they will be an excellent tool in providing precise source information since they will not undergo unknown numerous deflections caused by the magnetic fields of space. However, it is expected that the flux from any point source will be so much smaller than the diffuse background radiation that detection of any point source will be extremely difficult.

The two general objectives for gamma-ray astronomy are to obtain information sufficient to establish the source, its intensity and the mechanisms which produce the emissions; and to obtain data on gamma-rays emitted by supernova remnants and strong radio sources.

The manned space programme will also be instrumental in establishing large aperture radio telescopes in space for the astronomers. However, the magnitude of their antenna elements are such that large supporting structures will have to be erected and at this time there is little available capability to construct and maintain the necessary orbital locations for these telescopes. The radio astronomer will desire: a long wavelength (MHz to a few hundred KHz) antenna with a 20-kilometre aperture, and a millimetre wave dish of about 100 feet diameter (about three times our largest ground-based unit).

Space astronomy has a vast potential of increasing the scientific knowledge of the universe. To do this we must always be in a position to capitalize on the experiences of man in space with astronauts from the community of astronomers going into space to observe the universe.

F. **Bioscience Plans**

In Chapter 1 some achievements in space bioscience were discussed. These represent first steps towards the important goals of investigating extraterrestrial life and assessing its nature, origin and level of development; to studying the effects of space on Earth organisms and developing fundamental theories and models of the origin and development of life in the universe. The space bioscience programmes of the future will be essentially an evolution of work already begun. They will focus both on obtaining basic scientific knowledge and providing biological knowledge useful to man on Earth.

Extensive studies of the effects of space environments on Earth organisms will be pursued actively during the immediate future. The next Biosatellite will place a pig-tailed monkey into orbit for 30 days. Scientists will make detailed measurements of the central nervous system, including alertness, sleep-wakefulness cycles, decision making, performance and fatigue. They will evaluate the cardiovascular system including heart action, blood pressure and circulatory dynamics. Metabolic measurements will be made of food utilization, energy production and biochemical changes in tissues such as muscle, bone and kidney. Behavioural tests will also be performed to assess the effect of weightlessness on task performance.

Following the 30-day mission, there are plans for another Biosatellite mission lasting 21 days and involving several test species. Rat experiments will study gross changes such as alteration in size and shape of muscle, bone or other body tissues; and will also expand the general metabolic information obtained in earlier flights. Rats will also be used to study circadian biological rhythms, i.e., rhythms whose period is about an Earth day in length. Circadian rhythms are found in many plants and animals, including man. The fundamental question is: do these rhythms rise from some innate character of the organism or are they impressed upon the animal by periodic geophysical phenomena which have approximately 24-hour cycles? After changing the periodicity of exposure by going into Earth orbit, a comparison with controls should show if moving the animals into the new environment changes the rhythms.

During this 21-day flight, plants will, for the first time, be grown through an entire life cycle to detect effects of weightlessness on the development and function of a complex organism throughout its life span. Finally, isolated human liver cells will be aboard to permit initial evaluation of the impact of weightlessness on cellular and subcellular functions.

The search for extraterrestrial life will, during the immediate future, also draw support from a rapidly evolving laboratory programme. In particular, lunar landings will permit an examination of the lunar surface for the signs of life or prebiotic and postbiotic chemical residues.

Beyond these ongoing programmes lie numerous research options and flight mission approaches which can lead to attainment of space bioscience goals. Extension of the Biosatellite Programme and individual experiments carried aboard manned orbital missions could extend previous work to clarify unresolved questions or introduce totally new research approaches. A logical follow-up of early primate work might be the "Bio A" experiment, exposing four primates to weightlessness for up to 90 days. Measurements could include: total food and water intake, waste analyses, energy exchange, electrocardiogram, electroencephalogram, electro-oculogram, Electromyogram, temperature, blood pressure and flow, arterial and venous oxygen tension respiratory gas exchange blood volume and composition and psychomotor testing for performance and behaviour. The high level of experiment sophistication would lead to increased understanding of response mechanisms to weightlessness.

A potential extension of biorhythm studies could employ a Pioneer spacecraft to place such test subjects as the common potato or the cockroach in an orbit around the Sun to totally isolate them from

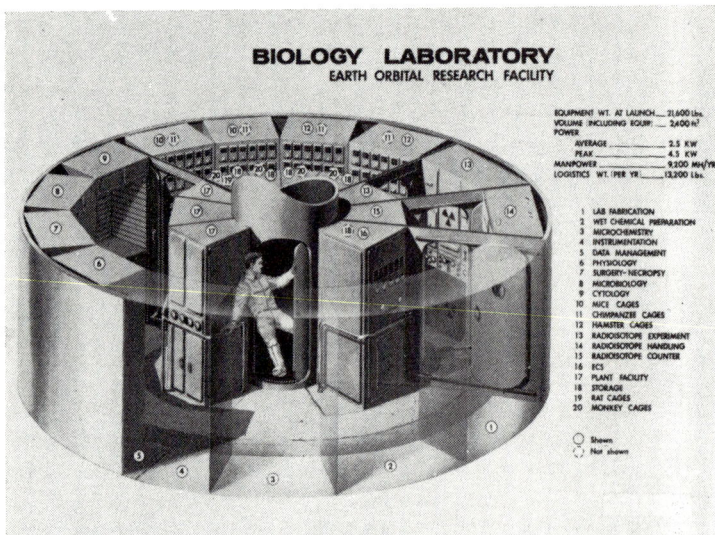

Figure 5-11.

geophysical cues, providing an independent line of evidence on the source of these rhythms.

Various degrees of man's participation in bioscience research in orbit may be used to supplement automated approaches. The crewmen's role might be to assist the ground-based researcher in monitoring experiment equipment, collecting records, preserving specimens and taking necessary maintenance or remedial actions. Eventually the researcher himself may become a crewman aboard an advanced workshop or definitive space station. In the more distant future the onboard researcher may have at his disposal an integrated life sciences laboratory (*Figure 5-11*). Such a facility would contain a wide array of common use and special purpose instrumentation and might house, simultaneously, a dozen or more species of test subjects. The research programme of a life sciences laboratory, extending over five or more years, could encompass hundreds of experiments across the fields of geosensitivity, genetics, morphogenesis, metabolism, biorhythms and behaviour.

Extraterrestrial studies beyond those discussed above must await the advent of further planetary missions. Planetary orbiting spacecraft will provide data on atmospheric and surface conditions which will permit clearer definition of experiments in planetary life detection and chemical evolution. The execution of these experiments will be deferred until soft-landing vehicles become available. Automated life-detection experiments will then give preliminary answers to some of the questions of extraterrestrial biology. However, the greatest return to bioscience from planetary exploration will come in the still distant future when man brings his research talents to bear on planetary research in person.

These plans, then, represent potential means by which NASA can approach fulfilment of its goals in bioscience.

G. Summary

It is less than 35 years ago that a man stood *alone* against the elements of the Antarctic winter to prove that it was possible to do useful scientific observation and research there. He, and those who followed him, were so successful in proving their point that today an arctic or antarctic tour of study is commonplace in completing the experience of a well-rounded ecologist or Earth sciences specialist.

Although it may be many years away, man's capability may eventually be expanded to accomplish full-fledged exploration and research in the new worlds of space, the Moon and the planets. The goal of this work is to squeeze from space the last drop of useful knowledge, not only for the intellectual satisfaction of man but to understand more clearly the environment on Earth and to modify it for the betterment of mankind.

Within these five chapters some few aspects of the U.S. space programme have been related. Some of man's experiences in space will be given in additional chapters. The accomplishment of the space programme and many related applications of hardware items required for space are being applied to man's better living every day. It is intriguing to consider what the next several years and decades will bring to mankind. The Apollo Programme is the next giant step across space to the next body of solid matter in our universe and in the future we hope to move even beyond the Moon to the planet Mars.

What does the vast area of space hold in the form of scientific knowledge that will add to our understanding? An encyclopaedia of questions could be written. We cannot speculate at this time on how quickly the answers will come. We can be sure, however, that answers we do find will lead to a series of new complexities and additional questions. We can, therefore, look forward with anticipation to the excitement of a vigorous search and the challenge of engineering to be done.

Introduction
To Space Flight

PART II

(Two Chapters)

by

DONALD K. SLAYTON

ALAN B. SHEPARD

L. GORDON COOPER

Donald K. Slayton.

Alan B. Shepard.

L. Gordon Cooper.

NASA astronauts.

CHAPTER 6

Manned Flight Operations

A. Project Mercury

Place man in orbit . . . return him safely to Earth . . . acquire and evaluate data to support long-duration space missions. What was once science fiction became, less than a decade ago, the true-to-life objectives of Project Mercury, America's first manned exploratory venture into space

Project Mercury, made possible by the far-sighted and creative efforts of members of the scientific and engineering communities, marked only the beginning of even more ambitious and more complex encounters with the unknown rigours of space. As with any undertaking of such magnitude, there were many unknowns confronting us when this venture began; but paramount, perhaps, were the monumental tasks of developing the systems and techniques to achieve these seemingly basic goals and proving man's ability to survive and function in these systems.

We knew that our approach to defining and solving the unique problems presented by this new environment would constitute the preliminary groundwork for all future manned space flight activity. Therefore, considerable thought and effort went into planning every phase of these early, pioneering flights. There were 25 major launches accomplished in Project Mercury between August, 1959, and May, 1963 — six of these were manned *(Figure 6-1)*.

Project Mercury Manned Flights

Flight	Crew	Launch date	Description	Duration, hr:min
MR-3	Shepard	May 5, 1961	Suborbital	0:15
MR-4	Grissom	July 21, 1961	Suborbital	0:15
MA-6	Glenn	Feb. 20, 1962	Orbital	4:56
MA-7	Carpenter	May 24, 1962	Orbital	4:56
MA-8	Schirra	Oct. 3, 1962	Orbital	9:14
MA-9	Cooper	May 15, 1963	Orbital	34:20

Figure 6-1.

On May 5, 1961, the United States man-in-space programme became a reality with the lift-off of Alan Shepard's "Freedom 7" spacecraft. This was a 15-minute suborbital flight over a ballistic trajectory through space — during which pilot and spacecraft travelled at a speed of 4,375 m.p.h. 302 miles down the Atlantic Missile Test Range, reached a maximum altitude of 116 statute miles, and experienced five minutes of weightlessness. A second suborbital flight, piloted by Virgil Grissom, was launched in July of that same year.

The original primary objective of Project Mercury was achieved successfully on February 20, 1962, when John Glenn, piloting the "Friendship 7" spacecraft completed a planned three-orbit mission with no adverse effects from weightlessness. On his first pass over Australia, Glenn observed the lights of Perth and conveyed his thanks to the citizens of the Australian coast for their thoughtful greeting. Three months later, Scott Carpenter made our second three-orbit flight, using a mission profile almost identical to that used by Glenn. Walter Schirra earned the title "textbook pilot" in October, 1962, with his flawless performance while completing six orbital passes of the Earth in his "Sigma 7" spacecraft. His flight provided information on extended exposure to the space environment, additional operational experience, and an opportunity to conduct experiments and measurements in space.

Project Mercury was concluded with the 22-orbit, 34-hour and 20-minute flight of Astronaut Gordon Cooper in May, 1963. The major objective of this mission was an evaluation of extended weightlessness effects, and Cooper was in excellent condition upon recovery in the Pacific Ocean. Other scientific experiments conducted on that final Mercury flight included aeromedical studies, radiation measurements, photographic studies, and visibility and communications tests.

The painstaking care exercised in Mercury to insure the ultimate in personnel safety and mission success enabled us to achieve all programmed objectives. But, additionally, we gained sufficient knowledge to permit us to embark on Project Gemini with a greater degree of confidence than originally anticipated. We had designed, built, and tested a spacecraft capable of sustaining human life; we had learned to prepare launch vehicles for safe and reliable manned flight; and we had succeeded in establishing an operational world-wide network of radio and radar tracking facilities which permitted extensive monitoring and communicating with the spacecraft and pilot. We had also reached a very important milestone — having developed and operated the life support and biomedical instrumentation systems that were considered essential for mission success. Finally, we had acquired valuable experience in large-scale programme management and systems engineering.

B. Project Gemini

With this newly acquired knowledge and experience, our efforts now turned to Gemini, the follow-on programme designed to subject two men and their necessary supporting equipment to long-duration flights in preparation for trips to the Moon or beyond. Our objectives: to formulate and refine rendezvous and docking techniques; to determine how astronauts could live and work safely in space for extended periods; to conduct experiments determining the feasibility of extravehicular activity (EVA); and to develop the systems and perfect the methods for controlled re-entry and landing at pre-selected sites.

These goals were accomplished on ten manned Gemini flights *(Figure 6-2)*, representing almost 2,000 man-hours in space. Five of the flights involved extravehicular activity, during which astronauts spent more than 12 hours outside the spacecraft. Rendezvous was accomplished on six of the flights, and five of these also included docking and multiple docking manoeuvres.

The primary purpose of Gemini III, the first manned flight in this series, was to verify the integrity and compatibility of the spacecraft with the crew. Astronauts Virgil Grissom and John Young served as command pilot and pilot, respectively, for this three-orbit mission. They performed the first in-flight manoeuvres to change the orbital path of a spacecraft. They are also credited with having performed the first manually-controlled spacecraft re-entry to an Atlantic splashdown, 60 nautical miles from the predicted landing point.

Gemini IV astronauts James A. McDivitt and Edward H. White first demonstrated the feasibility of extravehicular activity and evaluated a simple manoeuvring device. Following complete de-

Gemini Manned Space Flights

Gemini mission	Crew	Launch date	Description	Duration, day:hr:min
III	Grissom Young	Mar. 23, 1965	Three revolution manned test	0:04:52
IV	McDivitt White	June 3, 1965	First extended duration and extravehicular activity	4:00:56
V	Cooper Conrad	Aug. 21, 1965	First medium-duration flight	7:22:56
VII	Borman Lovell	Dec. 4, 1965	First long-duration flight	13:18:35
VI–A	Schirra Stafford	Dec. 15, 1965	First rendezvous flight	1:01:53
VIII	Armstrong Scott	Mar. 16, 1966	First rendezvous and docking flight	0:10:41
IX–A	Stafford Cernan	June 3, 1966	Second rendezvous and docking; first extended extravehicular activity	3:01:04
X	Young Collins	July 18, 1966	Third rendezvous and docking; 2 extravehicular activity periods; first docked target-vehicle-propelled high-apogee maneuver	2:22:46
XI	Conrad Gordon	Sept. 12, 1966	First rendezvous and docking initial orbit; 2 extravehicular activity periods; second docked target-vehicle-propelled high-apogee maneuver; tether exercise	2:23:17
XII	Lovell Aldrin	Nov. 11, 1966	Rendezvous and docking; umbilical and 2 standup extravehicular activity periods; tether exercise	3:22:37

Figure 6-2.

pressurization of the cabin, they opened the spacecraft hatch.   Edward White exited from the spacecraft to perform the first, powered extravehicular manoeuvring in history.   His life-support system consisted of a small chest pack (called the Ventilation Control Module) with oxygen supplied through a 25-foot umbilical hose assembly.   The hand held manoeuvring unit was a self-contained, cold-gas propulsion unit using two 1-pound tractor jets and one 2-pound pusher jet.   His specially designed space suit was worn with the extravehicular cover layer for micrometeorite and thermal protection.   He also wore a special sun visor to protect his vision while outside the spacecraft.

During his 20-minute EVA, White followed the time line illustrated in *Figure 6-3*.   *Figure 6-4* is one of the many photographs taken by Command Pilot James McDivitt and shows Edward White in the perfect posture for manoeuvring with a hand-held unit.   In recounting his experiences outside the spacecraft, White reported that he had perfect control of his movements using the hand-held

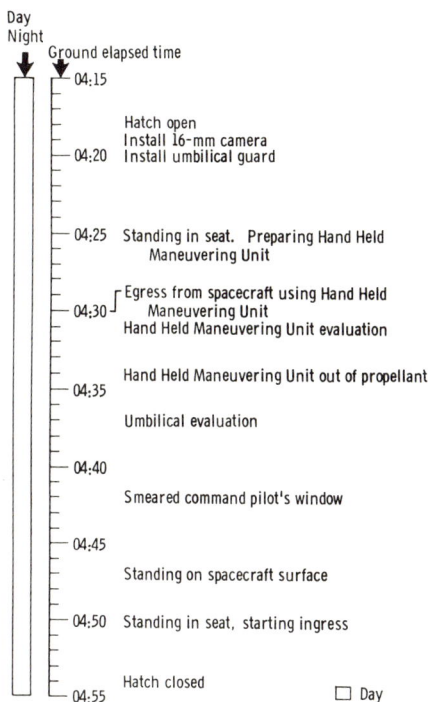

Figure 6-3. Gemini IV extravehicular time line.

manoeuvring unit (which he called "the gun"). He expressed some disappointment in not having more fuel to manoeuvre with the gun but found, after the gun's gas supply was exhausted, that he could still manoeuvre using the tether as a guide — although this method did have some limitations. He described his encounter with tether dynamics as follows:

"... You're actually like a weight on the end of a string. If you push out in one direction and you're at an angle from the perpendicular, when you reach the end of a tether, it neatly sends you in a long arc back in the opposite direction. ..."

He recalled absolutely no sensation of falling while out of the spacecraft — comparing the sensation he had to flying over the

Figure 6-4. Extravehicular activity during Gemini IV. Note classic posture of pilot for manoeuvring with hand-held manoeuvring unit.

Earth from about 20,000 feet — and he reported no difficulty in recontacting the spacecraft, particularly in trying to move back. He further explained that ". . . As long as the pushoffs are slow, there just isn't any tendency to get in an uncontrollable attitude."

Gemini IV proved that EVA was feasible but identified several areas where equipment performance needed improvement.

Gemini V, piloted by Gordon Cooper and Charles Conrad, marked the first flight on which fuel cells were used to provide electrical power for onboard systems. Its primary objectives were to evaluate performance of the rendezvous guidance and navigation system, using the radar evaluation pod (REP); to demonstrate and evaluate spacecraft performance for a period of eight days; and to evaluate the effects of prolonged exposure of the crew to the space environment.

In describing the activities during the first few orbits of this flight, Cooper said that the lift-off was very smooth and positive and that the trajectory was almost as perfect as a trajectory could be. He compared the Titan II launch vehicle as being considerably smoother and more solid than the Atlas he had ridden previously on his Mercury flight. He concluded that "the Gemini launch vehicle was really a Cadillac". He recalled the events surrounding the ejection of the REP by saying in part:

". . . it was ejected nominally, and very shortly after we ejected it, we were getting into the second night side, and as we began to turn around — yaw to the left — to get the REP in sight, we could see the light flashing very brightly on the nose of the spacecraft. It was quite near us and moving out in the scheduled manner away from us. We succeeded in getting a few pictures of it as it moved on out . . . We got radar lock on. The radar behaved ideally . . . We got radar range and range rates, and we got a great deal of data as the REP passed on out in this out-of-plane fashion in which we had ejected it. . . ."

Cooper praised the behaviour of the fuel cells, saying that they worked perfectly and were better than anyone had hoped they would be. He pointed out that they had powered one section of the cell down for long periods of time and it came back on strong and provided all the electrical power needed.

The Gemini V crew had taken about 350 photographs in addition to some 16-mm film, and Conrad pointed out that this material would provide "some pretty useful geological information". He and Cooper positioned tropical storm "Doreen" on two successive days of their flight, and they also observed hurricane "Betsy" during its early stages off the coast of Brazil.

Visual acuity was one of the six scientific experiments conducted on Gemini V. Its purpose was to measure man's visual acuity before, throughout, and after the flight to ascertain the effects of prolonged spacecraft environment on vision. Another objective was to test the use of basic visual acuity data to predict the limiting naked-eye visual capability to identify objects on the surface of the earth in daylight. To gain a better understanding of this particular experiment, let us go back for a moment to Cooper's first flight in Project Mercury and review some of the interesting visual sightings he reported:

". . . I could detect individual houses and streets in the low-humidity and cloudless areas such as the Himalaya mountain area, the Tibetan plain, and the south-western desert area of the United States. I saw several individual houses with smoke coming from the chimneys in the high country around the Himalayas . . . I could see fields, roads, streams, lakes. I saw what I took to be a vehicle along a road in the Himalaya area and in the Arizona-West Texas area. I could first see the dust blowing off the road, then could see the road clearly, and when the light was right, an object that was probably a vehicle.

"I saw a steam locomotive by seeing the smoke first; then I noted the object moving along what was apparently a track. This was in northern India. I also saw the wake of a boat in a large river in the Burma-India area . . . I saw the lights of Perth, Australia, and a bright orange light from the British oil refinery to the south of the city. If there is moonlight, then cloud layers and ground features can be seen. The moonlight was bright enough to detect motion on the ground. On several occasions I could see light from cities on the ground through the clouds."

Equipment used in the Gemini V experiment consisted of an in-flight device for testing visual acuity, a photometer to monitor the window of the spacecraft, two ground test sites, instruments to monitor atmospheric and light conditions at each site, and a training van. The experiment started several months before the flight and both pilots completed six sessions in the training van during this period. During the flight, in-flight vision tests were conducted each day by the Gemini V crew. Ground patterns were laid out near Laredo, Texas, and near Carnarvon, Australia. At the Laredo site, 12 background test areas were used with markings made of white gypsum. In Australia the markings were made of white shells. Use of these materials permitted changing the ground patterns as desired.

Preliminary results of the flight showed that visual performance was not degraded during the mission. Due to unfavourable cloud conditions only one quantitative reading was achieved of the ground markings at the Laredo site. Results of that observation indicate that the visual performance of the pilot during space flight was within the statistical range of his preflight visual performance and that laboratory visual acuity data can be combined with environmental optical data to predict correctly the astronaut's limiting visual capability to discriminate small objects on the surface of the Earth in daylight.

Before the first orbital space flight, the biomedical community expressed considerable concern for man's capability not only to perform in such an environment but even to survive in it. This concern resulted in a variety of predicted human responses to the space environment, and medical monitoring, therefore, became an important and inherent part of each manned mission.

The combined Mercury and Gemini space flights provided almost 2,000 man-hours of weightlessness, so that predicted effects of space flights could be evaluated against actual findings. As evidenced by the comparisons shown in *Figure 6-5*, many of the preflight predictions failed to materialize under actual space flight conditions. It was also disclosed that many of the responses observed were similar to reactions encountered in the Earth environment under comparable conditions of increased or modified human activity i.e., high heart rates, dehydration, loss of appetite, fatigue. Our experience emphasized that the causes of these various effects are not peculiar to the space environment alone but, rather, are dependent on a number of conditions which may or may not exist when these effects are observed. For instance, the use and overall performance of certain crew equipment was a contributing factor in the eye irritation experienced on two of the flights.

The extended duration of Gemini VII provided the best opportunity for extensive medical observation of a flight crew and was particularly helpful in arriving at our conclusions in this area. Frank Borman and Jim Lovell found that most of the problems, which existed on paper before the Gemini flights, concerning man's survival in space disappeared with the completion of their Gemini VII flight. Commenting on this, Jim Lovell reported:

"It is interesting to note that there were no body functions that we could observe that did not operate as well in a zero-g

Predicted	Observed
Dysbarism	None
Disruption of circadian rhythms	None
Decreased g-tolerance	None
Skin infections and breakdown	Dryness, including dandruff
Sleepiness and sleeplessness	Interference (minor)
Reduced visual acuity	None
(a)	Eye irritation
(a)	Nasal stuffiness and hoarseness
Disorientation and motion sickness	None
Pulmonary atelectasis	None
High heart rates	Launch, reentry, extravehicular activity
Cardiac arrhythmias	None
High blood pressure	None
Low blood pressure	None
Fainting postflight	None
Electromechanical delay in cardiac cycle	None
Reduced cardiovascular response to exercise	None
(a)	Absolute neutrophilia
Reduced blood volume	Moderate
Reduced plasma volume	Minimal
(a)	Decreased red-cell mass
Dehydration	Minimal
Weight loss	Variable
Bone demineralization	Minimal calcium loss
Loss of appetite	Varying caloric intake
Nausea	None
Renal stones	None
Urinary retention	None
Diuresis	None
Muscular incoordination	None
Muscular atrophy	None
(a)	Reduced exercise capacity
Hallucinations	None
Euphoria	None
Impaired psychomotor performance	None
Sedative need	None
Stimulant need	Occasionally before reentry
Infectious disease	None
Fatigue	Minimal

ª Not predicted.

Figure 6-5. Human response to space flight.

environment as they do on Earth. The only real problem that we encountered was the deterioration of the muscles due to lack of exercise. The absence of gravity made life very easy during the flight, and our muscles took full advantage of the vacation. When we landed and got out of the spacecraft, our legs felt somewhat heavy and stiff. This feeling of heaviness went away after several hours, but the leg muscles became sore after doing just some ordinary walking, very similar to doing strenuous exercise without first being trained. We, of course, exercised during the mission to help prevent this weakening of the muscles. We had an elastic-type cord which we could loop around our feet and then stretch with our arms to exercise our leg and arm muscles. We did this exercise very faithfully three times a day, and I am sure it helped in keeping our bodies in tune, but the small spacecraft volume prevented adequate body conditioning. We feel, however, that in a spacecraft with adequate volume man can survive the rigours of space much longer than previously anticipated."

Environmental hazards and the effects on man appear to be of less magnitude than originally anticipated. The principal physiologic changes noted were orthostatism for some 50 hours postflight, as measured with a tilt-table; reduced red-cell mass (5 to 20%); and reduced X-ray density (calcium) in the os calcis and the small finger. No abnormal psychological reactions have been observed, and no vestibular disturbances have occurred that were related to flight. Although much remains to be learned, it appears that if man is properly supported, his limitations will not be a barrier to the exploration of the universe.

Our first rendezvous mission, Gemini VI, was scheduled for launch on October 25, 1965, but was cancelled after a propulsion failure on the intended Gemini Agena Target Vehicle. The postponed flight, later redesignated as VI-A, was flown during the time Gemini VII was in space.

Astronauts Frank Borman and James A. Lovell were command pilot and pilot, respectively, on Gemini VII — the longest manned space flight in history through the close of the Gemini Programme. The 14-day mission was programmed to evaluate the effects of extended weightlessness on the crew. But, in addition, Gemini VII

Gemini Rendezvous Summary

Mission	Target	Approach	Separation altitude, n. mi.	Orbit travel, deg
VI–A	Gemini VII spacecraft	Below	15	130
VIII	Gemini VIII target vehicle..	Below	15	130
IX–A: Initial rendezvous	Augmented target dock-ing adapter.	Below	...............12.5	130
No. 1 re–rendezvous		Equiperiod	0	 80
No. 2 re–rendezvous		Above	...............7.5	130
X: Initial rendezvous	Gemini X target vehicle	Below	 15	130
Re–rendezvous	Gemini VIII target vehicle..	Below	 5	 80
XI: Initial rendezvous	Gemini XI target vehicle	Below	10	120
Re–rendezvous		Stable orbit	 0	292
XII	Gemini XII target vehicle....	Below	10	130

Figure 6-6.

served as a target for the subsequently launched Gemini VI-A which was piloted by Astronauts Walter Schirra and Thomas Stafford.

This historic first rendezvous occurred on December 15, 1965. Gemini VI-A was first seen as a small point of reflected light, coming straight up to the already orbiting Gemini VII crew. As Gemini VI-A came closer, the sun reflecting off its adapter section, it appeared as a half-moon shape. Stafford was heard to exclaim: "Commencing braking manoeuvres", and VI-A slowed to within 120 feet of Gemini VII. With great precision, VI-A moved in close to VII, and the two spacecraft remained in close proximity — sometimes as close as one foot — for about five orbits.

The success achieved on Gemini VI-A and VII tested our rendezvous and endurance capabilities in space — two areas essential for Project Apollo. To gain additional experience, rendezvous manoeuvres were scheduled and accomplished on all remaining Gemini flights — providing as broad a spectrum of terminal phase conditions as possible (*Figure 6-6*).

Having rendezvoused two manned spacecraft, the next planned mission was to rendezvous and dock with an unmanned Agena Target Vehicle. Gemini VIII was the first mission in which an

Figure 6-7. The Agena target vehicle, as it appeared to the Gemini VIII crew during the station keeping activity. The Agena was approximately 45 feet from the spacecraft at this time.

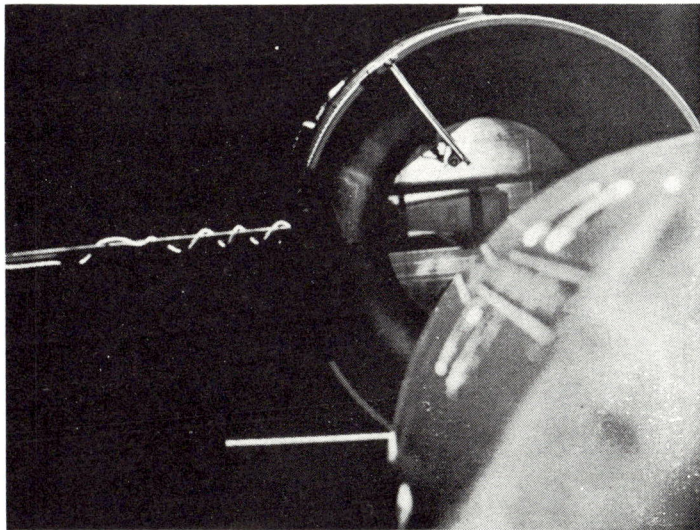

Figure 6-8. The Gemini VIII spacecraft is shown with its nose about two feet from the Agena target docking adapter just prior to accomplishing the first docking of two space vehicles in history.

Agena target was placed in orbit. The crew, Astronauts Neil Armstrong and David Scott, achieved rendezvous during the fourth orbit of the mission and performed the necessary station keeping manoeuvres to determine the status of the target vehicle (*Figure 6-7*). The spacecraft was manoeuvred to a position directly in line with the target docking adapter at a distance of about 3 feet. After inspecting the status panel, the docking cone, and the latches, the command pilot initiated the final approach by firing the aft-firing manoeuvre engines (*Figure 6-8*). Contact occurred with less than 2 inches of linear displacement, and very little angular misalignment, at a velocity of about ¾ ft/sec.

Gemini VIII and the Agena remained docked in a stable configuration for almost 41 minutes before a spacecraft attitude-control problem caused an unscheduled emergency undocking. Armstrong activated a secondary thruster system, which proved successful in stabilizing the spacecraft as it continued to roll at a rate of approximately one revolution per minute after the undocking sequence. The crew was then directed to terminate the mission and initiate re-entry procedures. (A more detailed description of this emergency is given in Chapter 7.)

This was our first in-flight emergency requiring the early termination of a mission. But, the effective implementation of emergency procedures confirmed that all the detailed planning and training which precede each flight are well worth the time and effort, and gave us increased confidence in the ground and recovery personnel who support each mission.

The primary objectives of the Gemini IX-A mission were rendezvous and docking with an Augmented Target Docking Adapter (ATDA) and the extravehicular activity of Astronaut Eugene Cernan. According to the mission plan, the spacecraft would rendezvous with the ATDA during the third revolution; to accomplish this, seven manoeuvres were scheduled.

The first manoeuvre involved a phase adjustment, initiated at 49 minutes and 3 seconds after lift-off, to attain the correct orbital catch-up rate for rendezvous with the ATDA in the third revolution of the spacecraft. Following completion of this manoeuvre, the Gemini IX-A orbited the Earth with an apogee of 147 nautical miles and a perigee of 124 nautical miles. Astronaut Thomas

Stafford executed a corrective combination manoeuvre beginning 1 hour, 55 minutes, and 17 seconds into the mission. This action adjusted the catch-up rate and the plane of the spacecraft orbit, bringing it closer to the ATDA's orbital plane.

An additional manoeuvre was performed as the spacecraft neared apogee on the second revolution. Termed a "coelliptic manoeuvre," the objective was to place the spacecraft into a circular orbit. Initiated 2 hours, 24 minutes, and 51 seconds into the flight, this action brought the spacecraft within 109 miles of the target, with a closure rate of about 126 feet per second, and about 12 miles below the target orbit. Pilot Cernan related that the spacecraft was about one mile from the target as they passed over New Guinea during the third revolution — at about 4 hours and 11 minutes of mission elapsed time.

As the spacecraft moved across the Pacific Ocean, the tracking station at Hawaii picked up the voice communication, and command pilot Stafford revealed that the shroud on the nose of the ATDA had not separated. In describing the situation, Stafford said:

"We have a weird-looking machine here. . . both the clam shells of the nose cone are still on, but they are open wide. The front release has let go and the back explosive bolts attached to the ATDA have both fired. . . . The jaws are like an alligator's jaw that's open at about 25 to 30 degrees and both the piston springs look like they are fully extended. . . . It looks like an angry alligator out here rotating around." (*Figure 6-9*.)

During the rest of that pass over Hawaii and continuing across the United States, flight controllers studied the situation and considered several possible means of breaking the shroud loose from the target. As one possibility, the ground would transmit several commands to the target, first to "rigidize," the second to "unrigidize" the docking cone. These actions were carried out, after the crew had backed to a safe distance away to observe the activity. The commands caused the shroud to move, and the "alligator's jaw" to close partially.

After determination that the shroud had not jettisoned, an alternate plan was worked out in Houston. At a time when docking had been planned, the Mission Control Centre in Houston gave Stafford instructions to align his spacecraft and to perform

Figure 6-9. The augmented target docking adapter as seen by the Gemini IX-A spacecraft after the rendezvous in space. Command pilot Tom Stafford referred to the ATDA and its still attached shroud as an "angry alligator".

a manoeuvre which would place the spacecraft into an orbit about 2½ miles above and 11 miles behind the target. About an hour and a half later, Stafford completed the equi-period rendezvous which had originally been programmed for the 28th hour of the flight. This manoeuvre, a completely onboard operation, used the computer and a handheld sextant to obtain guidance.

Stafford executed a separation manoeuvre at an elapsed flight time of 7 hours and 14 minutes. He later reported satisfaction with the results after tracking the ATDA on radar. Following this manoeuvre, the Gemini IX-A spacecraft orbited the earth at an apogee of 160 nautical miles an a perigee of 156 nautical miles; the ATDA remained in an orbit with an apogee of 161 nautical miles and a perigee of 159 nautical miles. Mission Control Centre, Houston, predicted that during the sleep period scheduled for the crew before the final rendezvous attempt, the spacecraft would move about 60 miles ahead of the target.

The third rendezvous, designed to simulate a lunar module rendezvous (rendezvous from above), would undertake to investigate possible conditions of a lunar rendezvous which might take place if the lunar excursion module had descended to the 50,000-foot level above the moon's surface. With the spacecraft about 80 miles ahead of the target, the first manoeuvre towards affecting the rendezvous was initiated at 18 hours, 23 minutes into the flight to adjust the spacecraft altitude. About 2 hours and 39 minutes later, the crew was preparing for the terminal phase initiate manoeuvre of this rendezvous. Stafford and Cernan then reported difficulty in visually acquiring the target using this mode of rendezvous but stated they had a solid radar lock-on. This difficulty resulted when they attempted to visually sight the target against the background of the Atlantic Ocean and the sand dunes of the Sahara Desert. The crew was not able to see the ATDA until they were within three miles of it. They said during a debriefing, later, that even after they had visual acquisition, the ATDA would be intermittently lost to sight against various terrain features.

During final rendezvous with the ATDA, the crew manoeuvred to within about three inches of the shroud to take closeup pictures of the shroud wires. Stafford graphically described this activity by saying:

"We kept clear of the dipole antenna, rolled the Gemini on its side, and rolled right up to where the X axis of the Gemini was 90 degrees to the X axis of the ATDA, and rolled into it, and snapped the pictures — making sure the alligator wouldn't bite us that way."

Stafford and Cernan reviewed the situation after the rendezvous. Stafford requested that any extravehicular activity be postponed until the following day because of crew fatigue. They had completed three rendezvous events in less than a day, all by different modes, and the crew was quite fatigued by the close attention demanded. Permission was granted and the flight controllers on the ground immediately started to revise the flight plan.

After completing their station keeping operation with the ATDA, Stafford and Cernan performed a separation manoeuvre high above the African continent. Several hours later, after another scheduled rest period, the crew performed several experiments. These consisted of zodiacal light photography, airglow horizon photography, and a communications system experiment to check operations through the ionosphere.

About 5.30 a.m., EST, the following day, some 45 hours into the flight, Stafford and Cernan began preparations for the extra-vehicular activity (EVA). According to plan, this operation would commence at the 49-hour, 26-minute point, just as the spacecraft entered the daylight portion of an orbital circuit. During the "making ready" period, ground stations held their communications with the spacecraft to a minimum, collecting only the essentials of information, and the crew advised on the status of the prepara-tions. Stafford said, near the 47-hour point, ". . . we've got the big snake out of the black box," meaning they had removed the 25-foot umbilical from its stowage place.

As the 49th hour drew near, Stafford told the ground stations that they were slightly ahead of schedule in their EVA preparations and that they were in the process of drinking a lot of water as the flight surgeon in the Houston Control Centre had advised. Shortly thereafter, the Carnarvon station relayed the word to them that they were "Go" for cabin depressurization, which took place between Canton Island and Hawaii. Stafford reported depressuriza-tion complete at 49 hours and 19 minutes. Three minutes later the crew opened the hatch. Then Cernan stood in the seat, retrieved the S-12 micrometeorite impact package, deployed the handrails, attached the docking bar mirror, and set up his EVA 16-mm camera. He seemed enthralled with the space view. He also noted that the tasks were somewhat difficult to accomplish in this weightless, suit-pressurized environment where all objects he worked with tried to float away.

At the 49-hour, 43-minute elapsed-time point, Cernan moved outside the spacecraft. He saw Los Angeles, he thought, and mentioned seeing Edwards Air Force Base. Cernan remarked about the difficulty in getting to desired vantage positions with the "snake" seemingly all over him, but he gradually worked his way toward the spacecraft's adapter section. He then returned to the hatch area, so that he and Stafford could change the film in the EVA camera and illuminate the EVA lights for the night orbital period. After this, Cernan returned to the adapter vicinity and reported there were no streamers or hanging straps as there

Figure 6-10. Astronaut manoeuvring unit as configured for Gemini IX-A. Extravehicular life support system also shown.

had been in the case of the VI-A and VII spacecraft. As IX-A came into the sunlight over Africa, Cernan moved into the adapter area where the Astronaut Manoeuvring Unit (AMU) was stowed (*Figure 6-10*). As he began to plug into the AMU circuits, Cernan noted that the exertion caused some fogging of his helmet visor. He also encountered difficulty in deploying the attitude control arms on the AMU, which ". . . presented far more difficulty to us in zero-g than they did in the simulation", Stafford said. In addition to problems in connecting the oxygen hoses and electrical circuits, Stafford described Cernan's communications through the AMU transreceiver as having a "lot of garble". The command pilot stated that Cernan's tasks seemed to be about four or five times harder to accomplish than had been anticipated. They were about ready to give a "no go" on the AMU if the visor continued fogging. Although Cernan rested, the fogging condition did not improve significantly. Stafford and Cernan evaluated the situation after sunrise. They felt continued fogging constituted a flight safety hazard and, after ground concurrence, the AMU experiment was scrubbed. Cernan continued to rest and gradually gained about 25% vision through the visor. After he had switched to the umbilical communications lead, voice contact immediately improved between command pilot and pilot. After more rest the fogging reduced to about 40%, but, as he began to retrieve the docking bar mirror, the fogging grew worse. Cernan also became quite warm. According to the flight plan he was scheduled to take photographs of the sunset but Stafford decided that Cernan should get back in the spacecraft before then.

At 51 hours and nine minutes, Stafford helped Cernan back into the seat. They spent the next 17 minutes stowing the umbilical, and secured the hatch at 51 hours and 30 minutes elapsed time. The crew had been in extravehicular conditions for two hours and five minutes and had accomplished cabin repressurization without incident.

One of the most exciting aspects of the entire Gemini Programme (and the primary reason for rendezvous and docking) was to use the target-vehicle propulsion systems to greatly increase the manoeuvring potential of the manned vehicle. The Gemini X crew, Astronauts John Young and Michael Collins, were first to test this manoeuvring capability.

Gemini X was launched exactly as planned and inserted into orbit with a perigee of 87 nautical miles and an apogee of 146 nautical miles. A half-hour after lift-off, the Gemini X spacecraft trailed the Agena X by 850 miles; and the Agena VIII passive target vehicle, which had been in orbit four months, trailed the Gemini X spacecraft by 500 miles.

A series of five major manoeuvres were performed by the Gemini X crew in preparation for the rendezvous with the Agena X target scheduled for the fourth revolution. Those manoeuvres were completed and, after five hours and 21 minutes of flight, the crew reported through the Tananarive, Malagasy, tracking station that they were 40 feet from the Agena X. Later, while the spacecraft was over the Coastal Sentry tracking ship, Young and Collins were given a "go" for docking and that phase of the mission was completed after five hours and 53 minutes of elapsed flight time.

During the early phases of the flight, before rendezvous, the crew accomplished sextant readings and performed other activities to attempt rendezvous without using ground-computed data. Although the computations were made, the crew decided to use the ground-computed data to accomplish the rendezvous. Fuel usage was high during the terminal phase of the first rendezvous.

The resultant shortage of fuel placed an additional constraint on most phases of the planned flight activities from that point. The flight controllers did, however, greatly alleviate the situation by determining that the Gemini X spacecraft should stay docked with the Agena X for almost 39 hours to get the maximum benefit from the propulsion system of the target vehicle. By using the spacecraft-target vehicle combination, most mission objectives were achieved. There were six major manoeuvres of the docked Gemini-Agena vehicles during this time period: three using the primary propulsion system of the Agena X and three using the Agena's secondary propulsion system.

The first manoeuvre produced an orbit with an apogee of 412·2 nautical miles and a perigee of 158·5 nautical miles. In describing his personal reaction to this manoeuvre, John Young used these words:

". . . . Mike threw the switch and a minute and 24 seconds later . . . it was really something. We had a negative 1-g and were driven forward in the cockpit . . . we got a tremendous thrill . . . on our way out to apogee and a new world's record for altitude. . . ."

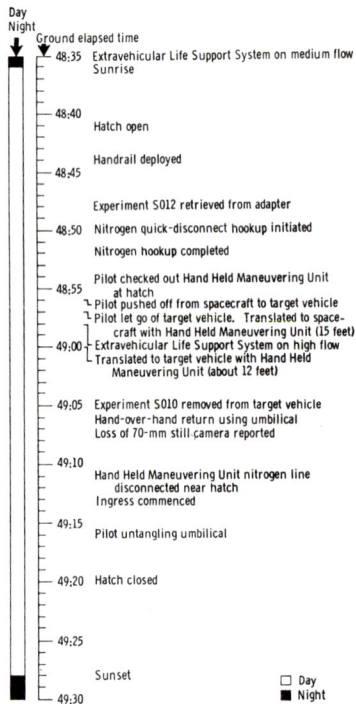

Figure 6-11. Gemini X umbilical extravehicular time line.

This manoeuvre was followed by one which adjusted the height and resulted in an orbit with an apogee of 205·8 nautical miles and a perigee of 158·4 nautical miles. The next manoeuvre was to make the orbit circular and was completed with an orbit which had an apogee of 208·7 nautical miles and a perigee of 203·9 nautical miles. Three smaller manoeuvres, all performed by using the secondary propulsion system of the Agena X, ended with the Gemini X spacecraft and the Agena X still docked and in an orbit with an apogee of 208·5 nautical miles and a perigee of 205·5 nautical miles.

The crew undocked the spacecraft from the Agena X after 44 hours and 40 minutes of flight. Shortly thereafter, a manoeuvre was performed permitting the spacecraft to rendezvous with the

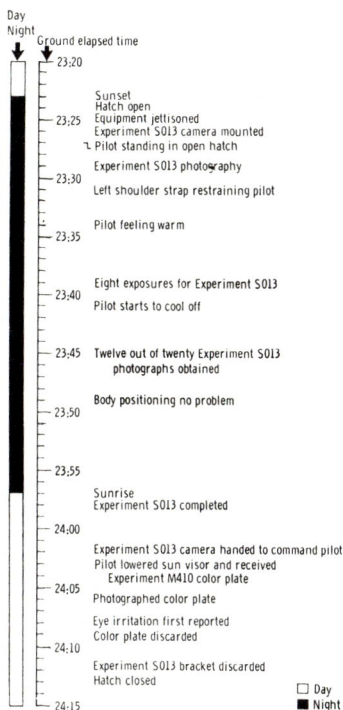

Day
Night
Ground elapsed time

23:20

Sunset
Hatch open
23:25   Equipment jettisoned
Experiment S013 camera mounted
Pilot standing in open hatch

Experiment S013 photography
23:30
Left shoulder strap restraining pilot

Pilot feeling warm
23:35

Eight exposures for Experiment S013
23:40
Pilot starts to cool off

23:45   Twelve out of twenty Experiment S013
photographs obtained

Body positioning no problem
23:50

23:55

Sunrise
Experiment S013 completed
24:00

Experiment S013 camera handed to command pilot
Pilot lowered sun visor and received
Experiment M410 color plate
24:05   Photographed color plate

Eye irritation first reported
Color plate discarded
24:10

Experiment S013 bracket discarded
Hatch closed   ☐ Day
24:15   ■ Night

Figure 6-12. Gemini X standup extravehicular time line.

Gemini VIII Agena. Three hours later, Young and Collins reported they were closing in on the Gemini VIII target vehicle. At 48 hours and 42 minutes (48 hours and 35 minutes ground elapsed time) into the flight, immediately after rendezvous with the passive Gemini VIII target vehicle, the Gemini X umbilical EVA was initiated. The sequence of events for the umbilical EVA is shown in *Figure 6-11*. (*Figure 6-12* gives the time line for the standup EVA performed earlier in the Gemini X mission.)

The Gemini X pilot was expected to perform an extensive evaluation of the hand-held manoeuvring unit, including precise angular attitude changes and translations. However, the flight plan for the umbilical EVA required a number of other activities prior to this evaluation. One activity was to translate (*Figure 6-13*) to

Figure 6-13. Beginning of the Gemini X extravehicular transfer for collection of experiment package.

the target vehicle at very short range, using manual forces alone, and retrieve the Experiment S010 Agena Micrometeorite Collection package attached near the docking cone. The pilot, Mike Collins, described his activities with the hand-held manoeuvring unit (HHMU) as follows:

"Okay, we're in this EVA. I got back and stood up in the hatch and checked out the gun and made sure it was squirting nitrogen. That's the only gun check-out I did. In the meantime, John manoeuvred the spacecraft over toward the end of the TDA [target docking adapter], just as we had planned. He got in such a position that my head was four to five feet from the docking cone. It was upward at about a 45° angle, just as we planned. I believe at one time there you had trouble seeing it, and I gave you [command pilot] some instructions about forward, forward, stop, stop. So I actually sort of talked John into position.

"I translated over by pushing off from the spacecraft. I floated forward and upward slowly and contacted the Agena. I grabbed hold of the docking cone as near as I can recall, at about the two o'clock position. If you call the location of the notch in it, the 12 o'clock, I was to the right of that — at about the two o'clock position and I started crawling around. No, I must have been more about the four o'clock position, because I started crawling around at the docking cone counter-clockwise, and the docking cone itself, the leading edge of the docking cone, which is very blunt, makes a very poor handhold in those pressure gloves. I had great difficulty in holding on to the thing. And, as a matter of fact, when I got over by the S010 (Agena micro-meteorite collection) package and tried to stop my motion, my inertia (the inertia of) my lower body, kept me right on moving and my hand slipped and I fell off the Agena. When I fell off, I figured I had either one of two things to do. I could either pull in on the umbilical and get back to the spacecraft, or I could use the gun. And I chose to use the gun. It was floating free at this time. It had come loose from the chestpack. So, I reached down to my left hip and found the nitrogen line and started pulling in on it and found the gun, and unfolded the arms of the gun and started looking around. I picked up the spacecraft in view. I was pointed roughly toward the spacecraft. The spacecraft was forward and below me on my left. The Agena was just about over my left shoulder and below me, or down on my left side and below me. I used the gun to translate back to the cockpit area. Now, I was trying to thrust in a straight line from where I was back to the cockpit, but in leaving the Agena I had developed some tangential velocity, which was bringing me out around the side and the rear of the Gemini. So what happened was, it was almost as if I was in an aeroplane on down wind for a landing, and in making a left-hand pattern I flew around and made a 180° left descending turn, and flew right into the cockpit. It was a combination of just luck, I think, being able to use the gun. At any rate, I did return to the cockpit in that manner, and John again manoeuvred the spacecraft. When I got to the cockpit, I stood up in the hatch and held on to the hatch. John manoeuvred the spacecraft again up next to the Agena. This time we were, I think, slightly farther away, because I felt that rather than trying to push off I would use the gun and translate over. And I did, in fact, squirt the gun up, depart the cockpit and translate over to the docking cone using the gun as a control device. The gun got me there. It wasn't extremely accurate. What happened was, as I was going over, I guess in leaving the cockpit, I somehow developed an inadvertent pitch-down moment, and when I corrected this out with the gun, I developed an upward translation as well as an upward pitching moment. So I did damp out the pitch. I converted that downward pitch moment into an upward pitching moment, and then I was able to stop my pitch entirely. But, in the process of doing that, I developed an inadvertent up translation, which nearly caused me to miss the Agena. As a matter of fact, I came very close to passing over the top of the Agena; and I was just barely able to pitch down with the gun and snag hold of the docking cone as I went by the second time."

A total of 14 experiments were scheduled for completion by the Gemini X crew. Here is a brief description of those experiments and the results achieved:

● The star occultation navigation experiment is designed to determine the value of star occulting measurements in developing a simple, accurate, and self-contained orbital navigational capability. On Gemini X, this experiment was accomplished with the Gemini-Agena target vehicle control system because it was necessary to remain docked. This experiment was terminated early because of concern over excessive use of propellant, but the crew obtained good data.

● The flight crew performed the ion-sensing attitude control experiment in an excellent manner, and, in addition to performing required operations, they obtained photographs of the attitude indicators while manoeuvring. The purpose of this experiment is to develop a navigation system which can sense vehicle attitude by using flow variations on a specially designed system.

● The colour patch photography experiment was terminated before completion when an eye irritation problem caused termination of the standup EVA; however, one of the three planned series of exposures was obtained. The purpose of this experiment was to determine if existing photographic materials could accurately reproduce the colour of objects photographed under the environmental conditions which exist in space.

● The fact that any data were obtained on the zodiacal light photography experiment is attributed to the crew's decision to combine this experiment with the ion-sensing attitude control experiment.

• No fuel was allocated for the synoptic terrain and synoptic weather photography experiments, but the crew obtained good photographic data while in drifting flight.

• The micrometeorite collection package was retrieved from the Gemini VIII Agena target vehicle. However, the new micrometeorite collection package was not put in place as planned because of hand-hold problems encountered by the pilot. The crew performed the micrometeorite collection experiment later than originally planned because of the early termination of the first EVA. The package was retrieved during the second EVA and handed into the spacecraft; subsequently, it floated out of the spacecraft and was lost in orbit.

• The decision to remain docked to the Agena target vehicle after the initial rendezvous placed a major constraint on the experiment concerning the ultraviolet astronomical camera. However, several exposures more than planned were obtained.

• The quantity of ion-wake measurement was decreased because of the deletion of docking practice. The crew performed all required operations for the completion of this experiment.

Three other experiments were conducted according to the flight plan. They concerned a Tri-Axis Magnetometer, a Beta Spectrometer, and a Bermsstrahlung Spectrometer.

Gemini XI, the ninth manned space flight of the programme, was launched September 12, 1966. The primary objective of the mission — to rendezvous and dock with the target vehicle during the "first revolution" — was accomplished . . . as were the following seven secondary objectives assigned to the flight:

Conduct docking practice;

Conduct extravehicular activity;

Conduct docked manoeuvres which included a high-apogee excursion;

Conduct a tethered vehicle test;

Demonstrate an automatic re-entry; and

Park the Agena target vehicle.

The initial orbit attained by Gemini XI was an apogee of 151 nautical miles and a perigee of 87 nautical miles. In achieving the "first revolution" rendezvous with the previously launched Agena, Gemini XI performed a minor plane change to the left shortly after 29 minutes into the flight, then made a terminal phase burn manoeuvre after 49 minutes and 58 seconds of the mission had elapsed. The latter manoeuvre, with minor midcourse corrections, placed the spacecraft in position to initiate the braking manoeuvre after about 1 hour and 18 minutes of the flight.

During the next several hours, the crew performed various sequences of the ion-wake measurement experiment, including one undocking and redocking. The first docked manoeuvre using the Agena's primary propulsion system was begun after four hours and 28 minutes of flight. Each crewman performed an additional docking practice before entering the first sleep period, which was scheduled to start about eight hours after lift-off.

Gordon opened his hatch and began the umbilical EVA at 24 hours and two minutes after lift-off. He set up a camera, retrieved an experiment package, then moved to the nose of the spacecraft and attached a tether from the Gemini Agena target vehicle to the docking bar (*Figure 6-14*). This operation proved more difficult and tiring than expected. As a result, Conrad and Gordon decided to terminate this phase of EVA because of pilot fatigue. The hatch was open 33 minutes.

After 25 hours and 37 minutes, ground-elapsed time, the pilot again opened the hatch and jettisoned equipment no longer needed for the mission. The other major activity during the second day of the flight was spent performing various sequences of the airglow horizon photography experiment. The second sleep period began after 31 hours and 30 minutes of flight and was completed at 39 hours.

The following day, the Agena primary propulsion system was used to place the docked Gemini-Agena configuration into an elliptical orbit, with an apogee of 741·5 nautical miles and a perigee of 156·3 nautical miles (*Figure 6-15*). This manoeuvre was initiated after 40 hours, 30 minutes and 15 seconds of flight. During the ensuing two revolutions, the crew took the photographs required for the synoptic terrain, synoptic weather, and the airglow

Figure 6-14. This EVA of astronaut Richard Gordon took place 160 miles above the Atlantic Ocean and prompted command pilot Charles Conrad to exclaim "Ride 'em, Cowboy!".

Figure 6-15. The northwest coast of Australia as seen from the Gemini XI spacecraft at an altitude of 740 miles. This is the most impressive view of the curvature of the Earth yet taken by man.

photography experiments. Three and a half hours later, a retrograde manoeuvre was performed which lowered the apogee of the docked vehicles to 164·2 nautical miles.

Conrad and Gordon then started preparations for the standup EVA which lasted two hours and eight minutes. (*Figure 6-16* shows the time line for this standup EVA.)

After 50 hours of flight, the spacecraft was undocked and the tether exercise began (*Figure 6-17*). The initial rotational rate achieved was 38 degrees per minute. Later the crew increased the rotational rate to about 55 degrees per minute, thus providing the first — although very small — artificial gravity field in space. This exercise was to evaluate the basic feasibility of rotating tethered-vehicle operations as the operations might apply to generating

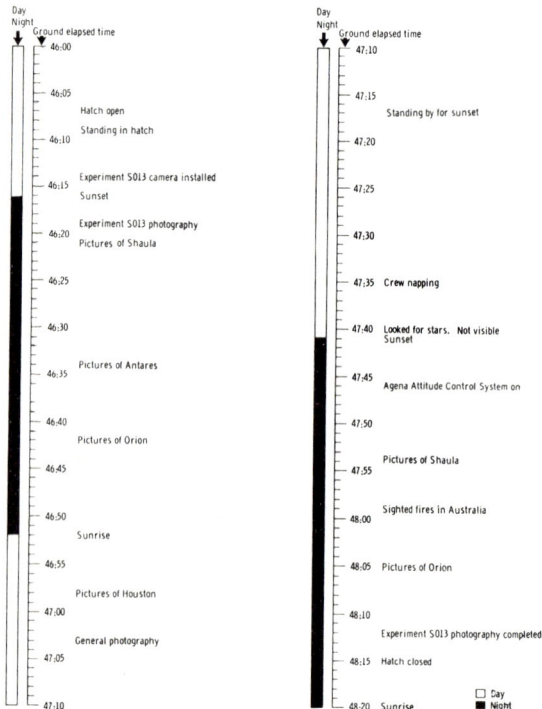

Figure 6-16. Gemini XI standup extravehicular time line.

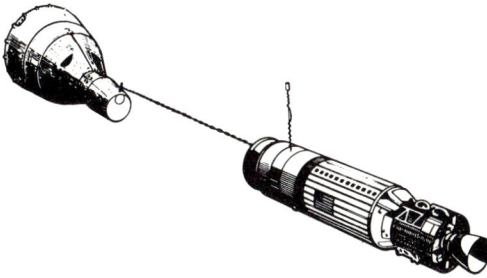

Figure 6-17. Gemini spacecraft/target-vehicle tethered configuration.

artificial gravity or to station keeping. The exercise lasted three hours and consisted of connecting the spacecraft and target vehicle with a 100-foot dacron tether, and then using the translation thrusting capability of the spacecraft propulsion system to induce a mutual rotation. The result of this mutual rotation was that the vehicles essentially maintained a constant separation at the ends of the tether.

The next day, a series of manoeuvres was initiated, starting at 65 hours and 27 minutes of flight, to achieve re-rendezvous with the target vehicle. Conrad and Gordon were station keeping with that target one hour and 13 minutes later.

Retrofire occurred over the Canton Island tracking station at an elapsed time of 70 hours, 41 minutes and 36 seconds. The crew performed all manual functions to prepare the spacecraft for re-entry. At 400,000 feet Conrad rolled the spacecraft to a backup angle of 44 degrees, and the computer commanded a bank angle for full lift and a right roll to recover from the backup bank angle.

At this time the crew agreed that the computer was operating properly and switched control to the automatic mode. Conrad followed all commands for control of the spacecraft with the attitude hand controller deactivated, so that, if a problem occurred, manual control of the re-entry could have been initiated in a minimum time.

The landing point achieved by the automatic re-entry was about one-and-a-half miles from the prime recovery ship, the USS Guam, after 71 hours, 17 minutes and 8 seconds of flight. After landing,

Conrad and Gordon decided to be retrieved by helicopter, and they were on the deck of the Guam 24 minutes after landing in the water. The spacecraft was picked up by the Guam 35 minutes later.

The final flight of the Gemini Programme began November 11, 1966, and ended four days later. On launch day the Atlas-Agena lift-off occurred at 2.07.59 p.m., one second earlier than planned. This launch was followed by the Gemini lift-off at 3.46.30 p.m., within a half-second of the planned time. Gemini XII was designed to investigate further the requirements for EVA; to obtain additionally desired experience in rendezvous, docking, and docked manoeuvring; and to perform a number of experiments. This final Gemini flight was an unqualified success.

In addition to achieving these and other objectives, command pilot James Lovell and pilot Edwin "Buzz" Aldrin set several individual space records. Lovell has logged more hours in space flight than any other man — 425 hours, 10 minutes and two seconds. Aldrin has logged more EVA time than any man — a two-hour, 27-minute standup EVA; a two-hour, eight-minute umbilical EVA; and another 51-minute standup EVA — a total of five hours and 26 minutes.

Lovell described the rendezvous of Gemini XII with the Agena target vehicle in these words:

"We were extremely fortunate because we turned the radar on early and we had a solid lock-on at 235 miles. We were led to expect, before the flight, that this range was highly improbable and we would have a much shorter range. You can imagine our confidence and elation as we waited for the rendezvous to take place. At the time for terminal phase initiation for the final rendezvous, Buzz noticed that the computer wasn't giving any change of range. I looked down at the little green light that tells us we had a radar lock-on, and it was off.

"We just looked at each other. We said, 'Oh, no, it can't happen to us. Anybody else or any other time but not this time.' Then it suddenly dawned on us that our radar had indeed failed. We went to the radar backup procedures, which we had practised quite a bit in preflight training but never really expected to use.

"The first thing on my list was to acquire the target visually. I looked up there and couldn't see a thing. Buzz took out his trusty sextant, which had an eight-power scope, and put it up to the window and spotted the target. I looked up again and that speck on the windshield turned out to be the Agena. So, we bore-sighted on the target and the rest of the rendezvous is history. It was successful and now I am sort of glad that we had a radar failure, because it gave us an opportunity to use the backup charts that all the crews had been practising with quite a bit but never really utilized."

Docking was subsequently accomplished, and the Agena primary propulsion system was used in docked manoeuvres to change orbit. During the tethered station keeping exercise which followed, the crew also performed a gravity-gradient exercise to study the feasibility of using gravity-gradient effects in the stabilization of manned spacecraft. The exercise consisted of tethering the orbiting vehicles together, then arranging the vehicles one above the other at the ends of the extended tether (that is, along the local vertical). By imparting the proper relative velocities to the vehicles in this arrangement, the vehicles would proceed into a constantly taut tether configuration and the tethered system would be captured by the gravity gradient. This captured behaviour would be manifested by oscillation of the system about the local vertical.

This sequence of the mission was described with these observations made by Lovell (*Figure 6-18* was taken during this exercise):

"It's actually a matter of station keeping without using any fuel.

"About this time we had a little thruster problem . . . the two and four thrusters were out and every time I wanted to pitch or yaw, I would roll. It really got to be quite frustrating. . . . Every time I wanted to do anything, I'd always roll. But we finally, through a learning curve, determined how to handle the situation by using a manoeuvre thruster — actually blipping it a little bit to bring it around and counteract this roll."

In practice they had decided that Aldrin would control attitudes and Lovell translational movements. The idea was to maintain

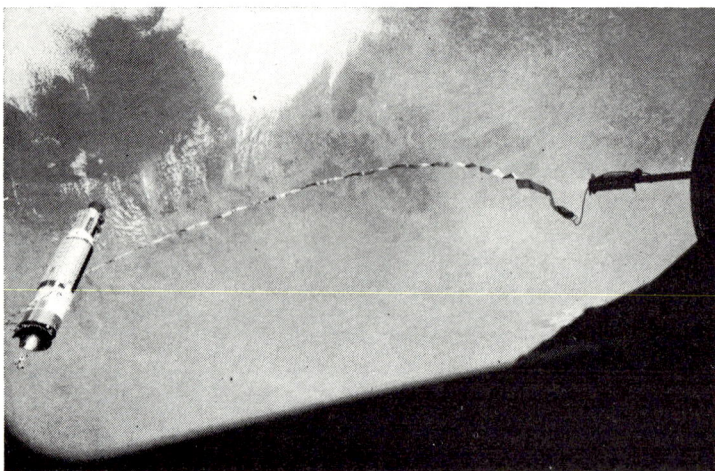

Figure 6-18. The Gemini XII crew photographed the Agena XII and its tether as the spacecraft and target passed over the Texas Gulf Coast during the 47th hour of the mission.

the position and then let go when all the rates had stopped to determine whether gravity had captured the spacecraft and was keeping it in position going around the earth. After encountering problems with attitude control, they decided to try to accomplish the goal by use of translational manoeuvres.

This continued through one night pass. During the next daylight pass, they tried again. Lovell recalls:

"Buzz got the slide rule out and made a few fast calculations, and we got above the Agena again, maintained this position, and it appeared to us then that our rates had indeed dampened. We let it go for the next two revolutions and finally we let the Agena go, too. There we were — two dead vehicles, captured by gravity in a vertical position going around the Earth."

There were three orbits allotted to the gravity-gradient tether exercise on the Gemini XII mission. Approximately half of this orbit time was used in establishing the starting conditions for the exercise. The remainder of the allotted time was spent observing the subsequent motion of the system.

Initiation of the system consisted of various translational and attitude thrusting manoeuvres by the spacecraft, and an active stabilization of the target vehicle using the target-vehicle control system. After the flight crew had ascertained that acceptable initial conditions had been achieved, the crew deactivated the target-vehicle control system and terminated all spacecraft thrusting. The resulting motion was one of limited amplitude oscillations relative to local vertical. It was evident that the system was indeed captured by the gravity gradient. After initial perturbations, the tether became constantly taut, and the attitude oscillations of the spacecraft were of sufficiently limited amplitude that the crew were able to view the target vehicle almost continuously. Under these conditions, the target vehicle was never observed to rise toward the horizon by more than approximately 60° from local vertical.

Initiation of the gravity-gradient exercise was greatly hampered because some spacecraft control thrusters were malfunctioning. Attitude control had degraded to the extent that the preflight planned procedures for setting up the gravity-gradient exercise could not be accomplished. Despite this handicap, the crew was able to devise a backup procedure consisting of judicious use of remaining thrust capability to provide initial conditions for a successful gravity-gradient capture.

The crew judged that the gravity-gradient exercise was more difficult in simulation training than during actual flight. With a properly functioning control system, the gravity-gradient-capture could have been accomplished with relative ease and certainty.

Another highly successful portion of the flight was the completion of three scheduled extravehicular activities. Gemini IX-A, X, and XI had shown that EVA (or "space walking" as it is more commonly known) was not as easy as Edward White's 20-minute venture on Gemini IV had led us to believe. Eugene Cernan, while attempting to fly the astronaut manoeuvring unit during Gemini IX-A, was forced to quit after his visor became fogged; the Gemini X EVA, performed by Michael Collins, was shorter than planned due to low spacecraft fuel; and Richard Gordon was able to complete only a portion of his umbilical EVA on Gemini XI before over-exertion and fatigue forced him to return to the spacecraft. In all of these cases, the crews had successfully simulated

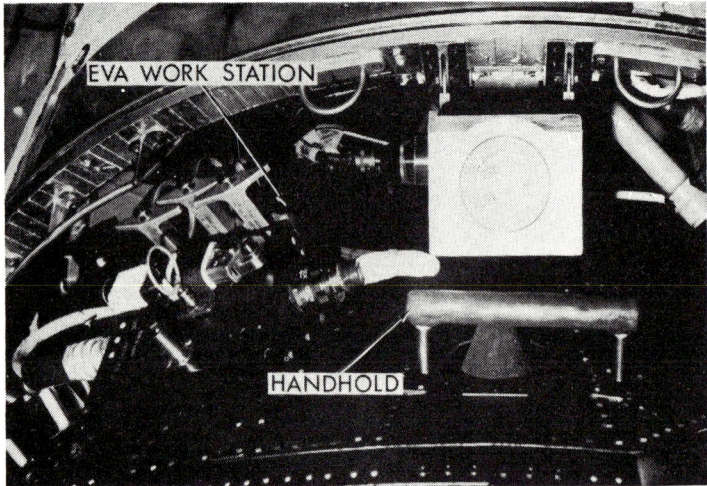

Figure 6-19. Gemini XII extravehicular work station on target docking adapter.

their assigned tasks on the ground and in the zero-g aeroplane, but they were unable to complete the tasks in flight.

The mission of Gemini XII was consequently changed to investigate the fundamentals of extravehicular activity — to define and solve the problems of doing simple tasks while outside the spacecraft in a pressurized suit.

Gemini XI mission results raised significant questions concerning man's ability to perform extravehicular activity satisfactorily with the existing knowledge of the tasks and environment. The Gemini X umbilical EVA results had established confidence in the understanding of extravehicular restraints and of workload; however, the Gemini XI results indicated the need for further investigation. And, so, the Gemini XII extravehicular activity was redirected from an evaluation of the astronaut manoeuvring unit to an evaluation of body restraints and extravehicular workload. Attaching the spacecraft/target-vehicle tether and obtaining ultraviolet stellar photographs were other objectives. The extravehicular equipment for the Gemini XII mission included a new work station in the spacecraft adapter, a new work station on the target docking

Figure 6-20. Gemini XII extravehicular system.

adapter (*Figure 6-19*), and several added body restraints and handholds. The pilot's personal EVA equipment (*Figure 6-20*) was nearly identical to that of the Gemini IX-A pilot.

The flight-crew training for the Gemini XII EVA was expanded to include two periods of underwater simulation and training (*Figure 6-21*). During these simulations, the pilot followed the intended flight procedures and duplicated the planned umbilical EVA. The procedures and times for each event were established

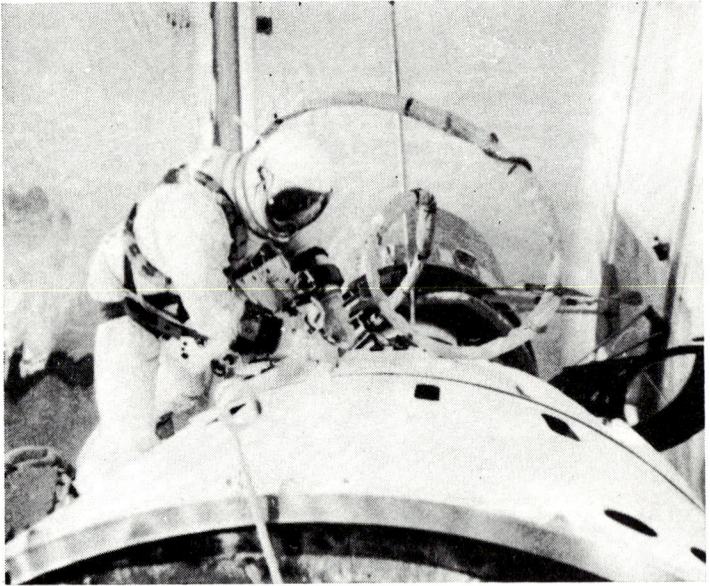

Figure 6-21. Underwater simulation of Gemini XII extravehicular activity.

and were used to schedule the final inflight task sequence. The underwater training supplemented extensive ground training and zero-g aircraft simulations.

Aldrin contracted the preparation for EVA with training required for the launch, rendezvous, and re-entry phases of a mission. In the latter phases, simulators have been designed which create the situation accurately. However, EVA is so much a factor of zero-g environment that it is impossible to establish completely accurate simulators on Earth. He pointed out that great success was achieved by underwater simulations in solving the EVA problems.

To increase the margin for success and to provide a suitable period of acclimatization to the environment before the performance of any critical tasks, the standup extravehicular activity was scheduled prior to the umbilical activity. The planned extravehicular activity time line was intentionally interspersed with two-minute rest periods. Procedures were also established for monitoring the heart rate and respiration rate of the EVA pilot; the crew was to be advised of any indications of a high rate of exertion before the condition became serious. Finally, the pilot was trained to operate at a moderate work rate, and flight and ground personnel were instructed in the importance of workload control.

The first standup EVA was very similar to that of the two previous missions. As indicated by the time line in *Figure 6-22*, the ultraviolet stellar and the synoptic terrain photography experiments were routinely accomplished. During the standup activity, the pilot performed several tasks designed for familiarization with the environment and for comparison of the standup and umbilical extravehicular activities. These tasks included mounting the extravehicular sequence camera and installing an extravehicular hand-

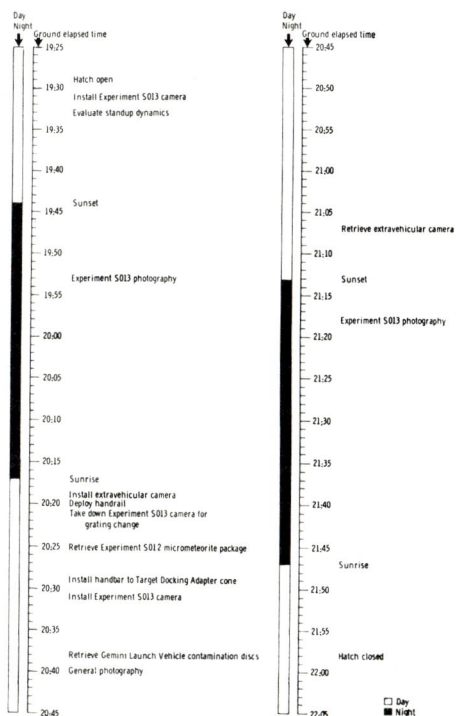

Figure 6-22. Gemini XII first standup extravehicular time line.

Day
Night
Ground elapsed time

42:35

42:40

42:45

Sunrise

42:50 — Hatch open. Extravehicular Life Support System in high flow
Standup familiarization
Rest

42:55 — Evaluate extravehicular camera installation
Rest

43:00 — Move to nose on handrail
Attach waist tether to handrail. Evaluate rest position
Hookup spacecraft/target-vehicle tether

43:05 — Tether hookup complete

43:10 — Activate Experiment S010 micrometeorite on target vehicle
Prepare Target Docking Adapter work station

43:15

43:20 — Observing hydrogen vent outlet
Return to hatch area and rest
Hand extravehicular camera to command pilot

43:25 — Pick up adapter camera from command pilot
Move to adapter section
Position feet in fixed foot restraints

43:30 — Install adapter extravehicular activity camera
Rest and general evaluation of foot restraints

43:35 — Unstow penlights

43:40

Sunset
43:45 — Torquing bolthead with torque wrench

Day
Night
Ground elapsed time

43:45

Disconnecting and connecting electrical connectors
43:50 — Rest

Removing cutters from pouch

Cutting wire strands and fluid hose
43:55

Loosening Saturn bolt
Removed feet from foot restraints.
44:00 — Evaluating waist tethers
Saturn bolt removed

44:05 — Saturn bolt installed
Saturn bolt tight
Evaluating hooks and rings

44:10 — Rest
Pulling Velcro strips
Connecting electrical connectors

44:15 — Feet in foot restraints
Retrieve adapter camera. Sunrise

44:20 — Move to cockpit
Install extravehicular camera
Move to Target Docking Adapter and hookup waist tethers

44:25 — Rest

44:30 — Disconnect and connect electrical and fluid connectors
Evaluate Apollo torque wrench

44:35 — Disconnect one waist tether and evaluate bolt torquing task
44:40 — Jettison waist tethers and handholds
Evaluate torquing task with no waist tethers

44:45 — Wiping command pilot's window
Return to cockpit
Observing thrusters firing

44:50 — Start ingress

□ Day
■ Night
44:55 — Hatch closed

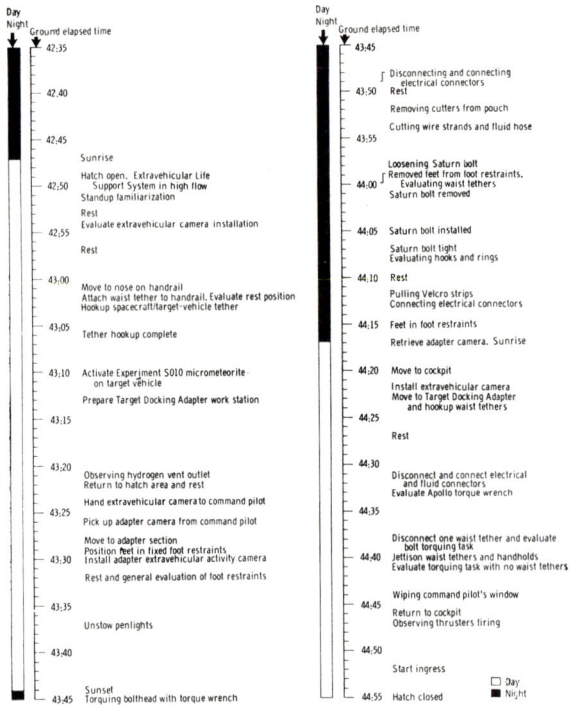

Figure 6-23. Gemini XII umbilical extravehicular time line.

rail from the cabin to the docking adapter on the target vehicle. The standup activity was completed without incident.

The umbilical EVA preparations proceeded smoothly, and the hatch was opened within two minutes of the planned time (*Figure 6-23*). The use of waist tethers during the initial tasks on the target docking adapter enabled the pilot to rest easily, to work without great effort, and to connect the spacecraft/target-vehicle tether in an expeditious manner. In addition, the pilot activated the Experiment S010 Agena Micrometeorite Collection package on the target vehicle for possible future retrieval. Before the end of the first daylight period, the pilot moved to the spacecraft adapter where he evaluated the work tasks of torquing bolts, making and breaking electrical and fluid connectors, cutting cables

Day
Night
Ground elapsed time

66:05   Hatch open

66:10   Equipment jettisoned

66:15   Sunset

66:20

         Ultraviolet photography of stars

66:25

66:30   Exercise

66:35

66:40

66:45   Ultraviolet photography of sunrise

66:50

66:55

67:00   Hatch closed

                         □ Day
                         ■ Night

67:05

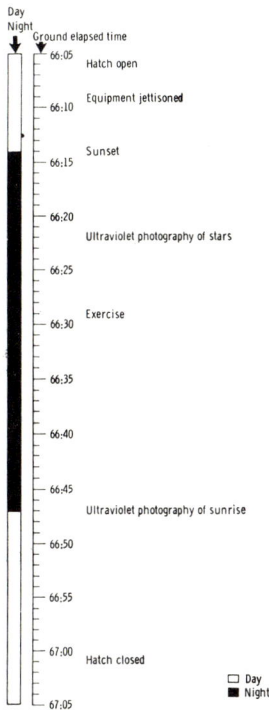

Figure 6-24. Gemini XII second standup extravehicular time line.

and fluid lines, hooking rings and hooks, and stripping patches of velcro. The tasks were accomplished using either the two foot restraints or the waist tethers. Both systems of restraint proved satisfactory.

During the second daylight period of the umbilical activity, the pilot returned to the target vehicle and performed tasks at a small work station on the outside of the docking cone. The tasks were similar to those in the spacecraft adapter and, in addition, included an Apollo torque wrench. The pilot further evaluated the use of two waist tethers, one waist tether, and no waist tether. At the end of the scheduled EVA, the pilot returned to the cabin and ingressed without difficulty.

A second standup EVA was conducted (*Figure 6-24*). Again, this activity was routine and without problems. The objectives were accomplished, and all tasks were satisfactorily completed.

The results of the Gemini XII EVA showed that all the tasks attempted were feasible when body restraints were used to maintain position. The results also showed that the extravehicular workload could be controlled within desired limits by proper procedures and indoctrination. The final, and perhaps most significant, result was confirmation that underwater simulation duplicated the actual extravehicular environment with a high degree of fidelity. It was concluded that any task which could be accomplished readily in a valid underwater simulation would have a high probability of success during actual extravehicular activity.

During the Gemini missions, a number of capabilities were demonstrated which met or exceeded the original objectives of extravehicular activity. Their basic feasibility of extravehicular activity was well established by 11 hatch openings and more than 12 hours of operations in the environment outside the spacecraft. The Gemini missions demonstrated the ability to control the extravehicular workload and to maintain the workload within the limits of the life-support system and the capabilities of the pilot. Standup and umbilical extravehicular operations were accomplished during eight separate night-time periods, confirming that EVA could be performed at night.

The need for handholds for transit over the exterior surface of the spacecraft was shown, and the use of several types of fixed and portable handholds and handrails was satisfactorily demonstrated.

The capability to perform tasks of varying complexity was demonstrated. The character of practical tasks was shown, and some of the factors limiting task complexity and difficulty were identified.

Several methods were demonstrated for crew transfer between two space vehicles. These include: (1) Surface transit while docked; (2) Free-floating transit between two undocked vehicles in close proximity; (3) Self-propulsion between two undocked vehicles; and, (4) Tether or umbilical pull-in from one undocked vehicle to another. All of these methods were accomplished within a maximum separation distance of 15 feet.

The hand-held manoeuvring unit was evaluated briefly, but successfully, on two different missions. When the unit was used, the extravehicular pilots accomplished the manoeuvres without feeling disoriented and without loss of control.

On four missions, equipment was retrieved from outside the spacecraft. One equipment retrieval was accomplished from an unstabilized passive target vehicle, which had been in orbit for more than four months.

Gemini X demonstrated that the command pilot was able to manoeuvre in close proximity to the target vehicle while the pilot was outside the spacecraft. The close-formation flying was successfully accomplished by co-ordinating the thruster firings of the command pilot with the extravehicular manoeuvres of the pilot. No damage nor indication of imminent hazard occurred during the operation.

Photography from outside the spacecraft was accomplished on each extravehicular mission. The most successful examples were the ultraviolet stellar spectral photographs taken during standup extravehicular activities on three missions and the extravehicular sequence photographs taken with the camera mounted outside the spacecraft cabin.

The dynamics of motion on a short tether were evaluated on two missions. The only tether capability that was demonstrated was for use as a distance-limiting device.

The requirements for body restraints were established, and the capabilities of foot restraints and waist tethers were demonstrated in considerable detail. The validity of underwater simulation in solving body restraint problems and in assessing workloads was demonstrated in flight and further confirmed by post-flight evaluation.

In summary, the Gemini missions demonstrated the basic techniques required for the productive use of extravehicular activity. Problem areas were defined sufficiently to indicate the preferred equipment and procedures for extravehicular activity in future space programmes.

Experiment	Missions Flown
Scientific	
Zodiacal light photography	V, VIII, IX-A, X
Sea urchin egg growth	III
Frog egg growth	VIII, XII
Radiation and zero-g on blood	III, XI
Synoptic terrain photography	IV, V, VI-A, VII, X, XI, XII
Synoptic weather photography	IV, V, VI-A, VII, X, XI, XII
Cloudtop spectrometer	V, VIII
Visual acuity	V, VII
Nuclear emulsion	VIII, XI
Agena micrometeorite collection	VIII, IX-A, X, XII
Airglow horizon photography	IX-A, XI, XII
Micrometeorite collection	IX-A, X, XII
Ultraviolet astronomical camera	X, XI, XII
Ion wake measurement	X, XI
Libration regions photographs	XII
Dim sky photographs orthicon	XI
Daytime sodium cloud photography	XII
Technological	
Reentry communications	III
Manual space navigation sighting	XII
Electrostatic charge	IV, V
Proton-electron spectrometer	IV, VII
Triaxis fluxgate magnetometer	IV, VII, X, XII
Optical communication	VII
Lunar ultraviolet spectral reflectance	X
Beta spectrometer	X, XII
Bremsstrahlung spectrometer	X, XII
Color patch photography	X
2-color Earth's limb photographs	IV
Landmark contrast measurements	VII, X
Basic object photography	V
Nearby object photography	V
Mass determination	VIII, XI
Celestial radiometry	V, VII
Star occultation navigation	VII, X
Surface photography	V
Space object radiometry	V, VII
Radiation in spacecraft	IV, VI-A
Simple navigation	IV, VII
Ion-sensing attitude control	X, XII
Astronaut maneuvering unit	IX-A
Astronaut visibility	V, VII
UHF-VHF polarization	VIII, IX-A
Night image intensification	VIII, XI
Power tool evaluation	VIII, XI
Medical	
Cardiovascular conditioning	V, VII
Inflight exerciser	IV, V, VII
Inflight phonocardiogram	IV, V, VII
Bioassays of body fluids	VII, VIII, IX-A
Bone demineralization	IV, V, VII
Calcium balance study	VII
Inflight sleep analysis	VII
Human otolith function	V, VII

Figure 6-25. Gemini experiment record.

The complement of experiments in the total Gemini Programme numbered 52. In general, each experiment was flown several times (*Figure 6-25*) to take advantage of varying flight conditions and resulted in 111 experiment missions — an average of 11 experiments per mission. The largest number of experiments, 20, was carried on the 14-day Gemini VII mission.

These experiments were divided into three categories: scientific, technological, and medical. There were 17 scientific experiments conducted during the programme. The 27 technological experiments were conducted in support of spacecraft development and operational techniques, and the eight medical experiments were directed toward determining more subtle effects than might be found during the regular operational medical measurements and preflight and post-flight examinations.

Experiment Performance Status

Gemini mission	Number of experiments	Experiments accom-plished [a]	Problems [b]
III	3	2	Experiment
IV	11	11	
V	17	16	Mission
VI-A	3	3	
VII	20	17	Experiment
VIII	10	1	Mission
IX-A	7	6	Mission
X	15	12	Mission
XI	11	10	Mission
XII	14	12	Experiment
Total	111	90	

[a] 80.3 per cent accomplished overall.

[b] 14.3 percent not accomplished due to primary mission problems; 5.4 percent not accomplished due to experiment equipment problems.

Figure 6-26.

The diversity of the experiments required considerable training by the crew. The training began with briefings by the experimenter to explain the experiment, the proposed method of operation, the probable training required, and the expected results. It was often determined in such briefings that various constraints would prevent the spacecraft and/or crew from accomplishing the experiment in the manner originally desired. In these situations, either the crew or the engineering and operational specialists could generally propose and develop alternate techniques which allowed accomplishment of the experiment objectives within the capabilities of the crew and the spacecraft.

The overall success of the Gemini Experiments Programme is indicated in numerical values in *Figure 6-26,* and is measured by the new or confirmed information provided for engineering, management, and scientific disciplines. If mission problems are not considered, a remarkable success is indicated. Experiment equipment problems affected only six of the 111 experiments performed on all missions. This performance was the result of the close teamwork of all participants as well as the capability to readily incorporate equipment and mission modifications up to launch time.

CHAPTER 7

Astronaut Training

Introduction

On December 12, 1965, Astronauts Walter Schirra and Thomas Stafford were strapped in a Gemini spacecraft atop a mighty Titan booster. The Gemini VI crew had been briefed that lift-off was characterized by noise, vibration, the event timer starting and the "feel" of lift-off. "How do you know we'll feel lift-off?" they asked. "Don't worry, when the timer starts you're on your way," they were told. On December 12 there *was* noise, there *was* vibration, and the timer did start, but the booster did not, and the crew did not. In the trade it was called a "Hold Kill".

Astronaut Schirra had only a few seconds for decision. Should he abort the mission by using the ejection seats, or should he stay with the spacecraft with the possibility of the Titan exploding as its engines shut down? He made the decision. They stayed with the spacecraft. The decision proved to be right. The Titan did not explode. Their previous training in this area of spaceflight paid off. A wrong decision — an abort — would have resulted in a long delay in preparing the spacecraft and booster again for flight, thus resulting in a long delay in flying the Gemini VI mission. The right decision allowed a short turn-around time before they were ready to try again.

On March 16, 1966, Astronauts Neil Armstrong and David Scott had completed the first docking in space as they brought their Gemini VIII spacecraft together with their Agena target vehicle. Approximately 27 minutes after docking, the spacecraft-target vehicle combination encountered greater than expected yaw and roll rates. Although their attitude control power switch was off, one of their orbit attitude and manoeuvre system attitude thrusters commenced firing. It fired intermittently at first — on for five seconds, then off for four; on for two seconds, then off for three; and then on. The position of the spacecraft attitude control power switch indicated that the problem was in the Agena target vehicle. They undocked, but continued to roll, showing that the rolling was being generated by the Gemini. A thorough set of control trouble-shooting techniques was completed by the crew to no avail! Roll and yaw rates had now increased to such an extent that the spacecraft was making about one full revolution per second. For approximately three minutes they tried to isolate the problem. Finally they decided to activate the re-entry control system to regain control of the spacecraft. The crew made the right decision — they solved a rough problem just in time to circumvent disaster.

How were they able to make the right decision? Their training paid off. They had studied the systems of their spacecraft. They knew what to look for, and they were thus able to decide what to do.

These two examples of skill and calm ability to isolate the cause of trouble in time of a crisis illustrate the necessity for the proper training and background knowledge. Prior to entering the space programme, many of the astronauts were test pilots. As such, they had learned to remain calm during crises.

The astronauts enter the space effort with an excellent academic and professional background. Thus each astronaut has already obtained a certain amount of required training prior to becoming a part of the astronaut team. After they are selected as astronauts, they are given even further specialized training, which is discussed later in the chapter. The training they bring with them to this programme can best be described by reviewing the criteria for astronaut selection.

A. Astronaut Selection Criteria

We presently have a group of 54 astronauts representing six selection processes (*Figures 7-1a* and *7-1b*). You will note that the seven original Mercury astronauts, selected back in 1959, were among other things, required to be graduates of a test pilot school, be rated as qualified jet pilots, and have at least 1500 hours of jet flying time. A bachelor of science degree in engineering, or its equivalent, was also required.

ASTRONAUT SELECTION CRITERIA

GROUP	NUMBER SELECTED	DATE OF SELECTION	BASIC REQUIREMENTS			
			GRADUATE OF TEST PILOT SCHOOL	JET PILOT RATING	FLYING TIME	EDUCATION
1	7	APR 59	REQ	REQ	1500 HRS	BS IN ENG OR EQUIVALENT
2	9	SEP 62	REQ	REQ	1500 HRS	BS IN PHYS, BIOLOGICAL OR ENG SCIENCES
3	14	OCT 63	OPT	REQ	1000 HRS	SAME AS FOR GP 2
4	6	JUN 65	OPT	OPT	OPT	GRADUATE WORK IN THE PHYS, BIOLOGICAL OR ENG SCIENCES
5	19	APR 66	OPT	REQ	1000 HRS OPT	SAME AS FOR GP 2
6	11	AUG 67	OPT	OPT	OPT	SAME AS FOR GP 4

Figure 7-1 (a).

COMPARISON OF ASTRONAUT GROUPS AT TIME OF SELECTION

	1959	1962	1963	1965	1966	1967
AGE	34.5	32.5	30.0	31.2	32.8	33.0
COLLEGE YEARS	4.3	4.6	5.6	8.0	5.8	8.3
FLIGHT HOURS	3500	2800	2315	*	2714	*

* PILOT EXPERIENCE NOT REQUIRED FOR SELECTION OF SCIENTIST ASTRONAUTS

Figure 7-1 (b).

With the second group of nine astronauts, added in 1962, we find the criteria basically the same; however, the bachelor of science equivalent was dropped and the earned degree in physical or biological sciences or in engineering became a mandatory requirement.

The third group of 14 astronauts, selected in 1963, also brought changes in the selection process. For this group, the requirement for test pilot certification was optional and could be substituted for the required jet pilot time which was reduced to 1000 hours from the previous 1500 hours. With this drop in actual flying requirements, increased emphasis was given to academic areas.

In 1965 six astronaut scientists were selected. For this group, flying status was desirable but not mandatory. As a result, three of the six entering the programme underwent flight training at Williams Air Force Base, Arizona. Although each of those selected was required to pass a Class I military flight status physical examination prior to acceptance, emphasis was on graduate work done in the natural sciences, such as medicine or engineering or comparable occupational experience.

The fifth group, consisting of 19 astronauts, came on board in 1966 with 1000 jet hours or certification from an Armed Forces test pilot school and a bachelor's degree in engineering, physical or biological sciences.

The latest recruiting effort, completed in 1967, yielded 11 astronaut scientists — two of whom are naturalized citizens of the United States, one having been born in Wales and the other in Australia. As the classification "scientist" implies, emphasis for the selection of this sixth group of astronauts was on scientific education rather than on flying experience (as was the case with the previously chosen fourth group). Following a brief period of general orientation activities at the Manned Spacecraft Centre, our newest astronauts began a programme of academic or "ground school" training. This training included orbital mechanics, computers, spacecraft orientation, and general math and physics refresher courses, as well as field trips for contractor facility orientation. In March, 1968, they began Air Force flight training to become qualified jet pilots.

Although crew qualifications may change and hardware become more refined as years go on, the importance of man as an integral part of the system will not change. The ability of man to assess a situation and provide the necessary input is as important today in space as it was during earlier explorations above and below the earth's surface. And in reflection, it is hard to single out a manned space flight during which the crew did not provide a contributing or decisive input. This pilot capability was especially proven during the Gemini programme. The importance of man to the mission and the importance of training to the man were vividly illustrated by the specific examples previously mentioned about the Gemini VI and Gemini VIII crews.

Each specific flight crew is trained in specific flight-oriented areas to fulfil a specific mission. However, prior to being assigned to a crew, the astronauts undergo a general training programme. This general training lays the foundation for specific crew training and, as a result, there is close inter-relationship between general training and crew training for a specific flight.

B. Objectives of Training

The astronauts' personal part in this space programme is to provide the crews needed to man the space flight vehicle and to provide an operational input to design. This input is based on individual judgment, past experience and education, engineering simulations, and actual space flight experience.

In Project Apollo, man for the first time will leave his orbit around the Earth and strike out for the Moon — first stepping stone in the exploration of the solar system. The navigation required for this voyage demands that the astronaut be skilful, not only as a spacecraft pilot and in carrying out the routine computations on digital computers in the spacecraft, but also in using other more complex equipment such as propulsion control systems and fuel cells for electric power generation. In addition to operating scientific equipment, the astronaut will be required to make meaningful observations — to select and interpret those phenomena which may be of significant scientific interest.

In order to equip the astronaut with an understanding of these space problems and the knowledge he will need to solve them, a continuing programme of astronaut training activities is conducted. The objectives of astronaut training then, can be summed up as follows:

1. To provide crew members prepared to operate the spacecraft in the best possible manner throughout both the normal flight phases and in emergency situations.
2. To provide crew members capable of competently accomplishing the scientific objectives of the flight.

Astronaut training can be broken down into four areas: academic, operational, contingency, and specific flight.

C. Types of Training

Academic

For the academic part of their training, the astronauts spend much of their time in classroom work to give them a thorough "grounding" in space-related subjects.

The academic curricula includes such subjects as geological processes, mineralogy and petrology, space science (physics of upper atmosphere and space), digital computers, astronomy, medical aspects of space flight, rocket propulsion, and communications. Most of the instructors are from various universities and government scientific research centres.

Just as the astronaut selection criteria has changed from group to group, so the scope of training has differed from group to group. For example, whereas the first groups had spent 250 classroom hours in academic courses after entering the astronaut training programme, the 11 astronaut scientists of Group Six devoted 330 classroom hours to scientific and technical lectures. The classroom work of Group Six was also unique in that many of the lectures were presented by the astronaut scientists themselves — each in his specialized field.

Although the subjects they studied were basically the same, the extent of coverage was broader. For example, the field of space science included solar physics (general structure, surface features, corona, chromosphere, cyclic behaviour, solar flares, babcock theory); solar wind (sources, quiescent structure, storm structure, Mariner and Explorer experiments); geomagnetic field (Hydronamic introduction, particle modulation, surface activities); moon and planets (atmospheric equilibrium, captured magnetic field/conductivity, shock structure); Jupiter (radio radiation, magnetic field; Aeronomy (neutral atmosphere, ionosphere, ion wake, and radio occultation experiments); and unmanned exploration of the planets.

Figure 7-2 shows Astronauts Frank Borman and Eugene Cernan performing classroom assignments and *Figure 7-3* shows Astronauts Russell Schweickart, Alan Bean and Walter Cunningham "moon viewing" at Kitt Peak National Observatory, Tucson, Arizona. The astronauts also attend astronomy courses at the Morehead Planetarium in Chapel Hill, North Carolina. This training further increases the crew capability to orient and control the spacecraft by use of celestial information, as well as to make various astronomical observations. At the planetarium they review the entire celestial sphere, which includes recognition of relative magnitudes, positions, distances and celestial co-ordinates of prominent stars and constellations near the planned orbital planes for the missions. They observe and identify constellations and star patterns as they drift across the window, using the flight plan and star charts for reference.

Since astronauts must be prepared to perform scientific experiments during space flight, they take additional courses to improve their ability as scientific observers. The geosciences — geology, geochemistry and geophysics — have been of primary interest in the preparation for lunar surface explorations. It is estimated that by the time the astronauts actually explore the surface of the moon, they will have worked their way well toward a degree in geology.

The astronauts make geology field trips to such places as: (1) West Texas; (2) Katmai, Alaska; (3) Valles Caldera, New Mexico; (4) Pinacate, Mexico; (5) The Volcanoes of Hawaii; (6) Grand Canyon, Arizona; (7) Flagstaff, Arizona; (8) Bend, Oregon; (9) Medicine Lake, California; and (10) The Icelandic Volcanoes. Courses emphasize vulcanism and impact geology and the related subjects of mineralogy and petrology. During field trips astronauts perform the function of field geologists, examining the surface and trying to determine its composition and origin.

Figure 7-4 shows Astronauts Donald Slayton and James McDivitt during geological training in Arizona. In *Figure 7-5* Astronaut Alan Shepard examines rock specimens during a geological field trip to the Marathon Basin Area of West Texas. In *Figure 7-6* Astronaut Walter Schirra rides a mule up the Grand Canyon Trail during geological training in Grand Canyon, Arizona.

To develop a trained eye in the field, the astronauts make many trips to view craters of all descriptions on the Earth's surface. Each group of astronauts has followed a different schedule for these field trips. Let's use the itinerary of the fifth group of 19 astronauts as an example. Their first geology trip was to the Grand Canyon. A team of skilled geologists helped them interpret the magnificent

Figure 7-2.   Astronauts Borman and Cernan — classroom work.

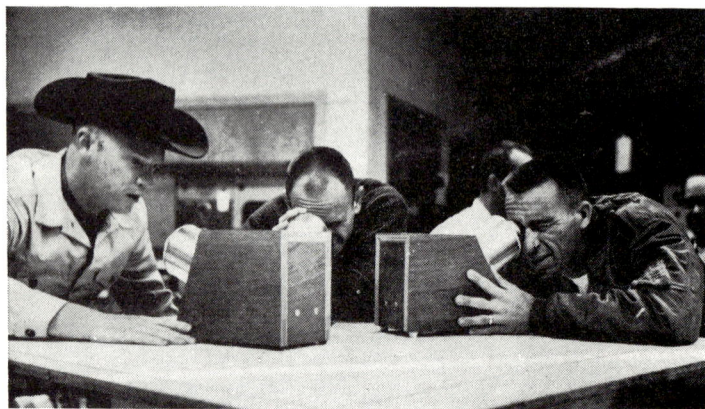

*Figure 7-3.   Astronauts Schweickart, Bean and Cunningham "moon viewing",
Kitt Peak National Observatory, Tucson, Arizona.*

Figure 7-4. *Astronauts McDivitt and Slayton, geological training, Arizona.*

Figure 7-5. *Astronaut Shepard studies composition of rocks in connection with geological field trip.*

Figure 7-6.   Astronaut Schirra rides mule up Grand Canyon for geology studies.

scenery for geological processes. This was just a two-day trip where they hiked down the Kaibab Trail and up the Bright Angel, camping out at the Phantom Ranch. Next they traversed the area around Bend, Oregon, where they studied volcanic craters, lava flows and cinder cones.

Their first big trip out of the continental United States was to Alaska. It was a one-week trip to the Katmai National Park on the Aleutian Chain. They flew into King Salmon and from there took a small float plane into Brooks Lake. They slept at a fishing camp. Army helicopters transported them to the area of volcanic activity for their geology studies.

Next, they made a short trip to the Valles Caldara, west of Los Alamos, New Mexico. A trip south of the border into Sonora, Mexico gave them the opportunity to study the Pinacates volcanic area.

The next trip out of the continental United States was to Hawaii, where they had a chance to study the immense volcanoes of Mauna Loa and Mauna Kea, very large lava flows, and craters of different types. They spent five days studying the features of the island. Then they made a short three-day trip to Arizona and New Mexico where they studied the Zuni Salt Lake near Springville, Hopi Buttes near the Utah border, and the famous Meteor Crater near Flagstaff.

Their final geology trip was to Iceland. Some of the terrain in Iceland is very similar to the photographs of the lunar surface as returned from Surveyor and Orbiter spacecraft.

Operational

Each astronaut is provided a reasonable exposure to each phase of flight environment, such as weightlessness, lunar gravity, vacuum in a pressurized suit, vibration and noise.

In our many simulators we can reproduce almost any aspect of space flight except weightlessness. The phenomenon of weightlessness (zero gravity) is explained in Chapter 4. The astronaut first encounters weightlessness in a KC-135 aircraft, which is a military version of the familiar Boeing 707 jet airliner (*Figure 7-7*). For about 30 seconds at about 35,000 feet of altitude, weightlessness can be produced aboard this aircraft flying a parabolic or Keplerian trajectory. During this type of training

Figure 7-7. Astronauts experiencing weightlessness in aircraft.

a spacecraft is often placed inside the aircraft so the astronaut can practise spacecraft egress and ingress for extravehicular activities (EVA) (*Figure 7-8*).

The air bearing Beta Trainer is another method used to simulate EVA (*Figure 7-9*). *Figure 7-10* illustrates how an astronaut "rides" a frictionless air bearing to simulate weightlessness in the horizontal plane. Using this device the astronaut becomes familiar with the hand-held manoeuvring unit, tether dynamics, and minimum reaction power tools which he will use during EVA. Each pilot spends approximately 15 hours on this platform practising extravehicular manoeuvring. This is done both in and out of a pressure suit and both pressurized and depressurized.

As discussed in Chapter 6, the experience gained during the EVAs performed on early Gemini flights proved that the brief periods of weightlessness produced while flying in the KC-135 aircraft were inadequate in preparing for the experiences encountered in actual space flight. To have sufficient time to perform tasks in a weightless condition NASA has learned that neutral buoyancy underwater is the best simulation of the zero gravity

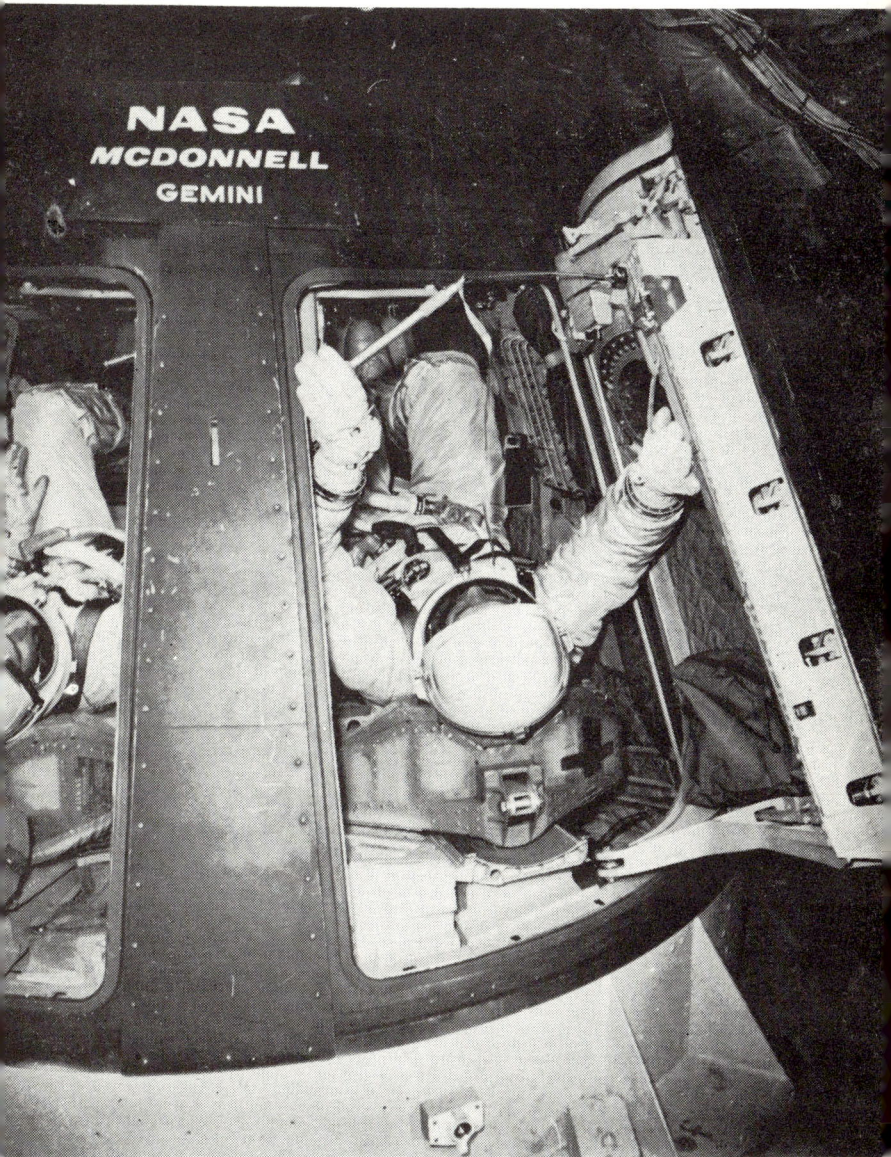

Figure 7-8. Weightless training with spacecraft in aircraft.

Figure 7-9. Beta Trainer.

conditions. Preparation for that underwater training begins with scuba diving at the Navy Underwater Swimmer School at Key West, Florida. The scuba course qualifies the astronauts for the work they must do in the water tank at the Manned Spacecraft Centre in Houston. (For further discussion on underwater EVA training see Chapter 6.)

Since lunar gravity is one-sixth that of earth gravity, the astronauts must learn to walk, run, jump, and work in an atmosphere where they will weigh one-sixth of their normal weight. By suspending an astronaut from cables against an inclined plane, lunar gravity can be simulated. This is done by using pulleys and weights and pneumatic devices to counter-balance five-sixths of the astronaut's weight (*Figure 7-11*).

To simulate the conditions of space, astronauts are trained inside vacuum chambers. These chambers are evacuated, and cooled with liquid nitrogen to a —300°F to simulate the ambient conditions of space. Here they learn to work in pressurized space suits, as well as in the simulated conditions of space. These chamber tests, conducted at the Manned Spacecraft Centre, take the astronauts to a simulated altitude of 180,000 feet. The importance of this training cannot be over-emphasized, since successful completion of the Apollo programme requires extra-vehicular activity. So, within the altitude chambers the crews learn to properly handle themselves during depressurization and

Figure 7-10. *EVA manoeuvring equipment as used on Beta Trainer.*

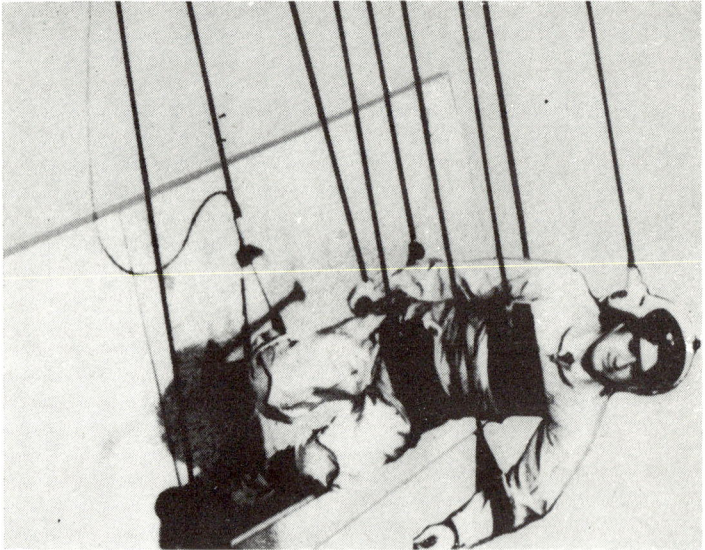

Figure 7-11. Partial gravity simulator.

pressurization cycles of their spacecraft, egress and ingress techniques, and the handling and operating of EVA support systems.

The astronauts also participate in tests of equipment to be used while in a pressurized suit during space flights. Astronaut Russell Schweickart wore a Gemini pressure suit continuously for several days to evaluate medical experiments planned during the Gemini flight. While wearing the suit he flew high performance aircraft, rode a centrifuge, and operated a Gemini mission simulator.

In tests over lava beds astronauts can get some feel for the difficulty which may be encountered while exploring the lunar surface. During such tests, the astronaut uses various pieces of prototype equipment to see how a particular pressure suit functions and how man can function with the suit. These experiments have disclosed tremendous differences between walking and operating in a pressurized and depressurized suit.

Since flying a jet approximates the situations found in "flying" a spacecraft, all astronauts must be competent jet pilots. As mentioned earlier, all astronauts were experienced jet pilots when they entered the programme, with the exception of members of the fourth and sixth groups.

The astronaut scientists with no pilot experience are given one year of Air Force flight training. During their flight training, 357 hours are spent in classrooms and 240 hours are spent flying various aircraft. Each day of the 53-week course is divided between actual flying time and classroom study. At the end of the 53 weeks, with training sessions averaging 10 to 12 hours a day, these astronaut-scientists earn their wings and are rated as qualified jet pilots.

Since hovering over the surface of the Moon in the Lunar Module in order to select a landing site is anticipated to be somewhat similar to flying a helicopter, all the astronauts are trained as helicopter pilots. This training is accomplished with the assistance of the Navy at Pensacola, Florida, and the Air Force at Ellington Air Force Base in Houston. Since flight skill is a requirement, each astronaut is required to maintain flying proficiency by logging approximately 110 hours each year in both jet aircraft and helicopters. Parachute experience prepares the crew member for an emergency which could require use of the ejection seat and personal parachute. The gliding parasail used can be towed aloft, thus permitting a number of descents in a short period of time (*Figure 7-12*). Training in this area includes parachute manoeuvring, surface obstacle avoidance, land and water landings, and the prevention of excessive drag after landing.

Another aspect of training astronauts for the Apollo missions involves a simulation of landing on the Moon. At the Langley Research Centre, the Lunar Landing Research Facility (LLRF) simulates landing a Lunar Module in one-sixth gravity by suspending the training vehicle from a cable which hydraulically balances five-sixths of the vehicle's weight. The LLRF has the same performance, control characteristics, and cockpit visibility as the Lunar Module. It can be flown at 17 miles per hour for 400 horizontal feet, 50 feet cross range, and 180 vertical feet.

The Lunar Landing Research Vehicle (LLRV) at the NASA Flight Research Centre, Edwards AFB, California (*Figure 7-13*) is another prototype of a trainer the astronauts use to practise

Figure 7-12. Astronaut Lovell, parasail training, Ellington AFB, Texas.

landing on the Moon. A jet engine supports five-sixths of the weight and an electronic control system compensates for the atmospheric effects, so that the astronaut can simulate actual landing under one-sixth gravity in a vacuum. The four welded aluminium alloy truss legs have a spread of 13 feet. The trainer uses a turbofan engine and hydrogen peroxide rockets. Actual Lunar Module instrument and control displays are used in conjunction with a variable stability auto-pilot system that causes the same basic reactions and sensations as the astronaut will experience on the Moon.

As you can see from *Figure 7-13*, the LLRV does not have a streamlined appearance, and you might wonder how it ever flies. But it is a very reliable training device and will be instrumental in providing the experience required for operating the Lunar Module to and from the surface of the Moon.

Astronauts also train for space flight in centrifuges (*Figure 7-14*).

The purpose of centrifuge training is to subject the astronauts to nominal launch and re-entry accelerations and to those "g" forces they might expect to encounter in the event of abort situations. The Navy's Johnsville facility has been used for most

LUNAR LANDING RESEARCH VEHICLE

Figure 7-13. Lunar landing research vehicle, Edwards AFB, California.

development and training programmes. The NASA Ames Research Centre centrifuge and the Manned Spacecraft Centre centrifuge are used for special projects. Accelerations up to 15 g's are experienced both suited and unsuited and in pressurized conditions.

In the Dynamic Procedures Trainer the astronauts spend a large percentage of their time going through manoeuvres and other phases of a mission where dynamic cues are important. A good portion of this training involves simulated launch aborts. Basically, for a crew, training consists of eight different groups of "runs" and is accomplished in six sessions of approximately 25 runs per session. During these runs the astronauts prepare themselves for failures associated with the launch vehicle or spacecraft. They also run the full gamut from a wind deviation problem through an electrical or sequential failure. These abort problems are fed into the system by the controllers seated at nearby consoles.

One of the examples of the value of this training was emphasized in the illustration used at the beginning of this chapter when, during preparation for the launch of Gemini VI, a split-second decision made by Astronaut Schirra averted a more serious problem.

Figure 7-14. Centrifuge.

During Project Gemini, the astronauts trained for rendezvous and docking manoeuvres in the Gemini Translation and Docking Simulator (*Figure 7-15*). In this trainer, the astronaut, seated in a Gemini spacecraft, could practise for the final phase of the rendezvous manoeuvres and actual docking of two spacecraft. Docking was a major portion of the Gemini programme and is essential to the Apollo programme, since the Lunar Module must dock with the Command Module in lunar orbit before return to Earth. Now that Project Gemini has ended, the Gemini spacecraft and Agena target vehicle are being replaced in this simulation by the Command Module and Lunar Module for Apollo docking practice (*Figure 7-16*).

The Aerospace Flight Simulator in Dallas, Texas, is frequently used by the astronauts for a variety of space programmes and manoeuvres to provide many of the sensations and visual scenes of actual space flight. Controlled by a complex of computers, the device makes it possible for the astronauts to work out procedures, solve problems, and simulate missions in real time with great accuracy. The astronaut rides in a spacecraft-like

Figure 7-15. Astronaut Gordon in Gemini translation and docking simulator, MSC, Houston, Texas.

Figure 7-16. Apollo docking simulator.

gondola which moves in roll, pitch, and yaw in response to his controls and accurate computer inputs. The astronaut makes his simulated flight in an inflated pressure suit and with the NASA developed extravehicular Life Support System chest pack which was designed for Gemini flights.

In *Figure 7-17*, Astronaut Neil Armstrong undergoes weight and balance tests in the Pyrotechnic Installation Building, Kennedy Space Centre, Florida.

The Apollo Mission Simulator (*Figure 7-18*) is the most complex simulator we have and the one in which the Apollo crews will spend the most time. In this device, the astronauts can simulate an Apollo flight from lift-off to within 500 feet of the lunar surface. The crew station of the simulator is identical to the Command Module, complete with switches, displays, communications, and controls. It contains realistic out-the-window television displays and instrument readings and displays for the entire lunar trip. It can reproduce every phase of a lunar flight, except weightlessness and lunar touch down. A crew using this simulator

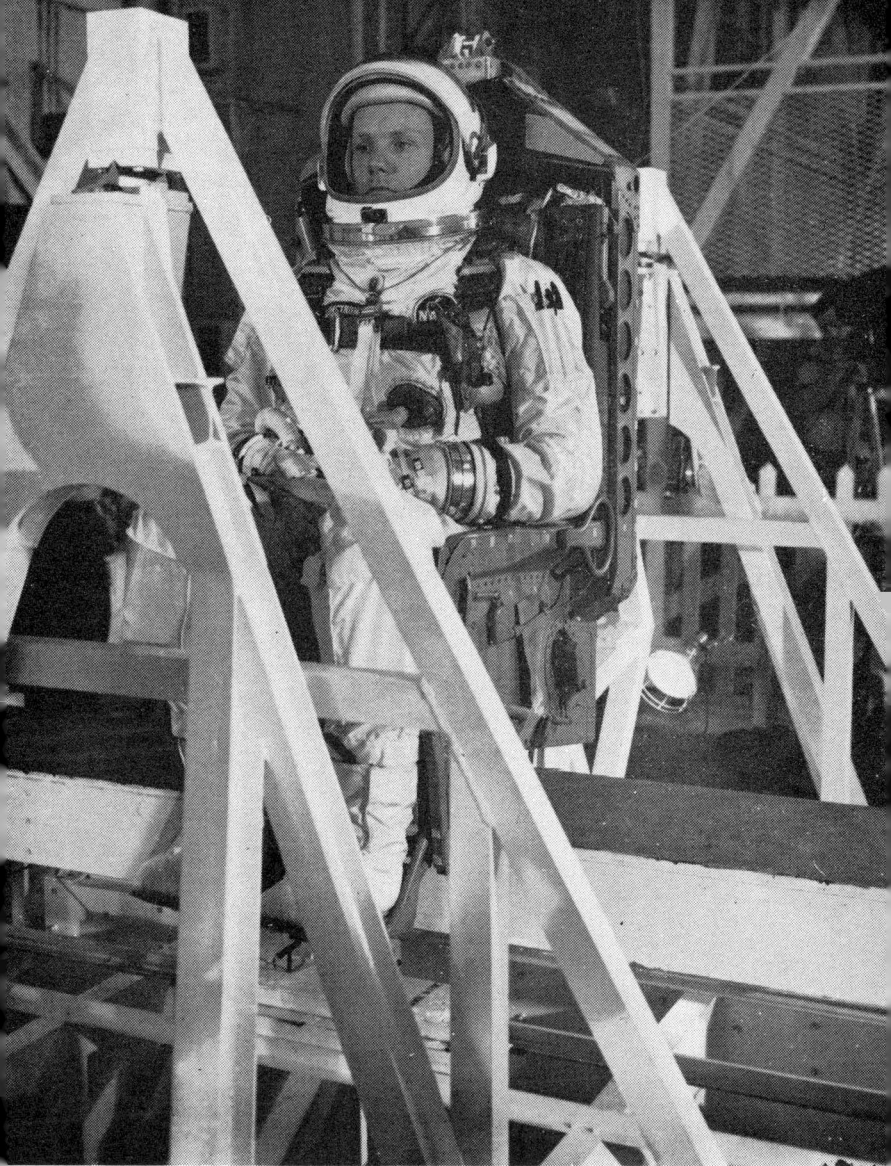

Figure 7-17. Astronaut Armstrong undergoes weight and balance tests at the Cape.

can complete the countdown and launch, go into earth orbit, be injected into translunar flight with course corrections, orbit the Moon down to an altitude of 500 feet, complete a rendezvous, and return to Earth, concluding with a re-entry and landing.

A miniature control centre is located next to the simulator. From this control centre, simulated flights are controlled, using associated digital computers which are capable of processing one million computations per second and which have an immediately accessible memory bank of 120,000 words. The spacecraft simulator responds to all signals and can execute the required corrections initiated by the crew.

A major portion of astronaut activity is their involvement in engineering development and testing of launch vehicles, spacecraft, and their various sub-systems. Although the astronauts do not actually design the systems, they monitor the design and assure that the designs reflect the crew's point of view. This is accomplished through design reviews, spacecraft and launch development tests, and individual engineering assignments.

For example, after the fifth group of 19 astronauts completed their basic space courses, they began studying Apollo design, which included the Command Module, Service Module, Lunar Module, and the launch vehicles. To help them acquire a better appreciation for operational aspects they toured the Marshall Space Flight Centre at Huntsville, Alabama, where the Saturn V is being developed. They also visited the plant at Michoud in Louisiana, where the first stage is built. Then they received a briefing at Cape Kennedy, where the final product is launched. This type of training, as well as experience in the various simulators mentioned previously, takes the astronauts to all NASA installations (see *Figure 7-19*) as well as various NASA contractor plants. Astronauts often gain much of their required flight experience by flying in NASA's T-38 jets on trips which may take them from Florida to California to Virginia and back to Texas in the same week.

As each astronaut carries out his all-encompassing training, he is assigned a specific engineering area and is responsible for keeping his associates informed on all significant developments in that area. For example, Scientist-Astronaut Joseph Kerwin and Astronaut Russell Schweickart helped solve a weight and stowage design

Figure 7-18. Apollo mission simulator.

problem of the Lunar Module. During stowage tests involving
the Lunar Module, these astronauts learned the importance of
weight inside the spacecraft and helped eliminate items from the
Lunar Module weighing as little as one ounce.

Survival

The long-duration earth-orbital flights include extensive periods
when the spacecraft is out of communication range of any of the
stations in the world-wide tracking network. An emergency
requiring an immediate return to Earth could result in a landing
in a remote area. So the astronauts must be prepared to land
in any part of the world and survive in all types of terrain —
water, desert and tropic. Thus the contingency portion of the
training programme includes water, desert and jungle survival
exercises. Water survival is conducted by NASA and the U.S.
Navy, while the desert and tropic training makes use of existing
U.S. Air Force schools.

NASA INSTALLATIONS

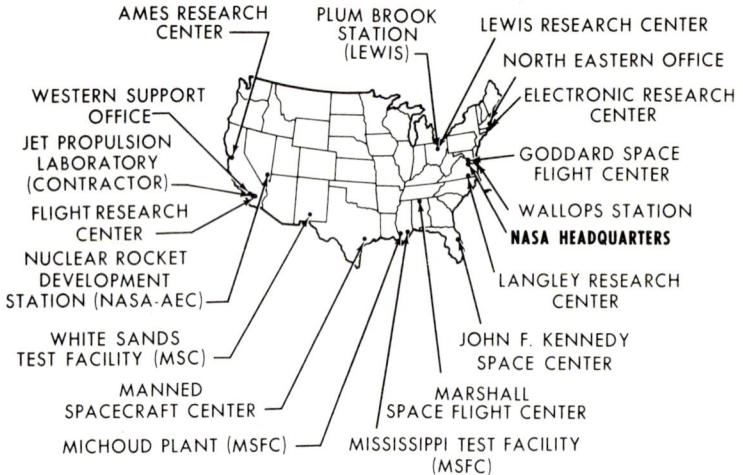

AMES RESEARCH CENTER —

PLUM BROOK STATION (LEWIS) —

LEWIS RESEARCH CENTER

NORTH EASTERN OFFICE

WESTERN SUPPORT OFFICE

ELECTRONIC RESEARCH CENTER

JET PROPULSION LABORATORY (CONTRACTOR) —

GODDARD SPACE FLIGHT CENTER

FLIGHT RESEARCH CENTER —

WALLOPS STATION

NASA HEADQUARTERS

NUCLEAR ROCKET DEVELOPMENT STATION (NASA-AEC) —

LANGLEY RESEARCH CENTER

WHITE SANDS TEST FACILITY (MSC) —

JOHN F. KENNEDY SPACE CENTER

MANNED SPACECRAFT CENTER —

MARSHALL SPACE FLIGHT CENTER

MICHOUD PLANT (MSFC) —

MISSISSIPPI TEST FACILITY (MSFC)

Figure 7-19.

Water survival training is conducted at the Pensacola Naval Air Station, Pensacola, Florida. Here the astronauts have the opportunity of riding the famous Dilbert Dunker, where they experience the difficulty of getting out of a spacecraft or aircraft if it should become inverted in the water. They also practise other water survival techniques, such as getting into a life raft (*Figure 7-20*).

For desert survival training, astronauts have visited the deserts of Nevada and Washington State during the month of August when temperatures are at their peak. In Washington State, the Air Force Survival School found some sand dunes in south-east Washington near Pasco. The astronauts spend two days in the classroom and then three days in the desert. On such an expedition they carry only survival kits, parachute, and $3\frac{1}{2}$ quarts of water for two days. They use a portion of a parachute to make a flowing robe and a burnoose for clothing. The rest of the parachute is used as a tent, with a life raft as the centre pole, providing a shelter against the 110°F desert heat.

Figure 7-20.   Astronaut Aldrin, water survival training.

For tropic survival training, the astronauts take a one-week trip to Panama. The first two days there are spent in the classroom learning the techniques needed to survive in the jungle. The highlights of the classroom training is the buffet on the last day, when they are served such gourmet delights as braised boa, iguana thermidor, fried rat, palm hearts, and tara root. In the morning of the third day they are taken into the jungle by helicopter. There they are formed into three-man teams like an Apollo crew. They hike until they find an ideal spot to construct a lean-to, which becomes their protection from the afternoon rains. In the heart of the jungle the astronauts live "off the land" for three days with nothing but a parachute and a spacecraft survival kit — which contains no food.

Flight Crew Assignments

The basic academic, operational, and contingency training requires over two years. The completion of this "basic" training is a happy time for the astronauts because they are then considered qualified for assignment to a specific space flight crew and can then begin training for a particular mission. Training for a specific flight begins at least six months prior to flight date.

During specific flight crew training, a typical Gemini crew spent approximately 160 hours in spacecraft checkout. About 40 of those hours were spent in the spacecraft itself, including 25 hours during pad tests at Cape Kennedy.

A variety of simulators were used to prepare the Gemini crews for a specific flight. The major facilities for such preparation are the mission simulators. As previously mentioned, these simulators can simulate a space flight from launch to landing. They permit development of the specific flight plan and familiarize the crew with the required timing, tasks and techniques required.

Again, during Gemini flight training, each crew member spent approximately 110 hours in the Gemini Mission Simulator. This simulator can partially duplicate a mission without leaving the ground. The mission simulator is primarily designed to perfect flight crews' knowledge in spacecraft systems and rehearse operations and actions to be taken during normal and emergency crew procedures (*Figure 7-21*). The basic systems training is performed at the Manned Spacecraft Centre. Actual flight plan practice with full gear is conducted at Cape Kennedy.

Generally, 13 hours per crew member were spent in egress and recovery training. Egress and recovery training begins with classroom instruction, progresses through simulations from an engineering mockup, and concludes with egress practice both from a spacecraft floating in a water tank and in the Gulf of Mexico, and both from water surface and underwater. However, underwater egress is restricted to the water tank. Final egress training is done with full gear (*Figure 7-22*).

The Gemini teams also practised helicopter recovery following water egress training in the Gulf of Mexico.

During this same period the crews attend briefings, technical meetings, evaluations and simulations. While all of this training is progressing, the crews are expected to maintain their flight proficiency in high performance aircraft. This is done primarily whenever some free time can be found and by using jet trainers for travel between activities.

The astronauts are also expected to keep themselves physically fit. There is no astronaut physical fitness programme in their regular training schedule — that is, in the sense that everyone

GENERALIZED MISSION SIMULATOR SCHEMATIC

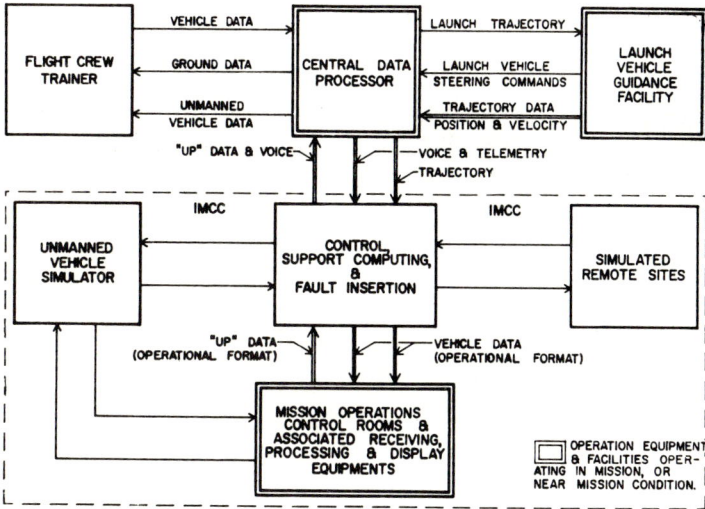

Figure 7-21.

does the same thing or follows a strict pattern. Facilities, equipment, and advice are readily available; but what they do and when they do it is left up to them. There is no regular period set aside each day for exercise and sports, and the only requirement is that they keep themselves in a constant state of readiness for space missions. That single requirement is the key to the astronauts' physical fitness. The space programme began with the promise that the men participating in it would have to be physically fit — and competition is the thing that keeps them that way. There are 54 astronauts today, and everyone of them who isn't going on the next mission would like to be. If they are in shape to fly when their turn comes, they can go, and if they aren't, they can't. It's that simple. There is always a qualified person ready to replace them. Some of the astronauts begin their day by running a mile or two. They all enjoy the facilities of the Astronaut Gymnasium located at the Manned Spacecraft Centre. In that facility their favourite activities include handball, squash, weight-lifting, and practice on the trampoline.

Figure 7-22. Astronauts Conrad and Cooper, water egress training.

The astronauts are not only trained to fly the spacecraft and perform the scientific experiments connected with each flight, they are also prepared to assist their fellow astronauts during flight. During an actual mission astronauts who remain on earth serve as Capsule Communicators (Cap Com) in various locations throughout the Worldwide Tracking Network. During a manned mission, it is an astronaut's voice that completes the final count-down for lift-off from the launch pad. *Figure 7-23* shows Astro-naut Donald Slayton during a manned Gemini mission watching his console in the Mission Control Centre (MCC) located at the Manned Spacecraft Centre in Houston, Texas. As the centre of the huge global network of tracking and communications stations, the MCC provides centralized control for the missions.

Conclusion

With their academic, operational, and contingency training, engineering participation, and specific missions, the astronauts' time is completely absorbed in preparing for the challenges of space flight. Thus, when our first astronauts land on the Moon, you can be sure they will be prepared to conduct a thorough

Figure 7-23. Astronaut Slayton in Mission Control Centre.

scientific survey. They will be prepared to handle any emergency that may occur. And, unlike in the case of unmanned spacecraft, astronauts will be able to return their manned craft to Earth, with its human "recorders" and precious lunar samples. Thus, they will be prepared not only for the lunar mission but also for future space adventures beyond the Moon!

What does all this mean to the young people of today? Simply this. We are faced with a challenge that will shape the destiny of our world for years to come. This is the challenge of space exploration, and we must do our utmost to insure that this challenge is adequately met. For some of our youth, the possibility to participate as an astronaut might exist. However, we must frankly admit that this number over the next few years will probably not exceed 100. So what opportunities exist for the bulk of our youth? How can they become involved in what we must agree is one of the most historic programmes of our time?

Our space programme is very broad and includes a great deal more than the training for space flight. For each astronaut, for example, there are thousands of others whose jobs are just as vital to the success of the manned space missions. These individuals represent almost every field of science and engineering, and they participate in an astonishing variety of areas of investigation. Medical personnel study not only the reactions of pilots but also racing drivers, bullfighters, and others under physical and psychological stress. Biologists, physiologists, radiologists and doctors must determine the physical condition necessary to sustain human life during lunar exploration. Chemists and chemical engineers develop new materials necessary to absorb tremendous heats of re-entry. Suit designers and engineers join with geologists to determine what the space suit should be like not only to sustain life but to perform the tasks required in space, while nutritionists must design condensed life-sustaining foods to be carried on extended missions. And so the list grows: aerodynamicists, structural engineers, electrical engineers, physicists, thermodynamicists, metallurgists, meteorologists, data analysts, and so on. Yes, the space programme offers a challenge to many types of trained individuals.

One way NASA extends this challenge to young people is by co-operating with universities in space research. Dealing this way serves a double purpose: NASA gets the product of many of the finest minds, and the universities use their NASA research work to train new scientists and engineers at the highest levels of ability and creativity. Many experiments carried aboard NASA satellites and deep space probes are conceived and designed by scientists and engineers within the colleges and universities.

Just as the astronaut team continues to study and prepare for the future, so young people of today must learn to live with education as a way of life, considering education not as an obligation but as a challenge.

We are entering into an exciting time in the history of man on this earth. We are really helping to make history, for as nations in this world co-operate with each other in space exploration, friendship and understanding will surely develop. Perhaps the co-operative exploration of outer space and the new knowledge it brings to us can lead us to this goal.

PLEASE
RETURN
TO
M. P. ATKINSON